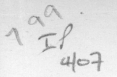

Critical acclaim f

'[A] rollicking good tale about an enduring ......
– the creation of a reference sequence of the human genome that
was ⌐he Great Genome Race of 2000 . . . Davies's book is a
straightforward account of extraordinary accomplishment . . . The
book deserve: he audience it will gar.. by w. .. ? the race to
cover The Race' Robert Cook-Deegan, *American S. ist*

'[The] race is over, and Davies was there, all along, providing the
running commentary – and there, too, at the finishing line. In *The
Sequence*, he hands out the prizes' Gail Vines, *Independent*

'Davies chronicles the events with verve and authority. He gives
many striking examples of what DNA sequences have already taught
us . . . An abiding impression is how much science has changed
since 1978 when Fred Sanger sequenced the first genome'
Walter Gratzer, *Times Higher Education Supplement*

'His book is consistently engaging, full of interesting perceptions and
overflowing with enthusiasm . . . Davies is able to give an insider's
point of view . . . a highly entertaining account of what has at times
seemed a rather undignified "race" to complete the sequence'
Steve King, *Spectator*

'The tale of the Tristanians is one of many good little stories in this
quick journalistic history of contemporary genomics . . . Davies has
done a nice job of assembling these stories and delivering them in
clear language that neatly demystifies many scientific complexities'
Paul Raeburn, *Business Week*

'Davies has done a superb job . . . a tantalizing glimpse of the ethical
perils and technological possibilities awaiting humanity'
Nick Owchar, *Los Angeles Times*

'Savvy . . . an admirable narration of how science and scientists work
in the real world' *Kirkus Reviews*

'Davies tells the story with verve . . . *The Sequence* is a good start'
Jan Witkowski, *Nature*

Kevin Davies was the founding editor of *Nature Genetics*, the world's premier genetics journal, and co-author of *Breakthrough: The Race for the Breast Cancer Gene*. He graduated from Oxford University with a BA in biochemistry, received a PhD in genetics from the University of London, and did post-doctoral work at MIT and Harvard Medical School before joining the editorial staff of *Nature* in 1990. He is now Editor-in-Chief of Cell Press in Cambridge, Massachusetts.

By Kevin Davies

Breakthrough: The Race for the Breast Cancer Gene
(with Michael White)
The Sequence

The Sequence

Inside the Race for the Human Genome

KEVIN DAVIES

PHOENIX

To Kyle & Fiona

A PHOENIX PAPERBACK

First published in Great Britain in 2001
by Weidenfeld & Nicolson
This paperback edition published in 2002
by Phoenix,
an imprint of Orion Books Ltd,
Orion House, 5 Upper Saint Martin's Lane,
London WC2H 9EA

First published in the USA in 2000
by The Free Press, Simon & Schuster, Inc.,
under the title *Cracking The Genome*

A CIP catalogue record for this book
is available from the British Library.

ISBN 0 75381 316 5

Printed and bound in Great Britain by
Clays Ltd, St Ives plc

Contents

The double helix is indeed a remarkable molecule. Modern man is perhaps 50,000 years old, civilization has existed for scarcely 10,000 years and the United States for only just over 200 years; but DNA and RNA have been around for at least several billion years. All that time the double helix has been there, and active, and yet we are the first creatures on Earth to become aware of its existence.

—FRANCIS CRICK

```
DEFINITION       Human Genome
LOCUS            HSGP010101        3.158 Gb          DNA
DATE             04-25-03
ACCESSION        Assembly 1.0
ORGANISM         "Homo sapiens"
   Eukaryota; Metazoa; Chordata; Craniata; Vertebrata; Euteleostomi;
   Mammalia; Eutheria; Primates; Catarrhini; Hominidae; Homo.

TITLE            The Sequence
/source          1,3150000000
/chromosome      "1-22,X,Y"
/note            "Book of Life, Holy Grail, Human Blueprint"

///...
ORIGIN
   1   agctcgctga gacttcctgg accccgcacc aggctgtggg gtttctcaga taactgggcc
  61   cctgcgctca ggaggccttc accctctgct ctgggtaaag ttcattggaa cagaaagaaa
 121   tggatttatc tgctcttcgc gttgaagaag tacaaaatgt cattaatgct atgcagaaaa
 181   tcttagagtg tcccatctgt ctggagttga tcaaggaacc tgtctccaca aagtgtgacc
 241   acatattttg caaattttgc atgctgaaac ttctcaacca gaagaaaggg ccttcacagt
 301   gtcctttatg taagaatgat ataaccaaaa ggagcctaca agaaagtacg agatttagtc
 361   aacttgttga agagctattg aaaatcattt gtgcttttca gcttgacaca ggtttggagt
 421   atgcaaacag ctataatttt gcaaaaaagg aaaataactc tcctgaacat ctaaaagatg
 481   aagtttctat catccaaagt atgggctaca gaaaccgtgc caaaagactt ctacagagtg
 541   aacccgaaaa tccttccttg caggaaacca gtctcagtgt ccaactctct aaccttggaa
 601   ctgtgagaac tctgaggaca aagcagcgga tacaacctca aaagacgtct gtctacattg
 661   aattgggatc tgattcttct gaagataccg ttaataaggc aacttattgc agtgtgggag
 721   atcaagaatt gttacaaatc acccctcaag gaaccaggga tgaaatcagt ttggattctg
 781   caaaaaaggc tgcttgtgaa ttttctgaga cggatgtaac aaatactgaa catcatcaac
 841   ccagtaataa tgatttgaac accactgaga agcgtgcagc tgagaggcat ccagaaaagt
 901   atcagggtag ttctgtttca aacttgcatg tggagccatg tggcacaaat actcatgcca
 961   gctcattaca gcatgagaac agcagtttat tactcactaa agacagaatg aatgtagaaa
1021   aggctgaatt ctgtaataaa agcaaacagc ctggcttagc aaggagccaa cataacagat
1081   gggctggaag taaggaaaca tgtaatgata ggcggactcc cagcacagaa aaaaaggtag
1141   atctgaatgc tgatcccctg tgtgagagaa aagaatggaa taagcagaaa ctgccatgct
1201   cagagaatcc tagagatact gaagatgttc cttggataac actaaatagc agcattcaga
1261   aagttaatga gtggtttttcc agaagtgatg aactgttagg ttctgatgac tcacatgatg
1321   gggagtctga atcaaatgcc aaagtagctg atgtattgga cgttctaaat gaggtagatg
1381   aatattctgg ttcttcagag aaaatagact tactggccag tgatcctcat gaggctttaa
1441   tatgtaaaag tgaaagagtt cactccaaat cagtagagag taatattgaa gacaaaatat
1501   ttgggaaaac ctatcggaag aaggcaagcc tccccaactt aagccatgta actgaaaatc
```

```
1561 taattatagg agcatttgtt actgagccac agataataca agagcgtccc ctcacaaata
1621 aattaaagcg taaaaggaga cctacatcag gccttcatcc tgaggatttt atcaagaaag
1681 cagatttggc agttcaaaag actcctgaaa tgataaatca gggaactaac caaacggagc
1741 agaatggtca agtgatgaat attactaata gtggtcatga gaataaaaca aaaggtgatt
1801 ctattcagaa tgagaaaaat cctaacccaa tagaatcact cgaaaaagaa tctgctttca
1861 aaacgaaagc tgaacctata agcagcagta taagcaatat ggaactcgaa ttaaatatcc
1921 acaattcaaa agcacctaaa aagaataggc tgaggaggaa gtcttctacc aggcatattc
1981 atgcgcttga actagtagtc agtagaaatc taagcccacc taattgtact gaattgcaaa
2041 ttgatagttg ttctagcagt gaagagataa agaaaaaaaa gtacaaccaa atgccagtca
```

continue . . .

The sequence is only the beginning.

—J. CRAIG VENTER

Preface to the paperback edition

THE QUEST TO DETERMINE the complete sequence of the human genome is, to all intents and purposes, over. For more than a decade, scientists around the world have toiled to assemble the order of more than three billion units of DNA that make up our genetic code. Buried within this text are tens of thousands of genes, the parts list of the human body and the unquestionable key to human health.

This Herculean task, popularly known as the Human Genome Project, was scheduled for completion in 2005, with a total price tag of some three billion dollars. And yet, here we are, years ahead of schedule, celebrating one of the greatest, most humbling scientific achievements in history. The accomplishment was officially sealed in February 2001 by the publication of two extraordinary issues of the esteemed scientific journals *Nature* and *Science*. Within a few dozen pages of these special issues, large teams of scientists deigned to describe the breathtaking landscape of the human genome. It was a triumph of technological prowess and endurance that President Clinton had hailed (somewhat prematurely) the previous summer as "the language in which God created life."

The quest for the sequence of the human genome would have been a compelling story under almost any circumstances. Since 1990, an international consortium of academic scientists, funded by governments and philanthropies, had made painstaking progress in its fifteen-year mission to decipher the complete DNA sequence of man. Most of the work was carried out by research centres in the United States, although fully one third of the sequence was the responsibility of the Sanger Centre near Cambridge, led by the respected geneticist Sir John Sulston.

But in May 1998, an increasingly pedestrian project, bogged down in federal bureaucracy and political correctness, was suddenly transformed into a gripping, no-holds-barred race to claim the human blueprint. A brash, brilliant American scientist named J. Craig Venter, no stranger to controversy over issues such as gene patenting, astounded the scientific community with his shocking decision to sequence the genome solo, backed by hundreds of millions of biotech dollars. Some scientists dismissed his ideas as delusional and his strategy flawed, but the majority fretted that he might indeed sequence the genome years ahead of the international project. Were he successful, Venter would not only embarrass the official government alliance, he would also lay claim to an invaluable genetic trove—the identity of thousands of human genes, and potentially the keys to find cures for cancer, Alzheimer's disease, diabetes, and other scourges of the modern world.

For two years, the public consortium and Venter's new company, Celera Genomics, located just outside Washington, D.C., waged an increasingly bitter struggle. It was a clash of ideologies and sequencing strategies, academia against big business, public ownership versus private entrepreneurship—some might even say, good versus evil. For while the consortium vowed to provide free and instant access to its DNA sequence data, Celera's ambition was to sell its sequence information to pharmaceutical clients. Adding insult to injury, Venter was perfectly able to supplement Celera's rapidly generated genome sequence with the latest public data deposited by his rivals.

Venter hired an outstanding team of molecular and computational biologists, including a Nobel laureate, but the unsung heroes were three hundred deluxe DNA sequencing machines. Venter surveyed their prolific output—a million letters of DNA per machine per day—from his control center on a pair of enormous video screens, which bore an uncanny resemblance to the bridge of the USS *Enterprise*. Meanwhile, just a few miles down Wisconsin Avenue, Francis Collins, Director of the National Human Genome Research Institute, marshaled the efforts of the five major sequencing centers charged with ramping up their output to keep pace with Celera.

In June 2000, Venter and Collins declared an uneasy truce and a hastily convened ceremony at the White House found the two rivals flanking President Bill Clinton (with Tony Blair participating via satellite) and sharing the credit for the sequencing of the genome.

THE SEQUENCE IS MY ACCOUNT of mankind's astonishing adventure to read and understand its own genetic blueprint. I began to think seriously about writing this book in the summer of 1998, shortly after Venter's stunning decision to take on the public project. Venter's tempestuous rivalry with Collins, Sulston, and the official project spiced up a potentially bland saga of reading the book of life. But the main purpose of the book is to highlight the uses of the human genome in improving our health and longevity, and understanding human history and evolution. The real hero of *The Sequence* is the sequence itself.

The Sequence went to press towards the end of 2000, after the White

House ceremony, while both teams were frantically analyzing their genome sequences prior to officially publishing their results. In early December, Venter sent off his genome report to *Science,* headquartered in Washington, D.C., while the public consortium sought out the editors of *Nature,* the chief rival to *Science.* Both papers were published two months later, the week of February 12, 2001. The timing was special for two reasons. Publication of the papers marked the centenary of a note published in the *Journal of the Horticultural Society* by Cambridge University botanist William Bateson, which accompanied the English translation of an obscure 1865 paper by Gregor Mendel that established the basic laws of heredity, only to fall into obscurity. It also fell fifty years after a precocious young American virologist named James Watson arrived in Cambridge with the naïve notion of discovering the structure of DNA.

THIS UPDATED VERSION OF *The Sequence* includes an epilogue that considers the historic accounts published in *Nature* and *Science.* Contrary to popular belief, the sequence is not yet complete. There remain thousands of gaps, inconsistencies, and plain errors that will need to be fixed over the next few years. Indeed, given the sheer repetitiveness of certain sections of the genome, which hampers DNA sequencing efforts, we may never have a truly complete version of the human genome.

For now, the hype associated with the human genome is gradually shifting to a new target, the proteome—the hundreds of thousands of discrete proteins, the products of our genes, and the targets of existing and future drugs. Identifying these proteins, determining their shapes, and elucidating their networks, is a monumental undertaking that is drawing intense interest from both the public and private sector. It will probably consume most of the twenty-first century.

Meanwhile, the quest to understand the genome sequence will last much longer than it took to obtain it. The genome sequence has not suddenly produced the cure for cancer or AIDS, but in time, that might change. Within weeks of the genome publications, scientists announced that a new drug called STI-571, better known as Gleevec, was proving extraordinarily effective in treating patients with chronic myelogenous leukaemia. This news is exciting because Gleevec was deliberately created to block the abnormal gene product that triggers the cancer. Even though we now know that some tumors acquire mutations that negate the drug's benefits, Gleevec demonstrates that the era of rational drug design, based on information gleaned from the human genome, is upon us.

Within a few years, scientists will be able to link specific pinpoint variations in your genome with a personal predisposition to cancer or heart disease or mental illness. In a decade or two, you will be able to send a cheek swab to a biotech company to produce a copy of your DNA on DVD. Most of your genome sequence will be identical to the consensus sequence reported in 2001. But about one letter in a thousand will be different, perhaps reading a G instead of a T, or a C in place of an A. All told, there are some three million of these single-letter differences between your genome and anyone else, differences that could explain why you might suffer from migraines or myopia, while your neighbour is susceptible to asthma or angina.

The sequence will yield clues to identifying, diagnosing and treating virtually every disease that affects humans. In time, it will reveal far more, such as the genetic components of human behavior and personality. We may still not know the identities of the "gay gene" or the "math gene," but scientists are discovering exciting genes linked to longevity, mental illness, and other intriguing facets of personality and behavior. Recently, for example, scientists in Oxford identified the mutated gene responsible for a profound language disorder. Another group showed how mutations in three discrete genes are co-inherited to give rise to a genetic form of obesity, illustrating how specific variations in the genome can shape susceptibility to common diseases.

Progress in understanding the sequence is closely accompanied by rapid progress in propagating and manipulating it. Remarkable advances in IVF (in vitro fertilization) can allow parents to screen embryos before implantation for a growing number of genetic traits. Fertility clinics could one day offer screening for personality traits in addition to disease genes. Even today, your genes convey information of immense interest to insurers and employers. Your chances of acquiring life insurance or health insurance or keeping a job could hinge on keeping your genetic information private. Without exception, we all carry dozens of misspellings in our genetic code that could have profound consequences for our health and disease susceptibility. Environmental factors indisputably play important roles in these disorders, but genetic factors unequivocally shape our susceptibility.

The genome also plays a vital role in understanding cloning and stem cell technology. The focus in this book is the linear sequence of three billion As, Cs, Ts and Gs—the currency of DNA—in the genome. But human development depends on the exquisitely orchestrated activation of genes to specify the hundreds of different cell types in the body. The potential therapeutic advances represented by stem cells represent a daunting medical and ethical challenge. It is the challenge of witnessing the sequence in action.

FOR ALL THE IMPORTANCE of the sequence, there is one unresolved question about the protagonists in the race for the genome that is summed up with two words: who "won"? In the past year, the principals—Venter, Collins, Sulston, and Lander—have all been feted with awards and prizes from learned societies and organizations. Whether any of these gentlemen (or their colleagues will) share the ultimate scientific honour only time will tell.

Both groups sequencing the genome greatly depended on each other: Celera made extensive use of the public sequence data to produce its genome assembly, while the public consortium only established the necessary efficiency after Venter launched his gambit back in 1998. It also relied heavily on the DNA sequencing machines supplied by Venter's sister company, Applied Biosystems.

Scientists who have had the opportunity to compare the public sequence with Celera's database say that, while there are important differences between the two groups, there is no outright winner. Celera gains credit for having assembled longer contiguous stretches of NA and supplementing its human sequence with mouse genome data, produced during the first half of 2001, which helps reveal the locations of human genes based on regions of closest similarity. The public sequence is praised for greater reliability in interpreting more complex regions of sequence, and of course for insisting on free and unlimited access to the data. In other words, both sequences complement each other.

While access to the Celera database is restricted, it is now possible, thanks to a graduate student at the University of California Santa Cruz, to view the sequence of the human genome in all its glory. Back in 2000, Jim Kent met the challenge of writing a computer program to place in order the hundreds of thousands of DNA fragments that constituted the public alliance's dataset. He finished the assembly just three days before the joint White House celebration in June 2000. But Kent did not stop there: he also developed a web browser to display the entire genome sequence via the internet (http://genome.ucsc.edu). Starting with a display of any human chromosome, it is possible, with just a few keystrokes, to navigate its length, zooming in on a region of interest until individual genes come into focus. Further magnification reveals the naked DNA sequence. The process is so effortless that it is easy to forget the years of toil, the hundreds of millions of pounds spent, and the thousands of man-years sacrificed to reach this historic goal.

Kevin Davies, Cambridge, Massachusetts, October 2001

Introduction

We are digital archives of the African Pliocene, even of Devonian seas; walking repositories of wisdom out of the old days. You could spend a lifetime reading in this ancient library and die unsated by the wonder of it.

—RICHARD DAWKINS

O_N APRIL 1, 1993, two hundred guests gathered in a ballroom at an upscale Washington, D.C., hotel for an important birthday party. As the nervous host, I hoped everything would go smoothly: I had mailed invitations around the world, hung banners from the walls, and booked the best entertainment I could find. As the guests arrived, they paid compliments on how well the baby looked. How could I disagree? After a nerve-wracking twelve months as the sole editor of a fledgling science magazine called *Nature Genetics,* it was only natural to mark its first anniversary with a party.

Nature Genetics was a spinoff from the prestigious British journal *Nature,* which, since its inception in 1869, halfway through the reign of Queen Victoria, has been "nature's finest midwife, interpreter and namesake," in the words of Stephen Jay Gould. Another commentator called *Nature* "the chic place for scientists to disport themselves." Many of the science breakthroughs reported in the newspapers or on television every week are recapitulations of reports first published in *Nature.*

I joined *Nature* in 1990, leaving behind a less-than-auspicious career as a molecular geneticist, hunting for genes that cause terrible human diseases such as cystic fibrosis and muscular dystrophy. Determined to make an impression my first day at *Nature*'s editorial headquarters, just off The Strand in London, I arrived dressed in an exquisite Italian double-breasted suit, only to discover a Dickensian office populated by a motley collection of disheveled journalists barely visible behind towering stacks of newspapers and magazines. The superficial air of civility could be shattered at any moment by an ugly fracas over the phone between an editor and the aggrieved author of a spurned manuscript. The editor, Sir John Maddox, was mostly sequestered in his office, protected by a staunch secretary and an impenetrable veil of cigarette smoke, although he would assuredly emerge late on a Monday to purchase a bottle of Bordeaux and two packets of cigarettes to help meet the weekly deadline.

As with any other magazine, *Nature* endured a few unfortunate incidents, notably the time that Maddox traveled to Paris in the company of his friend, the illusionist the Amazing Randi, to investigate the astounding claims of Jacques Benveniste that antibodies could leave a ghostly imprint in water. But despite such episodes, *Nature's* reputation remained secure. Among the thousands of reports that have graced its pages, including some of the most celebrated discoveries of the past century, one stands head and shoulders above the rest—indeed above the entire rest of the pantheon of scientific literature. In the spring of 1953, two precociously gifted scientists working in Cambridge, England, mailed a brief manuscript to the editor. As President Clinton told one of the authors almost fifty years later, the opening lines contained "one of the great understatements of all time." Then again, a certain sense of humility is in order when you have solved one of the mysteries of life. The letter began:

> We wish to suggest a structure for the salt of deoxyribose nucleic acid (D.N.A.). This structure has novel features which are of considerable biological interest.

Thus did James Watson and Francis Crick introduce the most celebrated scientific discovery of the twentieth century. The two-page letter contained just one drawing, a simple black-and-white figure for "purely diagrammatic" purposes. It was the first glimpse of the double helix, the defining scientific icon of the age, rivaled only by a sheep named Dolly.

Four decades after the discovery of the structure of DNA, human genetics research was enjoying a boom period. Researchers launched the first gene therapy trial, arranged the birth of the first genetically screened human embryo, and navigated their way across the twenty-three pairs of human chromosomes in search of genes that, when mutated, cause cancer and other diseases. A procession of exciting reports poured into the *Nature* office, and many worthy findings had to be turned down. On one occasion, Maddox questioned his biology editors for rejecting a paper reporting the mapping of the gene for an inherited form of Lou Gehrig's disease. "We must remember 'the David Niven factor,' " he said (the debonair British film actor had been another celebrity victim of motor neuron disease), because, after all, *Nature* was "in the business of selling magazines."

But the most important indication of the flourishing state of human genetics at the beginning of the 1990s was the start of an ambitious international program—the $3 billion Human Genome Project—to determine the complete instruction manual of humans by reading the precise sequence of the 3 billion chemical bases (A, C, G, and T) in human DNA. The smart editors at *Nature,* recognizing the significance of these developments, launched *Nature Genetics* in April 1992, with two noble goals in mind: first, to satisfy the prodigious output of human geneticists, and second, to sell more magazines.

THE WASHINGTON FIRST ANNIVERSARY PARTY was spread over two days, and the entertainment consisted of twenty of the most inspiring scientists that I knew of, ready to present the latest in cutting-edge research. As Maddox introduced the proceedings, I glanced at the program for reassurance. The meeting title, "Mapping the Future," was deliberately vague so that I could include speakers from all areas of genetics research. Of all those invited, only two people turned me down. Peter Goodfellow, the irrepressible chair of genetics at Cambridge University, would be missed, for he would typically dispense with slides altogether, drag a chair to the middle of the stage, and regale the audience with anecdotes. I was also sorry to lose the services of Bert Vogelstein, the brilliant Johns Hopkins University oncologist, but it was asking too much to postpone his son's bar mitzvah in Israel.

These absences notwithstanding, seated in the front rows of the audience was a veritable all-star cast. It included Berkeley geneticist Mary-Claire King talking about mapping the first breast cancer gene; an excited if exhausted Canadian, Marcy McDonald, representing the team that had just identified the mutation that causes Huntington's disease; Robin Lovell-Badge, the handsome heartthrob of my female production staff, who had identified the male "sex-determining gene"; Mark Hughes, who was helping to revolutionize genetic diagnosis; and the loquacious Ron Crystal, who would showcase the potential of gene therapy to treat diseases such as cystic fibrosis.

Just as baseball and football managers typically write in the name of their most indispensable player before all others when they pick their team, I had built the program around two figures with unrivaled drawing potential. The first was Francis Collins, a tall, lean figure from the University of Michigan, who had the distinction of publishing the first article in *Nature Genetics*—and for good reason. Collins had enjoyed a spectacular run of success since 1989. Working with a Canadian team, Collins identified the gene for cystic fibrosis, one of the most common genetic diseases among Europeans. Two years later, he isolated the gene for a cancer syndrome called neurofibromatosis (sometimes likened to the Elephant Man disease). He had also collaborated in the discovery of the Huntington's disease gene. Now he was hoping to go four for four by teaming up with King to snare the breast cancer gene—a glamorous, high-profile collaboration between two genetics superstars.

Collins's research record alone justified his prominent position in the program, as did his rare talent for public speaking. But there was one more reason, based entirely on rumors that had been swirling for the best part of a year. As Collins took the podium, he smiled and confirmed the worst-kept secret in town. He had agreed to accept an offer from NIH director Bernadine Healy to become the new chief of the Human Genome Project—the most ambitious, expensive, controversial project in the history of biology to sequence the complete human genetic code. (A formal press conference was held a few days later.) Collins would succeed James Watson, the founding director of the genome program, who had stepped down the previous year. For Collins, the opportunity to be entrusted with such a historic enterprise could not be passed up.

A successful conference needs to end on a high note, and the other name I had penciled into the agenda before anyone else was the closing speaker. J. Craig Venter had enjoyed an equally meteoric rise to the science stratosphere. While Collins was celebrating his discovery of the cystic fibrosis gene, Venter was toiling away at the NIH, a respected scientist but hardly a household name. That would change virtually overnight in the summer of 1991, when he described a revolutionary method for identifying the thousands of genes expressed in different tissues of the human body. Aided by one of the first commercially available DNA sequencing machines, Venter's laboratory found a way to produce reams of

DNA sequence data on hundreds of genes simultaneously, when other labs could study only one gene at a time. His method effectively bypassed the 95 percent of the human genome that has no known function, widely dismissed as junk DNA, to zero in on the most important DNA sequences: the genes, which carry the instructions to make the thousands of proteins in the human body. Venter had inspired a revolution in gene sequencing that brought him great fame, wealth, and no small measure of controversy. He was now the president of a nonprofit DNA sequencing institute, bankrolled by a venture capitalist to the tune of $70 million.

By the end of the conference, Collins and Venter had left nobody in any doubt that they would play dominant—perhaps *the* dominant—roles in the quest to decipher the riches inscribed in the human genetic code. Collins was effectively the leader of an international army of researchers on a quest for biology's holy grail, in the process tracking down the flaws in our DNA that cause thousands of inherited diseases. Venter's powerful new approach to DNA sequencing and gene identification would lead to the rapid identification of the majority of human genes in a few years, and perhaps provide the ideal complement to the task of sequencing 3 billion letters of DNA on twenty-three pairs of chromosomes.

Venter was already happily ensconced in his new institute, a few miles away from NIH, when Collins moved his laboratory to Bethesda. (In fact, Collins was scheduled to move temporarily into Venter's old laboratory.) What nobody knew was that these two doyens of DNA were on a collision course.

FIVE YEARS LATER, IN MAY 1998, Venter dramatically changed the course of the Human Genome Project as he informed Collins of his plans to form a new company that would sequence the entire human genome years ahead of the established deadline of 2005. He would use a simple sequencing strategy he had perfected on the genomes of bacteria, hundreds of the most sophisticated DNA sequencing machines, and one of the largest civilian supercomputers in the world. The *New York Times* trumpeted the news of Venter's brazen attempt to claim what many considered to be the human birthright, and with it the unthinkable prospect that the public human genome project was on the verge of extinction.

The Wellcome Trust, a British medical charity, quickly reassured the genetics community by doubling its funding of the Sanger Centre, the premier British DNA sequencing institute. Its goal was to guarantee production of a definitive sequence of the human genome that would stand the test of time. Collins followed suit, boosting support for the most productive sequencing centers in the United States. By all appearances, the slow march to decode the sequence of the human genome had been transformed into an epic battle between two sides with vastly different strategies and agendas. Venter's intent was for his company, Celera Genomics, to sequence the human genome years before expectation (leaving thousands of gaps if need be), to be able to patent hundreds of genes and sell precious information about the genome sequence for a gene king's ransom to the pharmaceutical industry. Collins's task was to kick-start an unwieldy federal program to keep pace with Venter's private effort and deliver the complete, gold-standard sequence years earlier than projected, all the while releasing its DNA data every night to make the human genome unpatentable.

For two years, these teams traded insults and accusations in the press while feverishly racing to sequence the human genome. The stakes were enormous; prestige and priority were on the line. Collins and his allies doggedly insisted that they were not racing with Venter, but that the increased pace of sequencing was part of their original strategy. Venter convened press conferences to mark major milestones in the human genome sequence, and at a congressional hearing in April 2000, claimed he had finished sequencing the DNA of a human being. The public sparring threatened to sully what by rights should be one of the most dignified and welcomed accomplishments in human history. "Intense competitors sometimes trade a little trash talk, and the media love it," commented Donald Kennedy, editor-in-chief of *Science* magazine. "The emphasis on the race may have the effect of obscuring the real story here, which is a magnificent scientific achievement."

Suddenly, in June 2000, the "race" was declared over, and Collins and Venter agreed to bury their differences to restore a measure of dignity to the quest for the human genome. The hastily organized victory announcement was premature in many ways: Collins's consortium had not quite reached their stated target of a "rough draft" of 90 percent of the sequence, but had made all their data publicly available. By contrast, Venter said his sequence was 99 percent complete, but only a handful of subscribers to his database were in a position to verify that claim. However, these were mere technicalities. In a June 26 ceremony at the White House, Collins and Venter stood proudly beside President Clinton as he proclaimed, "Today, we are learning the language in which God created life."

Decoding the human genome is a staggering achievement, one that has been compared favorably to every major technological achievement, from the invention of the wheel to the landing on the moon. We are the first species with the intelligence to be able to read the text of life (and as one wag put it a few years back, stupid enough to pay for it). But just what does cracking the human genome mean? How can we put this achievement into its proper perspective?

"This is just halftime for genetics," said Eric Lander, the director of the flagship American genome center at the Whitehead Institute, shortly after the White House ceremony. "It started around 1900, and the really interesting second half of the game is about to begin." The game indeed began a hundred years ago, when three plant breeders discovered the forgotten work of Gregor Mendel, a Bohemian monk. Mendel had established the fundamental rules of play by demonstrating that the inheritance of traits (in Mendel's case, the color and shape of peas) was determined by pairs of factors, later termed genes, which could be dominant or recessive. Shortly after the rediscovery of Mendel's work, Sir Archibald Garrod proposed that a disease called alkaptonuria was caused by the inheritance of a recessive gene—the first human "inborn error of metabolism."

That genes were composed of DNA was all but established in 1944, but the discovery did not catch the popular imagination until the seminal discovery of Watson and Crick in 1953. Monochrome photographs show the two young scientists staring in awe at their model of the spiraling ladder of DNA. (The Apple computer company adapted a photo of Watson for its "Think Different" advertising campaign.) The helical structure provided the secret of DNA's passage from generation to generation, whereas the rungs of the ladder, made up of four simple letters, held the key to the code of life. As Crick and others deduced a decade later, the sequence of these bases literally spells out the instructions for the synthesis of the proteins in our bodies. The 1970s gave rise to the genetic engineering revolution, as scientists devised ways to manipulate and sequence DNA and began sampling human genes.

In the mid 1980s, a group of scientists began to formulate a plan to assemble the complete sequence of all 3 billion letters of human DNA. Harvard University's Walter Gilbert, who shared the Nobel Prize for DNA sequencing, hailed the challenge as nothing less than biology's quest for the holy grail. After years of argument about the cost and wisdom of systematically procuring the sequence, the Human Genome Project finally got underway in 1990, with a scheduled completion date of 2005.

Early progress was rapid, highlighted by the identification of many genes that cause devastating diseases, including muscular dystrophy, Alzheimer's disease, and cancer. But the technology for painstakingly sequencing all 3 billion units of DNA moved more slowly. By the halfway mark of the project's 15-year mandate, only 3 percent of the human genome had been sequenced, raising doubts as to whether it would be finished on time. Venter seized this window of opportunity by coupling state-of-the-art DNA sequencing and computing technology with a daring DNA sequencing strategy. The test case was the genome of the fruit fly, one of the classic model organisms in biology, which Celera triumphantly completed in just four months in 1999. From that moment, there was little doubt that Venter would make good on his promise to sequence the human genome, years ahead of the original schedule.

REFERRING TO THE UNIVERSE, Galileo wrote, "This book is written in mathematical language and its characters are triangles, circles and other geometrical figures, without whose help . . . one wanders in vain through a dark labyrinth." As we prepare for the second half of genetics, we know virtually the entire text of the human genome, a string of 3 billion letters—about 750 megabytes of digitized information—that would fill about 5,000 books like this, and yet fit onto a single DVD.

If the first half was eventful, the second half promises to be spectacular. In the next few years, scientists aided by powerful computer algorithms will sift through the human DNA lexicon to identify all of the human genes. How many they will find it is too early to say—estimates have ranged wildly from 40,000 to over 100,000. The immediate challenge is to learn what these genes do and to divine links between the millions of pinpoint variations in our DNA sequence and our susceptibility to countless diseases. These advances will enable doctors to screen an individual's genome to produce a personalized scorecard of risks for common diseases including heart disease, diabetes, and mental illness, as well as recommending the most effective treatments for these conditions.

Within a decade or two, we may be carrying this information on our own personal DNA DVD, replete with information on our genetic susceptibility to disease and our tolerance to drugs. Clinics increasingly will be able to select genetic traits in human embryos by screening DNA before implantation and employ novel gene-based therapies to replace or repair faulty genes to cure inborn illnesses and cancer. And by the end of the game, we may know even enough about the secrets of our own genome to associate genes with elements of human character. I haven't even mentioned xenotransplantation, stem cells, and cloning.

The human genome indubitably holds the key to our future, but perhaps even more significantly, it also carries the secrets of our past. Studies of the variations in the genome sequence between humans and primates will reveal our evolutionary journey over the past 5 million years. Genome studies also shed light on the movement of populations out of Africa and across the globe over the past 100,000 years, revealing hidden truths about our identity as a population and as a species. DNA sequence variations also provide a unique molecular fingerprint of the living and the dead. Such studies have added important chapters to American and Russian political history, and DNA fingerprinting is playing an indispensable role in the legal system.

What I hope this book offers is a view of genetics as we momentarily regroup at halftime. It is the story of the people who are responsible for what is, at the very least, an extraordinary technological achievement, and is at best perhaps the defining moment in the evolution of mankind. It looks back at the highlights of the first half and looks ahead to the rest of the game. This book is not intended to be the definitive record of the politics of the genome project, nor is it an anthropological exercise designed to reveal hidden truths about the process of science. Rather, my goal has been to capture the excitement, intrigue, mystery, and majesty of the quest for biology's holy grail.

It is impossible to predict the final result of the game with so much left to play for, but there can be no doubt that this treasure of genetic information will irrevocably change our view of our place in the world. Our children will be diagnosed for diseases they have not even developed and treated with drugs that match their body chemistry. Our grandchildren may be plucked from a pool of cells bathing in a petri dish after being screened for hidden flaws in their DNA. And our great-grandchildren will have dominion over the generations to come, with the capability to engineer traits into the genetic material as easily as sewing a button on a shirt.

If the double helix is the prevailing image of the twentieth century, just as the steam engine signified the nineteenth century, then the sequence—the vast expanse of 3 billion As, Cs, Gs, and Ts—is destined to define the century to come. DNA is essentially digital information, a 3-billion-year-old Fortran code. Now that we have cracked the genome, we face the ultimate challenge of understanding what the sequence means and what it can teach us. We have the awesome potential—should we so desire—of rewriting the language of God and the responsibility of harnessing the genome to improve the human condition in an equitable and ethical manner. The childhood of the human race is about to come to an end.

Halftime is over.

CHAPTER ONE

Knights of the Double Helix

*It seems almost a miracle to me that 50 years ago we could have been
so ignorant of the genetic material and now can imagine that we will
have the complete genetic blueprint of man.*

—JAMES WATSON

IN THE MARCH 7, 1986, ISSUE of *Science* magazine, the president of the
Salk Institute, the Italian Nobel laureate Renato Dulbecco, advocated the
launch of a mammoth biology project to sequence the complete genome
of an organism so as to understand the genetic changes that lead to cancer.
Dulbecco argued that a large-scale program, rather than a piecemeal approach, was the best way to make progress in the war against cancer,
launched by President Nixon in 1971. A project to sequence a complete
genome would be important for the study of all disease and development,
not just cancer, and the obvious place to start was with the human
genome. By systematically sequencing the 3 billion letters of human
DNA, scientists would learn the identity of all 100,000 human genes, the
portions of the genetic code that carry the instructions to make the proteins of the body. Dulbecco's concluding remarks still resonate:

> Its significance would be comparable to that of the effort that
> led to the conquest of space, and it should be carried out with
> the same spirit. Even more appealing would be to make it an
> international undertaking, because the sequence of the human
> DNA is the reality of our species, and everything that happens
> in the world depends on those sequences.

Dulbecco's article was the most powerful pronouncement to that point of a groundswell of support for what another Nobelist, Harvard's Walter Gilbert, infamously called the quest for biology's "holy grail." Gilbert had made his declaration at a symposium a few months earlier at the University of California, Santa Cruz, organized by chancellor Robert Sinsheimer, which was held to consider the construction of a DNA sequencing institute. Recapping the meeting, Gilbert wrote, "The total human sequence is the grail of human genetics . . . an incomparable tool for the investigation of every aspect of human function."

Many branches of modern science have adopted the grail as their metaphor of choice. For physicists, the grail is the identification of one of the fundamental building blocks of matter. Biophysicists consider the grail to be the prediction of the complete three-dimensional structure of a protein from its one-dimensional chain of building blocks. And chemists invoke the sacred chalice in the quest to create a highly stable artificial element.

The grail has also become a popular, radiant symbol for the Human Genome Project, much to the bemusement of critics such as Harvard geneticist Richard Lewontin, who observed in 1992, "It is a sure sign of their alienation from revealed religion that a scientific community with a high concentration of Eastern European Jews and atheists has chosen for its central metaphor the most mystery-laden object of medieval Christianity." Lewontin castigated those who fetishize DNA and what he called the "evangelical enthusiasm of the modern Knights Templar and the innocence of the journalistic acolytes whom they have catechized." In Lewontin's opinion, this has resulted in the metamorphosis of DNA from an inert molecule to the blueprint of biological determinism, the manual of humans. Indeed, *human blueprint* is a common, if slightly more mundane, metaphor for the genome project, but it has its share of critics too, such as Ian Stewart, author of *Life's Other Secret:* "Genes are not like engineering blueprints, they are more like recipes in a cookbook. They tell us what ingredients to use, in what quantities, and in what order—but they do not provide a complete, accurate plan of the final result."

Matt Ridley, in his book *Genome,* also refuses to sanction the term *blueprint* to describe the human genome, in part because a two-dimensional map is a poor analogy for a one-dimensional digital code. Rather, the genome is "an immense book, a recipe of extravagant length."

• • •

PERHAPS THE BEST METAPHOR for the genome project, as suggested by the Whitehead Institute's Eric Lander, one of the leading figures in the Human Genome Project, is the prime organizer of the chemical elements: the periodic table. The principal cartographer of the chemical kingdom was Dmitry Ivanovitch Mendeleyev, a brilliant Russian chemist born the youngest of fourteen impoverished children in Siberia. As a youth, Mendeleyev was obsessed with understanding the physical properties of the elements. By his early thirties, the tall, stooped man with a stark resemblance to Rasputin was professor of inorganic chemistry at the University of St. Petersburg, determined to understand the atomic relationships among groups of elements. Some progress had already been made. English chemist John Newlands, for example, had noticed that different elements could be classified into eight discrete groups based on their chemical and physical properties, with the first resembling the eighth, the second the ninth, and so on.

Newlands's "Law of Octaves" was scornfully received—one contemporary suggested he would have more luck organizing the elements alphabetically—but Mendeleyev had noticed a similar pattern. A devotee of solitaire, Mendeleyev wrote out the names and properties of the sixty-one known elements on white cards. These he would arrange in rows and columns based on the elements' atomic weight, the results of his own experiments, calculations, letters from colleagues, and a voracious appetite for publications in five different languages. One night in February 1869, Mendeleyev had a dream in which he saw the alignment of the known elements in a single table.

In March 1869, Mendeleyev's paper, "The Dependence Between the Properties and the Atomic Weights of the Elements," was presented at a meeting of the Russian Chemical Society, and a short time later, he traveled to England to deliver the prestigious Faraday Lecture to the Chemical Society. After receiving his honorarium in a small silk purse decorated in the Russian national colors, Mendeleyev tumbled the sovereigns onto the table, saying, "I cannot contemplate accepting payment for a work of love, in the hallways made sacred by the memory of Michael Faraday." (The society sent a pair of engraved vases to Russia instead.)

Although Mendeleyev was guided by the known atomic weights of the elements, he possessed the confidence to digress when his instincts suggested an alternative order of elements. More impressive still, he left gaps in his table where he sensed that new elements lurked undiscovered. Sure enough, three elements—gallium, germanium, and scandium—

were later identified with exactly the properties Mendeleyev had foreseen, convincing his peers that his system was a genuine picture of chemical nature.

In 1955, almost fifty years after his death, Mendeleyev received the ultimate honor for a chemist (arguably greater than the Nobel Prize) when scientists artificially created the 101st element of the periodic table and named the short-lived radioactive isotope mendelevium. The Russian thus joined the select company of Albert Einstein, Alfred Nobel, and Enrico Fermi to have an element named after him.

Mendeleyev's periodic table has become the most important icon in chemistry, and arguably the most important tool of the industrial age. Perhaps it is not unreasonable, then, to seek comparisons with the foundation of biology, for after all, as the Oxford University chemist R. J. P. Williams put it, "Biology is the search for the chemistry that works." Lander has eloquently written of the comparison: "The Human Genome Project aims to produce biology's periodic table—not 100 elements, but 100,000 genes; not a rectangle reflecting electron valences, but a tree structure, depicting ancestral and functional affinities among the human genes."

The complete human genetic sequence will reveal the fundamental properties of all human genes, allowing their functions and interactions to be integrated into a miraculously complete picture of human biology and evolution. In the same way as the building blocks of chemistry were rendered finite 130 years ago thanks to the work of Mendeleyev, Newlands, and others, biology too is on the verge of becoming finite. Just as chemists can recognize atoms by their distinctive mass and charge, biologists will use gene chips and other new technologies to recognize each gene from thousands of alternatives.

But as Francis Crick has pointed out, there is one potentially important difference. The principles enshrined in the periodic table are truly universal, signifying the invariant properties of chemical elements dispersed throughout the universe. But if life exists on other planets, there is little reason to believe that the genetic code adheres to the same pattern as it does on earth, for chance played a major part in the origin of life as we know it.

Comparing the sequence to the periodic table of elements is an elegant analogy but is of limited everyday relevance. As mundane as it is, the human genome is essentially one huge parts list. A Boeing 777 contains about 100,000 parts, but as Eric Lander points out, "having a parts list doesn't tell you how to put it together." The sequence of the human genome will eventually reveal the identity of 50,000–100,000 key components of the human genome and provide an enormous advance in the practice of medicine. However, identifying all the genes will not, in and of themselves, explain how the human mind and body work.

DULBECCO'S 1986 commentary in *Science* catapulted the concept of a Human Genome Project into the scientific mainstream, but he was not the first to hit on the idea. Charles DeLisi, director of the Office of Health and Environmental Research at the Department of Energy, had been exploring the feasibility of such a project for six months. The Department of Energy had a long-standing interest in the effects of radiation on mutation rates. One major project was to investigate the *hibakashu,* the Japanese survivors of the atomic bombs dropped at the end of World War II, for possible increased rates of mutation frequency and birth defects.

In March 1986, just a few days before the publication of Dulbecco's commentary, DeLisi hosted a small workshop in Santa Fe to discuss the idea of sequencing the human genome under the auspices of the Department of Energy. Most of the participants agreed that sequencing the human genome was feasible, but that it should be done only after a physical map of the genome had been assembled. Some questioned the Department of Energy's qualifications to run such an operation. The garrulous David Botstein, the Stanford University geneticist who first proposed a genetic map of human DNA in 1980, chastised the initiative as "DOE's program for unemployed bomb-makers."

Two months after Dulbecco's controversial call for a "Big Science" genome project, legions of DNA dignitaries gathered at the Cold Spring Harbor Laboratory on Long Island, New York, for a meeting entitled "The Molecular Biology of *Homo sapiens,*" organized by the laboratory's director, James Watson. Rumors were spreading by now that the Department of Energy was contemplating a program to sequence the human genome, and a special session at the meeting, chaired by Nobel laureate Paul Berg, was devoted to the idea.

The most ardent supporter of Dulbecco's proposal was fellow Nobelist Walter Gilbert, who proposed a "blind sequencing" project, randomly cutting the genome, sequencing the DNA fragments, and reassembling them. Gilbert provocatively wrote the estimated cost on the blackboard: $3,000,000,000. His vision of sequencing the entire human genome drew an openly hostile response, particularly among younger re-

searchers. Even when defrayed over fifteen years, the cost of $3 billion—
$1 for every letter of DNA—would necessarily deny funds for many
other worthy projects. Indeed, during the late 1980s, the proportion of
grants funded by the National Institutes of Health (NIH) fell from 40 percent to less than 25 percent.

Gilbert was so convinced of the opportunity for a private venture to
sequence the human genome that he crafted plans to build his own company, to be called the Genome Corporation. His idea was to have hundreds of scientists housed under one roof dedicated to sequencing the
entire human genome over ten years for about $300 million. The se-
quence would then be released for a price. In retrospect, his plans were
not so different from Craig Venter's a decade later. However, experts
blasted Gilbert's planned privatization of the human genome as obscene
and in violation of the traditional spirit of cooperation that underlies scientific research. For reasons that included a difficult tenure as the chief executive at Biogen and the stock market collapse of 1987, Gilbert was
unable to raise sufficient venture capital, and his dream of Genome Corp.
collapsed. Gilbert had much more success a few years later, when he cofounded Myriad Genetics, the Salt Lake City company that identified the
first breast cancer gene, *BRCA1,* in 1994.

A more philosophical argument against the systematic sequencing of
the human genome went like this: What was the purpose of methodically
sequencing all twenty-three pairs of chromosomes in their entirety when
only 5 percent or so of human DNA actually codes for genes? Many of
the distinguished scientists at the Cold Spring Harbor meeting ridiculed
Gilbert's idea of sequencing every morsel of DNA, preferring instead a
more directed effort to identify and sequence the genes themselves. The
South African Sydney Brenner, one of the founding fathers of the modern genetics era, had argued from the beginning that DNA sequencing
should focus first on the small percentage of the genome that contains
genes, the far less interesting "junk DNA" later. He stressed that he was
not against the project in principle. In characteristically humorous vein,
he wrote:

> I am not one who believes that mapping and sequencing the
> human genome is a boring, thankless task, suitable perhaps only
> for a penal colony where transgressing molecular biologists
> might serve sentences of up to 20 [million] bases. On the con-
> trary, I think that it is the most important, the most interesting

and the most challenging scientific project that we have, and that it will come to attract the best minds in biological research.

During 1987, a special National Research Council panel, chaired by Bruce Alberts, deliberated on the wisdom of the Human Genome Project. Its final report advocated an international program led by the United States to sequence human DNA, and it laid out guidelines for how the project should be organized. With the cost of DNA sequencing still obscenely high (several dollars per base), the panel recommended postponing DNA sequencing until improved technology had driven down the cost—probably in five years. Instead, early efforts should focus on mapping the human genome and characterizing the genomes of other organisms, such as the mouse, the fruit fly *Drosophila melanogaster,* as well as certain species of yeast and bacteria, which would be indispensable for interpreting the function of human genes. Improving DNA sequencing technology should also be given priority. The committee recommended government funding of $200 million per year for up to fifteen years, thus endorsing Gilbert's earlier estimate. About the only thing it refused to voice an opinion on was whether the genome project should be administered by the Department of Energy or the NIH.

By this time, the Department of Energy was pushing ahead with its own genome initiative. The plan survived a congressional hearing in March 1987, and DeLisi submitted his first budget, for $12 million, in 1988. But James Wyngaarden, the director of the NIH, was also seeking to boost NIH genome programs, telling Congress that $50 million would be required for a legitimate genome program. He proposed establishing a new Office of Human Genome Research, which would eventually be given authority to distribute its own grants. Watson urged the NIH director to nominate an active scientist who "would simultaneously reassure Congress, the general public, and the scientific community that scientific reasoning, not the pork barrel, would be the fundamental principle in allocating the soon-to-be-large genome monies."

It was sound advice, but to many observers, there was only one person who measured up to Watson's imperial ideals: Watson himself.

IN 1928, THE YEAR THAT JAMES DEWEY WATSON was born in Chicago, a painfully shy British Ministry of Health researcher published the results of an experiment that "no sane person would have contemplated." But this study would have a direct bearing on the infant Watson's crowning achievement a mere twenty-five years later. Fred Griffith was studying the virulence of *Streptococcus pneumoniae* bacteria, which he had shown could exist in two very different forms: the virulent form had a smooth appearance, whereas a nonvirulent form was called rough. For his classic experiment, he injected both forms into mice: live nonvirulent rough cells and dead, smooth, virulent bacteria. Within two days, many of the mice were dead, and from the animals' blood, Griffith recovered live bacteria with the rough appearance. Something had transformed the nonvirulent bacteria into the virulent strain, and the short odds were on protein, because DNA was too monotonous and carbohydrate too implausible.

Griffith's results caught the attention of Oswald Avery, a well-known immunologist at Rockefeller University in New York, who spent the better part of ten years trying to identify the substance responsible. In a heroic series of experiments, Avery and first Colin MacLeod, then Maclyn McCarty, cultured gallons of infected cells and, by a process of elimination, characterized the "transforming principle." A battery of chemical, enzymological, and immunological tests all indicated the unthinkable: the transforming principle was not protein, but a substance that had been virtually ignored since Swiss biochemist Johann Miescher first isolated a "nuclein" from the pus of discarded surgical bandages seventy-five years earlier: deoxyribonucleic acid.

In the autumn of 1943, just as the fifteen-year-old Watson was enrolling at the University of Chicago, Avery wrote a paper describing his results and on November 1, hand-delivered the manuscript to Peyton Rous, the esteemed editor of the *Journal of Experimental Medicine,* published by the Rockefeller University Press. Avery's exhaustive report all but proved that genes were made of DNA, but his circumspect conclusion drained the excitement:

> The inducing substance, on the basis of its chemical and physical properties, appears to be a highly polymerized and viscous form of sodium desoxyribonucleate [DNA]. . . . If the results of the present study are confirmed, then nucleic acids must be regarded as possessing biological specificity the chemical basis of which is as yet undetermined.

Only once did Avery allow himself the indulgence of unbridled speculation, when he included a line written by J. B. Leathes in a 1926 paper in *Science*. Leathes had shown that chromosomes were composed of almost equal amounts of protein and nucleic acid, causing him to question the conventional wisdom that the genetic material must be protein. It was a chilling quote: "[If the chromosomes] are packed from the beginning all that preordains, if not our fate and fortunes, at least our bodily characteristics down to the color of our eyelashes, it becomes a question whether the virtues of the nucleic acids may not rival those of amino acid chains in their importance." But when Rous returned the edited manuscript to Avery two weeks later with his hand-written comments, this passage had been crossed out for insufficient evidence. In less formal settings, however, Avery was less restrained. In a letter written to his brother, a bacteriologist, he excitedly described his group's breakthrough in identifying the material that instilled the nonvirulent bacteria with the "aristocratic distinctions" of the smooth, virulent variety, tempered with the instinctive need to rule out all other explanations. "If we are right," Avery wrote,

> then it means that nucleic acids are not merely structurally important but functionally active substances in determining the biochemical activities and specific characteristics of cells—and that by means of a known chemical substance it is possible to induce *predictable* and *hereditary* changes in cells. . . . It's lots of fun to blow bubbles—but it's wiser to prick them yourself before someone else tries to.

Shortly before the paper was published, Avery gave a valedictory seminar in which he concluded: "The evidence presented supports the belief that [DNA] is the fundamental unit of the transforming principle of pneumococcus type III." McCarty recalled there was a thunderous round of applause, followed by a deafening silence as the audience grappled with the implications of the work.

The paper by Avery, McCarty, and MacLeod is rightly considered a classic of the twentieth century (although sadly Griffith was not alive to appreciate it, having been killed in an air raid during the London blitz in 1941). The Nobel laureate Sir Peter Medawar hailed the paper as "the most interesting and portentous biological experiment of the twentieth century." But at the time, the paper was widely unappreciated and ignored. Avery's choice of journal was popular among the immunology crowd, but was not widely read by geneticists and general biologists, who were skeptical about the applicability of work on bacteria. The timing too was poor. The paper was published in February 1944, near the height of American involvement in World War II, so it was seen by only a fraction of the journal's normal American readership and none working abroad. (Word slowly spread after the war, such that Avery had to order three hundred extra reprints.)

Another reason was the influential criticism of fellow Rockefeller University professor Alfred Mirsky, who was convinced that Avery's preparations were contaminated with traces of residual protein and that this might still be the transforming substance. McCarty believed that Mirsky's criticism so swayed the Nobel Prize Committee in Sweden that Avery's nomination was postponed. Widespread acceptance of Avery's finding did not come until 1952, when Al Hershey and Martha Chase showed that genetic information was carried by viral DNA, not the protein coat. (The key experiment involved shaking off the viral shells using a kitchen blender and showing that this did not diminish infection.) However, Avery's death in 1955 denied him the chance of a belated Nobel Prize.

Among those most influenced by Avery's work was Erwin Chargaff, an Austrian biochemist who moved to the United States in 1928. Together with Ernst Vischer, he began studying the chemical composition of DNA from various sources, crudely separating and quantifying the four constituent bases: adenine (A), cytosine (C), guanine (G), and thymine (T). In 1949, they disproved the idea that there were equal amounts of all four bases. But Chargaff noticed something else—a "striking, but perhaps meaningless" trend. Regardless of the source of the DNA, the amount of A roughly corresponded to T, as did the quantities of G and C. Chargaff suspected that these 1:1 ratios might merely be a coincidence, but their profound significance was about to become apparent.

NO OTHER ACTOR PLAYS THE ROLE of the brash, opinionated, self-absorbed scientist quite like Jeff Goldblum. He has delivered variations on this theme as teleportation scientist Seth Brundle in *The Fly*, chaos expert Ian Malcolm in *Jurassic Park* and *The Lost World*, and computer geek David Levinson in *Independence Day*. Less well known is his virtuoso perform-ance as James Watson in the 1987 BBC docudrama, *The Race for the Dou-ble Helix* (also known as *Life Story*), about the dramatic events leading up to the discovery of the structure of DNA. (The BBC, perhaps wisely, ignored suggestions that Watson be played by Woody Allen and Crick by Peter O'Toole.)

In one of my favorite scenes, set in November 1951, a gumsnapping, thoroughly disinterested Watson is seated at the back of a freezing, sparsely populated auditorium at King's College in London while Rosalind Franklin, a talented crystallographer, delivers a formal lecture about her DNA crystals. After momentarily fantasizing about how Rosy (as Watson referred to her) would look if she dispensed with her glasses and fixed her hair, Watson loses interest in the talk and brazenly spreads the *Times* on the seat next to him. Suddenly his attention is captured by a dramatic slide showing the pattern of X rays diffracted through a DNA crystal. Watson is a novice in crystallography, but he knows enough to realize that the image is consistent with a helical structure. Back in Cambridge, he tries to recall the numbers Franklin presented for "unit cells" and water content, upon which he and Crick would build a three-chain model of DNA. But when Franklin travels up to Cambridge with her boss, Maurice Wilkins, to view the model, her reaction is one of ridicule. It turns out that Watson hopelessly underestimated the water content in her crystals by a factor of ten. The model was a fiasco.

In Crick's view, Goldblum's characterization of Watson was too "manic," and the constant chewing gum was a predictable American stereotype. But Goldblum effortlessly captured the way the scrawny, precocious American stood out among the British old boys. The twenty-two-year-old Dr. Watson had arrived at the Cavendish Laboratory in Cambridge in September 1951. One eminent geneticist described Watson as "tall, gawky, scraggly . . . shirttails flying, knees in the air, socks down around his ankles . . . his eyes always bulging, his mouth always open . . . a surprising mixture of awkwardness and shrewdness." Watson was twelve years younger than Crick (who had still not earned his Ph.D.), and unlike his smartly dressed, well-spoken colleague, Watson walked around like a tramp, given to muttering punctuated by characteristic snorts (a curious mannerism in effect to this day). But they complemented each other perfectly: Crick was knowledgeable in physics and crystallography; Watson was consumed with finding the properties of the gene. Both shared the conviction that DNA was more important than protein, and as their colleague Max Perutz observed, "they shared the sublime arrogance of men who had rarely met their intellectual equals."

After the model fiasco, Crick was forbidden from working on DNA and told to concentrate on working for his doctorate. They met Chargaff, who contemptuously dismissed them as "two pitchmen in search of a helix." At the beginning of 1953, Crick and Watson feared that their dreams of success had been thwarted by the California chemist Linus Pauling, who had sent details of his own DNA model to his son Peter in Cambridge. Pauling went on to win the Nobel Prize in chemistry and in peace, but his chance of a third was gone. Incredibly, Pauling's model, with the bases on the outside, bore a vague resemblance to Crick and Watson's own botched attempt. They promptly adjourned to their local pub, the Eagle, and toasted Pauling's mistakes. "Though the odds still appeared against us," Watson wrote, "Linus had not yet won his Nobel."

A few days later, while visiting Wilkins to share the news of Pauling's blunder, Watson was shown a new X-ray photograph that Franklin had taken of DNA—Photograph 51. The X rays formed the shape of a dark cross, which Watson recognized as the signature of a helical molecule. The moment he saw it, his jaw dropped, his mind racing to consider the implications, none more so than the possibility that DNA might have only two chains, not three. The journey home to Cambridge convinced him this was so: "Thus by the time I had cycled back to the gate, I had decided to build two-chain models. Francis would have to agree. Even though he was a physicist, he knew that important biological objects come in pairs."

PHOTOGRAPH 51 WAS A CRUCIAL EVENT in the determination of the structure of DNA, but by itself it was not enough. Watson finally gleaned the precise details of Franklin's DNA crystals—which he had misheard fourteen months earlier—from an unpublished report. Crick then suggested he arrange two intertwining backbones on the outside of the helix, but Watson could not ascertain the arrangement of the bases inside. Impatiently, he cut out cardboard models of the bases, using a configuration of each molecule suggested by his American colleague Jerry Donahue. As he arranged them in different permutations, like a child playing with a jigsaw puzzle, he saw that A paired neatly with T, just as C bonded with G. Each pair could be linked by two weak chemical bonds and took up about the same space. In a single stroke, Watson had solved the mystery of Chargaff's 1:1 ratios (A+G = T+C). The bases fit snugly inside the twisting metal backbones of the double helix model and raised the tantalizing prospect that the sequence of one strand could form a template for the synthesis of a new strand to replicate the genetic material. It is no wonder, then, that at lunchtime, Crick "winged into the Eagle to tell everyone within hearing distance that we had found the secret of life." It was the last day of February 1953. Among those who made the pilgrimage to Cambridge to view the double helix was a young graduate student at Oxford named Sydney Brenner, who recalled that it was "the most exciting day of my life . . . a revelatory experience."

The final version of Watson and Crick's 900-word paper announcing their DNA model was typed by Watson's sister Elizabeth, persuaded by the brotherly advice that "she was participating in perhaps the most famous event in biology since Darwin's book." The order of the authors' names on the greatest scientific discovery of the century was decided, of all things, by a coin toss.

In a scanty two pages in the April 25, 1953, issue of *Nature,* Watson and Crick launched the era of molecular biology, almost 500 years to the day that the fall of Constantinople to the Turks marked the dawn of the Renaissance. The brief text—so brief they failed to mention Avery's seminal 1944 paper—was accompanied by a simple sketch of the double helix. In the following pages were two related reports, one from Wilkins, the other from Franklin, presenting their X-ray diffraction patterns that supported the helix model, including Photograph 51. Emboldened by these reports, Crick and Watson immediately wrote a second, more elaborate analysis of the double helix, which appeared in *Nature* five weeks later. They noted that "any sequence of the pairs of bases can fit into the structure . . . the precise sequence of the bases is the code which carries the genetical information." The basis of mutations, they suggested, could be an alteration in the form or order of the sequence of bases.

Fifty years on, it is surprising to learn that Watson and Crick's prescient discovery received a frosty reception from many in the establishment, notably Chargaff, who resented the way they had grasped the significance of his own measurements. But the more enlightened recognized that Watson and Crick had made what Peter Medawar called "the greatest achievement of science in the twentieth century." Among those inspired by the double helix was the flamboyant Spanish artist Salvador Dalí, who said "this for me is the real proof of the existence of God." Dalí gave one of his paintings, *Galacidalacidesoxiribunucleicacid,* the subtitle

Homage to Crick and Watson. It depicts three parts of existence—life, death, and the afterlife—represented by the DNA, a cubic molecule, and the figure of God reaching down to resurrect the spirit of Christ, respectively.

WHEN WATSON WAS ASKED RECENTLY what his greatest accomplishment was, his answer was not, as one would have expected, the discovery of the double helix, but the authorship of *The Double Helix,* his shockingly candid 1968 account of the race to discover the structure of DNA. Citing objections from, among others, Crick and Wilkins, the president of Harvard University (where Watson was a professor) ordered Harvard University Press not to publish the book (which Watson had wanted to call *Honest Jim).The Double Helix* was eventually published by Atheneum in 1968 and became an instant best-seller.

Reaction to the Pepys diary of modern science was bitterly divided. Many readers relished the unbridled honesty and wicked humor with which Watson laid bare his passions (tennis, girls), dislikes (English food, weather), and ambitions (the Nobel Prize). He revealed scientists to be vain, arrogant, and, above all, human. (Crick took exception to this, saying that he had never been in doubt on the matter.) Two best-selling authors warmly welcomed Watson into their ranks: Jacob Bronowski *(The Ascent of Man)* suggested Watson's writing had "a quality of innocence and absurdity that children have when they tell a fairy story." Alex Comfort *(The Joy of Sex)* likened Watson to Spike Milligan from *The Goons,* and so admired his panache for storytelling that "we could do worse than give him a second Nobel gong for literature."

But many scientists were appalled by Watson's naked ambitions, not to mention his supercilious attitude to his peers and predecessors. The worst offense in the minds of many was his disdain for Rosalind Franklin, who tragically died of cancer in 1958 at the age of thirty-seven. Franklin was thus denied a share of the Nobel Prize in Medicine that was awarded to Watson, Crick, and Wilkins in 1962. The author Brenda Maddox calls Franklin "the Sylvia Plath of molecular biology, a genius whose gifts were sacrificed to the greater glory of the male." Her reputation has been restored, thanks in part to Watson's generous testimonial in the epilogue to *The Double Helix*. In March 2000, King's College, London, dedicated the Franklin-Wilkins building, opened by Princess Anne.

To Watson's initial surprise, Franklin readily accepted the double helix when she saw it, because it fit with her own growing suspicions (from her own experiments) that DNA had to be helical and that the bases somehow complemented each other in pairs. Indeed, there is good reason to believe that had Watson and Crick not elucidated the structure, Franklin or Wilkins or Pauling would have succeeded within a year or two. Of course, it is inconceivable that a more protracted route to the discovery of the structure of DNA could have matched the drama and impact of the events that led up to the Watson-Crick model and the iconic status of DNA in the past half-century.

It took twenty years for the Watson-Crick structure of the double helix to be visualized directly. In 1973, Alexander Rich, a structural biologist at MIT, produced atomic-resolution crystal structures of DNA. After seeing the pictures, Watson phoned Rich to thank him, because "I've just had the first good night's sleep in 20 years."

IN THE AFTERMATH of discovering the double helix, Crick and Watson were inundated with requests to perform various services, so much so that Crick drafted a form letter to handle the deluge:

Dr Crick thanks you for your letter but regrets that he is unable to accept your kind invitation to:

send an autograph	speak after dinner	attend a conference
provide a photograph	give a testimonial	act as chairman
cure your disease	help you in your project	become an editor
be interviewed	read your manuscript	write a book
talk on the radio	deliver a lecture	accept an honorary degree
appear on TV		

Crick was not trying to be antisocial, for there was still the small matter of finishing his Ph.D., which he was finally awarded in 1954 for his thesis, "X-ray Diffraction: Polypeptides and Proteins." He was also preoccupied with a new challenge: to break the genetic code, the molecular lexicon that somehow interprets and translates a humdrum string of four bases in

DNA into the twenty different amino acid building blocks of proteins. The double helix was the final proof that DNA was the genetic material and showed how these instructions could be copied from cell to cell, but it left scientists none the wiser as it produced the great diversity of proteins that make up the human body.

Two years before Crick and Watson's breakthrough, the Cambridge University chemist Fred Sanger had described a unique sequence of amino acids in part of insulin, a finding that demanded the existence of a heritable code that could exist only in DNA. The only variable portion of the double helix was the order of four different bases stacked in pairs like rungs of a ladder inside the two twisting backbones of the helix, ten rungs for each complete revolution of the backbone. Somehow, combinations of these bases had to distinguish twenty different kinds of amino acids, but if the code consisted of different pairs of four possible letters, there could only be 16 (4^2) possible combinations. That suggested that the code consisted of triplets, where there could be 64 (4^3) different combinations, more than sufficient to encode the diversity of amino acids.

A couple of months after the revelation of the double helix, Crick and Watson received a letter from the irrepressible Russian-born cosmologist George Gamow. The name was familiar to them. In 1948, Gamow had written a famous paper with his student, Ralph Alpher, predicting that traces of the incandescent origin of the universe should be detectable as background radiation, tangible evidence of the Big Bang. Typical of his puckish sense of humor, he persuaded another physicist, Hans Berthe, to add his name to the paper, so the list of authors read "Alpher, Berthe and Gamow," a riff on alpha, beta, and gamma, the first three letters of the Greek alphabet. Such radiation was found two decades later, although Gamow, wrongfully perhaps, was not granted a share of the ensuing Nobel Prize.

Gamow's letter contained word of a radical solution for the genetic code: he suggested that the bases of DNA formed a series of diamond-shaped spaces, into which specific amino acids could be inserted to form proteins. "The sequence of bases determines in a unique way the sequence of diamonds," Gamow wrote in a *Nature* paper in February 1954, and by his calculations, the number of possible permutations was exactly twenty, corresponding to the number of amino acids. The idea was for an overlapping code. In the sequence GCAT, for example, GCA would code for one amino acid, CAT for the next, and so on, which would be physically assembled on the surface of the double helix. However, Sydney

Brenner perceptively showed that the overlapping scheme could not account for the highly variable assortment of protein building blocks evident in the early sequences of Sanger and others.

Crick also had a more deep-seated reason to discount Gamow's suggestion. It was an instinctive hunch that predated even the double helix, which he called "The Sequence Hypothesis." He later wrote that "the specificity of a piece of nucleic acid is expressed solely by the sequence of its bases . . . this sequence is a (simple) code for the amino acid sequence of a particular protein." In other words, Crick believed that genes contained all the information necessary to specify the building of a protein.

In place of Gamow's overlapping code, Crick made two extraordinarily brilliant predictions about the genetic code; unfortunately, only one of them proved to be correct. The first stemmed from Crick's conviction that amino acids did not physically attach to DNA. Instead, he suggested that they must be linked to adapters that ferry the building blocks to the template, which was later shown to be correct.

His other scheme, even more exquisite, was dubbed the "comma-free code." The idea was that for every triplet of bases, only one permutation of letters could encode an amino acid. For example, if the triplet CAT was meaningful, then ATC and TCA were not. In this way, there was no need to invoke a type of comma separating each triplet, telling the cell where a gene began. Intriguingly, this scheme allowed for precisely twenty amino acids. "It seemed so pretty, almost elegant," Crick later reminisced. "You fed in the magic numbers 4 (the 4 bases) and 3 (the triplets) and out came the magic number 20, the number of amino acids."

Crick eventually published a formal account of his comma-free code in the *Proceedings of the National Academy of Sciences* in 1957. But it turned out that Crick had outsmarted himself. His theory proved so spectacularly flawed that the great historian of molecular biology, Horace Judson, praised it as "the most elegant biological theory ever to be proposed and proved wrong."

Around this time, Brenner and others recognized that the instructions in the genetic code were carried out of the cell nucleus by a transient strand of ribonucleic acid (RNA), a faithful facsimile of one strand of DNA, appropriately named messenger RNA. The identification of the RNA intermediary provided the key to solving the puzzle of the genetic code. The locksmith was a young, relatively unknown NIH scientist named Marshall Nirenberg, who devised a method of synthesizing trace amounts of protein in a test tube by adding artificial RNA messages with a known base sequence. Nirenberg's first synthetic RNA, made entirely of a base called uracil (U) that is used in RNA wherever T is found in DNA, yielded a protein composed entirely of one amino acid, phenylalanine. It was solid proof that the UUU triplet coded for phenylalanine. Nirenberg first presented his results in August 1961 at a sparsely attended session at the International Congress of Biochemistry in Moscow. Word of his breakthrough reached Crick, who invited Nirenberg to repeat his talk before hundreds of delegates in the plenary session at the same meeting. Crick considered it an epoch-making discovery, one for which Nirenberg shared the 1968 Nobel Prize in physiology or medicine.

In a stunning series of bacterial experiments, Crick and Brenner proved that the code was written in triplets. They generated mutations that added or removed one base at a time, thereby knocking the sequence of a gene out of phase, until the addition or removal of three bases restored the phase of the code. Although it was not as famous as Crick and Watson's 1953 masterpiece, Sir John Maddox considered it possibly "the most elegant paper *Nature* ever published." By the end of 1966, the entire genetic code had been cracked: sixty-four possible triplets of a four-letter alphabet held the instructions for a mere twenty amino acids, as well as the signals to start and stop protein synthesis.

In 1976, Crick took up a professorship at the Salk Institute in California where, at the age of sixty, he embarked on another major challenge in biology: human consciousness. By his own admission a clumsy experimentalist, Crick's contributions have been largely theoretical, but he has characteristically left his mark on the field, publishing influential papers on memory and other aspects of cognitive science.

The same year as the publication of *The Double Helix,* Watson left Harvard to become director of the Cold Spring Harbor Laboratory, rapidly turning it into one of the country's leading research centers. "Honest Jim" thrives in the rarified surroundings of the north shore of Long Island, presiding over important scientific meetings and fraternizing with New York high society, and living up to a description of him as an "impresario of molecular biology." A couple of years ago, Watson obligingly took me on a tour of the surroundings in his trusty Volvo station wagon, gleefully pointing out the palatial homes of his wealthy neighbors, such as Charles Wang of Computer Associates. Prominently on display in his custom-built home on the fringe of Cold Spring Harbor are countless mementos of his remarkable career. But on this day, the fate of the Human Genome Project takes a back seat to a more pressing problem: the height of the hedgerow that is partially obstructing the priceless view of the harbor from the house.

IN MAY 1988, THE NIH DIRECTOR, James Wyngaarden, offered Watson the position of associate director for human genome research. Despite his commitments at Cold Spring Harbor, Watson had few doubts about what he should do. "Only once," he later recalled, "would I have the opportunity to let my scientific life encompass the path from double helix to the 3 billion steps of the human genome." It was time to address the sequence of DNA itself.

Watson skillfully deflected scientific criticism of the project while searching for political support. One of his first and most important decisions—a spontaneous announcement during a press conference—was to devote 5 percent of the budget toward the study of the ethical, legal, and social implications of the genome project. It was a sincere effort to ensure that society was prepared for the tidal wave of information on the horizon. The prospect of rapid advances in the understanding of major genetic diseases also raised serious issues regarding genetic discrimination, inadequate treatment options, and eugenics. Recalling the Nazi atrocities against Jews, gypsies, and the mentally ill, Watson solemnly wrote: "We need no more vivid reminders that science in the wrong hands can do incalculable harm."

In October 1989, Watson's unit received its new status as the National Center for Human Genome Research and a budget for fiscal year 1990 of $60 million, which was roughly twice that of the Department of Energy component. The genome project officially kicked off in October 1990, but the cost of the fledgling program remained a sensitive issue. When *Nature* erroneously reported in 1991 that the Human Genome Project was in line for a $334 million increase, Watson wrote back sarcastically: "The uninformed readers could get a false and exaggerated impression about the size of the US budget for genome research from your article, creating unnecessary concern."

Although Watson's peerless stature and the promise of exciting medical breakthroughs that would inevitably spin off the genome project helped garner support for the program, many scientists still harbored serious reservations. Leading the opposition was Bernard D. Davis, a geneticist at the Harvard Medical School. In July 1990, Davis and two dozen fellow faculty members denounced the "politically unstoppable" initiative, arguing that just as the known DNA sequences of a few viruses had not had a profound effect on understanding viral biology, neither would the complete sequence of the human genome transform human biology. "The magnification is wrong," Davis complained, "like viewing a painting through a microscope." He concluded: "Our fundamental goal is to understand the human genome and its products, and not to sequence the genome because it is there." Similar concerns were voiced by British geneticist and author Steve Jones, who felt that piecing together the sequence would put man "in the position of a nonmusician faced with the score of Wagner's *Ring* cycle: many pieces of information, apparently making no sense at all, but in fact containing an amazing tale—if only we knew what it meant."

Other scientists were genuinely worried about the potential misuse of genetic information, such as Watson's former Ph.D adviser, Salvador Luria:

> Will the Nazi program to eradicate Jewish or otherwise "inferior" genes by mass murder be transformed here into a kinder, gentler program to "perfect" human individuals by "correcting" their genomes in conformity, perhaps, to an ideal, "white, Judeo-Christian, economically successful" genotype?

Luria's dig at President Bush's vision of a "kinder, gentler" society probably went unnoticed in the West Wing. During a White House ceremony in 1989 to award the National Medal of Honor to, among others, Stanley Cohen and Herbert Boyer, the founders of recombinant DNA technology, President George Bush described the government's foray into Big Biology as the "Human *Gnome* Initiative."

The Human Genome Project was designed to be a worldwide effort, with about two-thirds of the work to be handled by university and government groups in the United States, the remainder by the United Kingdom, France, Germany, and Japan. But the outspokenness that made Watson such a favorite with the politicians threatened to cause an international incident when Watson rebuked a leading Japanese scientist for his country's paltry investment in its national genome program. Responding to charges of "Japan bashing," Watson fired back, "You don't get anywhere by being a wimp."

But Watson's real troubles were closer to home, in the form of the new NIH director, Bernadine Healy, who was appointed by President

Bush in April 1991. In early 1992, the pair became embroiled in a series of increasingly bitter and public spats over several issues. One was the matter of Watson's private stock portfolio, with holdings in several major pharmaceutical and biotechnology companies, which might give the appearance of a conflict of interest—even though Watson publicly declared his financial interests every year. A second was Watson's blunt criticism of Frederick Bourke, a wealthy entrepreneur who was trying to lure two leading DNA sequencers to head a new private genome institute.

But a deeper, more philosophical issue concerned the issue of gene patenting. Healy strongly supported a controversial NIH decision to seek patents for hundreds of gene fragments identified by NIH scientist Craig Venter, if for no other reason than to obtain clarification from the Patent Office on the legitimacy of patenting genes of no known function. Watson was critical of Venter's research and bitterly angry with Healy's decision to go ahead with the patent application. To add insult to injury, Healy asked Venter to consult on the future of human genome research at the NIH while instructing Watson not to go public with any further criticisms.

Calling his position "untenable," Watson abruptly quit on April 10, 1992. It was a potentially devastating blow for the Human Genome Project, for there was no other scientist who could combine Watson's scientific statesmanship and political savvy. Watson fired one last shot as he returned to Cold Spring Harbor, telling *Science,* "I don't know how to get anyone to succeed me. I don't know anyone who doesn't have stocks. And I don't know anyone who would want to live with my boss."

Early speculation on Watson's successor centered on another Nobel laureate, the late Daniel Nathans (although he scoffed at the suggestion). However, the early success of the Human Genome Project was creating a new generation of leaders—not the pioneers of the genetics revolution in the 1970s but the cartographers of the human genome during the 1980s. This was a new breed of geneticist who could map disease genes and navigate vast stretches of a chromosome to find their prize, the most publicly recognized rationale for spending $3 billion on the project.

While Healy set her sights on recruiting Francis Collins, one of the most respected members of the new DNA detectives, to succeed Watson, the NIH scientist who had unwittingly contributed to Watson's resignation was considering an extravagant offer to leave.

• • •

NOTWITHSTANDING THE ABJECT FAILURE OF HIS Genome Corporation, the DNA sequencing company that was years ahead of its time, Walter Gilbert had an almost preternatural vision of biological research in general and the Human Genome Project in particular. Gilbert urged biologists to become computer literate and accept the DNA sequence *tsunami* hurtling toward them as a new reagent for research. How quickly would this onslaught arrive? "I expect that sequence data for all of the model organisms and half of the total knowledge of the human organism will be available in five to seven years, and all of it by the end of the decade."

It was January 10, 1991.

CHAPTER TWO

Reading the Book of Life

[The science of life] is a superb and dazzlingly lighted hall which may be reached only by passing through a long and ghastly kitchen.
—CLAUDE BERNARD

THE HUMAN GENOME—also known as the Book of Life, the Manual of Man, the Code of Codes—contains riches of almost inestimable value. But it must be said that it is one utterly boring read. Then again, more than 3 billion letters will do that to you. If the sequence were written in the same font as this book, each page would contain about 3,000 characters. An average gene would run about 5 pages. The DNA of an average chromosome would require about 200 books of the size of this one, and the entire human genome would necessitate a library of 4,000 copies. Put another way, it would take an entire lifetime to listen to an unexpurgated recital of the genome sequence.

Besides its daunting length, there is no perceivable plot, most of the text is utter gibberish, and hundreds of pages seem to be made up of the same sequence repeated ad infinitum. And when the words do make sense, the chapters frequently leave the reader totally confused as to the meaning. Even the authors of a popular genetics textbook felt obliged to ask, "Could such a collection of unbelievably dull reading ever be of any value other than as a lending library for insomniacs?"

The DNA strands in each cell are tightly coiled and condensed thousands of times inside the nucleus. The space between each letter in the genetic code (each rung in the double helix) is 0.34 nanometer, less than one billionth of a meter. The DNA in a human cell is squeezed into a nucleus about 0.005 millimeter in diameter, and yet fully extended, the DNA of a single cell would stretch to about 2 meters, or 6 feet. The period at the end of this sentence would encompass about 200 cells, or 400 meters of DNA. The total amount of DNA in the 100 trillion cells in the human body laid end to end would run to the sun and back about twenty times.

Perhaps an analogy would be helpful here. Suppose we magnified a typical human cell about 300,000 times, so that it is now the size of a generously proportioned living room. Now we'll assume that there is a Volkswagen Beetle parked in the corner, conveniently posing as the cell nucleus. On this scale, the DNA molecule of a chromosome would be represented by a strand of cotton thread several miles in length, wrapped and twisted and coiled into one of forty-six bundles filling the car's interior.

WE ALL CONTAIN TWENTY-THREE PAIRS OF CHROMOSOMES, one set passed down from each of our parents. Twenty-two pairs are conveniently numbered 1 through 22, and are known as the autosomes. The twenty-third pair is the sex chromosomes: females have two X chromosomes, and males have an X and a Y chromosome. Each chromosome is a bundle of DNA and protein: the DNA molecule is tightly wound and folded on a protein core. Most of the genome is junk DNA—repetitive sequences that have no known function, and are probably vestiges of viral assaults on the genome over hundreds of millions of years. There is one gene every 40,000 bases on average.

The sequence of human DNA can best be thought of as a giant encyclopedia. This encyclopedia has twenty-four volumes (chromosomes): twenty-two autosomes and the X and Y chromosomes. Each volume includes on average several thousand terms (genes), although the length of each entry varies considerably. Each definition is written in a bizarre language: each word contains just three letters (codon), and the alphabet of the entire encyclopedia uses just four letters. Some entries run just a few lines; others extend for pages. Each entry is interrupted by several paragraphs of gibberish, while other junk inserts can run for pages between definitions.

The four letters of the DNA alphabet are simple chemicals, or bases: adenine, cytosine, guanine, and thymine, referred to, respectively, as A, C, G, and T. Each letter pairs with a complementary letter on the opposite strand of DNA. Thus, C always pairs with G, and A with T. These "base pairs" form the rungs of the DNA spiral, the famed double helix. This exclusive pairing was a crucial factor in the discovery of the double helix and has extraordinary biological significance. Before a cell can divide, its genetic instructions must be copied so that a complete instruction manual can pass into each daughter cell. As the double helix unzips, each daughter strand serves as a template to generate a new complementary strand identical to the original (although as we shall see, mistakes do occur).

Each triplet of letters, or codon, of DNA within a gene carries the instruction for one of twenty different amino acids. For example, CAG codes for glutamine, TGC a cysteine, and so on. These amino acids are linked in an order determined by the base sequence of a gene, to make one of thousands of proteins that constitute the building blocks of the human body.

The center of each chromosome, called the centromere, plays a vital role in aligning the twenty-three chromosome pairs before cell division. At the ends of chromosomes are special structures called telomeres, discovered by the Nobel laureate Hermann Muller in the late 1930s, who proposed that telomeres "seal" the ends of chromosomes. Telomeres are made up of thousands of identical DNA repeats strung together. In all mammals, including humans, this repeat signature consists of six letters, TTAGGG. Telomeres act in a manner not unlike aglets, the plastic caps that prevent shoelaces from becoming frayed at the ends: they help the cell's DNA repair machinery distinguish broken chromosomes from the mere ends. As cells divide, the telomeres get progressively shorter, apparently acting as a counting mechanism to measure the age of cells. A hallmark of cancers, however, is the activation of an enzyme called telomerase, which maintains telomeres at their full length. Scientists at Geron, a biotechnology company in California that owns the patent rights on telomerase, as well as Dolly the sheep, have found that switching on telomerase in cells growing in culture renders them immortal.

The aglet at the end of a human chromosome looks like this:

```
GGGATTGGGATTGGGATTGGGATTGGGATT—(5')
CCCTAACCCTAACCCTAACCCTAA—(3')
```

The double-stranded DNA does not terminate neatly, but contains a staggered end. The strand that contains the TTAGGG repeats overhangs the complementary strand of DNA. But in a recent surprising discovery, it turns out that the tip of the telomere folds back and tucks into the chromosome, providing a loop that physically seals the end of the chromosome. (For trivia buffs, examination of the sequence above reveals that the first letter in the book of life is a G.)

THE IMPORTANCE OF DETERMINING THE PRIMARY STRUCTURE of the molecules of life is evidenced by the fact that Fred Sanger, the pioneer of both protein and DNA sequencing, is one of only four scientists to win the Nobel Prize twice (the others are Marie Curie, Linus Pauling, and physicist John Bardeen). Modest to a fault, he keeps his medals in the bank, the certificates in the attic, turned down a knighthood, and stubbornly insists that he earned his place at Cambridge University only because his parents were wealthy. The Sanger Centre, Britain's flagship DNA sequencing center, in the quiet village of Hinxton just a few miles south of Cambridge, is named in his honor.

For ten years, beginning around 1945, Sanger labored in the basement of the biochemistry department at Cambridge in what many regarded as a foolhardy attempt to determine the sequence of the building blocks that make up the hormone insulin. His raw material was about 10 grams (or 120 cows' worth) of insulin crystals obtained from the Boots Pure Drug Company. From time to time, he would notice two figures walk past his department, "deep in conversation, battering away at each other, rather a crazy couple. They used to walk backwards and forwards, getting so excited about something." Sanger recognized them as Crick and Watson, of course, although he later admitted that he could not quite grasp the excitement when he went to view their model of the double helix in the spring of 1953, being much more interested in proteins than in nucleic acids. But he did not have to apologize, for it was Sanger who was awarded the Nobel Prize first, in 1958, just three years after he completely solved the order of the fifty-one amino acid building blocks in the two chains of insulin.

Sanger spent much of the 1960s developing methods to sequence short RNA molecules, using a similar divide-and-conquer strategy that worked for proteins. But the challenge posed by DNA was immensely more difficult. Not only did DNA molecules consist of thousands of letters (compared to merely dozens in RNA), but also they were double stranded, whereas RNA molecules were composed of a single strand. The breakthrough came when Sanger had "the best idea I have ever had, being original and ultimately successful." It was called the "chain termination" (or dideoxy) method for DNA sequencing, which remains the chemical basis for modern automated DNA sequencing technology.

Sanger's ingenious idea involved synthesizing a new strand of DNA that was complementary to a single-stranded template in such a way that the identity of the last letter was marked. The order of bases in the new strand is determined by the template: G always pairs up with C, and A bonds exclusively with T. Along with the bases and enzyme to make the new strand, Sanger included a modified dideoxy nucleotide. This dideoxy base is incorporated into the growing DNA strand as normal, but cannot form a chemical bond with the subsequent base in the strand, which subsequently stalls. The same procedure is repeated for all four dideoxy nucleotides, producing a mixture of DNA strands of varying lengths. These fragments are then separated by size on a gel and the sequence read from a ladder of radioactive bands.

At around the same time, Walter Gilbert and Allan Maxam developed an alternative method for sequencing DNA, and Gilbert shared Sanger's Nobel Prize, awarded in 1980. But Sanger's method proved easier and more popular. In 1977, Sanger and colleagues proved just how easy by sequencing the 5,375 bases of a virus called øX174—the first complete genome of a DNA virus and the largest sequence determined to date. It was a remarkable result, not only for its technological prowess but also because it revealed that the sequences of different genes need not necessarily sit as discrete beads on a string but could actually overlap.

The Princeton physicist Freeman Dyson compares this finding to an exceptional Mozart duet for violins where the players stand facing each other with the music laying flat between them. One violinist plays from the top down in the regular manner, while the other plays upside down from bottom up, even playing sharp signs on different notes (flats could not be used because they lack rotational symmetry). "I like to call Fred Sanger's virus the Mozart virus," wrote Dyson. "It shows that nature can compose a genome as cleverly as Mozart could compose a duet."

Sanger's recipe for how to sequence DNA by the chain termination method became one of the most referenced scientific papers of all time. He retired at the age of sixty-five. It was no surprise that when Sanger was invited to write a brief memoir several years ago, he entitled his reflections "Sequences, sequences, and sequences."

• • •

WHEN STEPHEN HAWKING HAD WRITTEN the first draft of *A Brief History of Time,* his editor warned him that the manuscript was far too technical. "Every equation will halve sales," Hawking was told, so he begrudgingly removed every line of mathematical code except one: $E = mc^2$. The same judgment would appear to hold for the inclusion of raw DNA sequence, although there have been exceptions.

Jurassic Park, Michael Crichton's best-selling novel about the re-creation of dinosaurs, featured an excerpt of sequence from putative dino DNA extracted from amber. Mark Boguski, a bioinformatics expert at the NIH, checked Crichton's sequence and was disappointed to find that the source of the DNA was not an evolutionary cousin of the dinosaur such as a bird or a reptile, but a bacteria. "My respect for [*Jurassic Park* scientist] Dr. Wu's scientific ability vanished," joked Boguski. "After all, he was unable to determine with three Crays what it took me two minutes on a Macintosh to discover." Much to Boguski's surprise, Crichton invited him to contribute a segment of DNA for the sequel, *The Lost World.* Boguski duly submitted a sequence composed of chicken and frog DNA, but unbeknownst to the author, also included a hidden message that was revealed when the genetic code was translated into the corresponding amino acid symbols: "MARK WAS HERE."

Let's briefly dust off one of the volumes of the encyclopedia *Homo sapiens* and sample an extract. Most striking is the amount of junk that almost obscures the thousands of definitions, or genes. Surprisingly, only about 3 percent of each volume is composed of genes and distinguishing them from the dross requires an expert eye.

One of the well-thumbed pages in the eleventh volume of the encyclopedia reads in part as follows:

```
...ACTAGCAACCTCAAACAGACACCATGGTG
CACCTGACTCCTGTGGAGAAGTCTGCCGTT
ACTGCCCTGTGGGGCAAGGTGAACGTGGAT
GAAGTTGGTGGTGAGGCCCTGGGCAGGCTG
CTGGTGGTCTACCCTTGGACCCAGAGGTTC
TTTGAGTCCTTTGGGGATCTGTCCACTCCT
GATGCAGTTATGGGCAACCCTAAGGTGAAG
GCTCATGGCAAGAAAGTGCTCGGTGCCTTT
AGTGATGGCCTGGCTCACCTGGACAACCTC
AAGGGCACCTTTGCCACACTGAGTGAGCTG
CACTGTGACAAGCTGCACGTGGATCCTGAG
```

```
AACTTCAGGCTCCTGGGCAACGTGCTGGTC
TGTGTGCTGGCCCATCACTTTGGCAAAGAA
TTCACCCCACCAGTGCAGGCTGCCTATCAG
AAAGTGGTGGCTGGTGTGGCTAATGCCCTG
GCCCACAAGTATCACTAAGCTCGCTTTCTT
GCTGTCCAATTTCTATTAAAGGTTCCTTTG
TTCCCTAAGTCCAACTACTAAACTGGGGGA
TATTATGAAGGGCCTTGAGCATCTGGATTC
```

This is the DNA sequence that codes for one of the chains that make up hemoglobin, the protein in red blood cells that transports oxygen in the body. All genes use the triplet ATG (in bold type) to serve as the marker to "start" protein synthesis. Similarly, the TAA codon marks the end of the coding sequence, and indicates "stop."

Here is the first line of the beta-globin gene once more, setting off each triplet of DNA bases. Each triplet provides the code that specifies a particular amino acid building block in the protein:

ATG GTG CAC CTG ACT CCT G**A**G GAG AAG TCT

The sixth triplet after the start codon, GAG, codes for a glutamic acid. If the middle letter, A, is mutated to a T, the resulting triplet, GTG, codes for a different residue, valine (identical to the first triplet after the ATG). This corrupt code produces a malformed hemoglobin protein that causes sickle cell anemia, a devastating disease affecting millions of West Africans and African Americans.

Genes are often depicted as beads on a string, but keep in mind that DNA is not a single linear strand of information, but a double helix consisting of two complementary strands of bases. The sequence that comprises a gene can actually reside on either strand of DNA, as if our encyclopedia was written simultaneously in English (reading left to right) and Japanese (right to left). In some cases, small genes lie within the introns of larger genes. Many genes give rise to multiple proteins with subtly varying functions depending on which introns are spliced out of the gene—just as words in a dictionary can have a range of subtly (or even radically) different definitions.

The role of the vast majority of the genes in the human genome is to store the instructions to make the thousands of proteins that build our bodies and sustain life. The decoding of DNA into protein by the cell uses an intermediary—a close relative of DNA called RNA, but with a couple of key differences from DNA. RNA is a single-stranded molecule, not a twin-strand polymer like DNA. It is also a lot less stable, which is important given its chief role as a short-lived messenger.

The very specialization of the dozens of different cell types in our body stems from the exquisitely regulated expression of subsets of the 100,000 genes in the nucleus. The signals that tell different cells which genes to activate are found in special instructions, like page cues in the encyclopedia. First, the gene must become disentangled from the coiled tangle of fibers in the nucleus. Next, an enzyme called RNA polymerase attaches in front of the gene. As the double helix unwinds, the polymerase proceeds along the gene, building a molecule of messenger RNA that faithfully complements the gene.

After the RNA is edited to remove any intron sequences, it leaves the nucleus and moves to the body of the cell, where it is seized upon by large protein machines called ribosomes. These ribosomes read the RNA message like a laser scanning a bar code. As each triplet is read, the appropriate amino acid is drawn to the ribosome and chemically linked to its predecessor, gradually assembling the new protein. When the ribosome reaches the "stop" signal in the RNA, it falls off, and the nascent protein chain folds into a complex three-dimensional shape and is escorted to its appropriate destination inside or out of the cell.

IT SHOULD NOT BE SURPRISING that scientists disagree over the precise number of genes there are on our twenty-three pairs of chromosomes. But what is surprising is just how diverse these predictions are. In the summer of 2000, before the full sequence was available, estimates varied from about 30,000 to well over 120,000, illustrating the difficulty of analyzing the genome sequence for signs of genes amid the vast expanse of junk DNA. British computational biologist Ewan Birney has started an international betting contest, one dollar per bid, to see who can correctly predict the final tally, with the result likely to fall somewhere between 50,000 and 100,000.

Despite the lingering uncertainty over the number of genes in the human genome, that number is the same for everyone. The position of every gene is fixed along the chromosomes. Thus, the gene for the clotting factor that is mutated in the most common form of hemophilia sits near the tip of the long arm of the X chromosome. The *p53* gene, the most commonly mutated cancer gene, resides on the short arm of chromosome 17.

The key to understanding the extraordinary diversity of the human population is not so much the identity or location of the 50,000–100,000 genes, but the subtle variation within them. Eric Lander, director of the Whitehead Institute Center for Genome Research, is fond of showing a slide of two men standing side by side, both immaculately and identically dressed in white tuxedos and tails. Both are icons of American sporting life. On the left is Willie Shoemaker, champion jockey, winner of nearly nine thousand horse races. On the right is the late Wilt Chamberlain, the legendary Los Angeles Lakers basketball star who once scored 100 points in a game. It takes a few moments to recognize what you are seeing. Shoemaker is less than five feet tall and weighs all of ninety-six pounds soaking wet. Chamberlain stood over seven feet tall and weighed more than three hundred pounds. Shoemaker's head barely reaches up to Chamberlain's navel. So marked are the differences in their physiques that you barely notice that Shoemaker is white, Chamberlain is black. These are two wildly extreme interpretations of the book of life.

Now let's pretend that we could compare the genomic DNA sequence of Messrs. Chamberlain and Shoemaker, aligned side by side, just as in the famous picture. Remarkably, we find that for long stretches of sequence, the letters are identical. Indeed, we have to scan several hundred letters before we encounter a single spelling discrepancy between the two sequences. For any two unrelated people, there is roughly one alteration every 1,000 bases.

This sounds trivial in some respects, but spread over the entire genome, there are about 3 million sequence variations between any two people, and several functionally significant spelling differences per gene. These can have profound effects on gene function (as in the case of sickle cell anemia) or far more subtle influences on activity. The sites that vary most commonly between different people are known as single nucleotide polymorphisms (or SNPs, affectionately known as "snips"), and are going to have a critically important role in the future of medicine. With as many as 3 million variations scattered over 3 billion possible characters, including as many as 100,000 genes, one begins to appreciate the awesome, virtually unlimited reservoir of variation enshrined in the human genome.

The profound effect of SNPs is vividly illustrated by skin color. As humans spread out of Africa about 100,000 years ago, the protective role of the dark melanin pigment became less important than the need to take advantage of the reduced ultraviolet light further north to synthesize vita-min D in the skin. Thus, European people have a much fairer complex-ion, although the corollary is a greatly increased risk of sunburn and skin cancer, as seen in Australia, where two out of three people develop skin cancer during their lifetimes.

A few years ago, a group from the University of Newcastle Upon Tyne published a remarkable study of one of the important proteins in the pathway that synthesizes melanin, the melanocortin receptor. The New-castle group checked the sequence of this gene in dozens of fair-skinned, freckled, red-haired individuals and found a striking pattern of SNPs, single-letter changes in the gene sequence that would diminish the protein's activity. Although many other factors are involved in skin pig-mentation, these results suggest that changes in a single gene have a pro-found effect on skin and hair color. "If true," says Mary-Claire King, "variation at this locus, which encodes evolutionarily important but su-perficial traits, has been the cause of enormous suffering."

Except for the X chromosome in males, we have two copies of each chromosome, and hence two sets of each gene that function with equal efficiency under most circumstances. However, there is an intriguing group of "imprinted" genes whose activity depends crucially on whether they were inherited from the mother or the father. Some of these genes are active only on the paternally derived chromosome, while the maternal copy sits silent. For other genes, the situation is reversed, and only the ma-ternal gene is expressed. Evolutionary biologists are searching for clues to explain this curious phenomenon. Perhaps genes that are paternally ex-pressed favor the growth of the embryo, because males are interested in having as many offspring as possible. By contrast, imprinted genes that are maternally expressed, are likely to hold embryonic development in check to allow the mother to conserve resources.

WE OWE OUR VERY EXISTENCE to mutations in our DNA. Mutations in-troduce variations into the DNA sequence of all organisms, the sheer essence of evolution. Damage to the integrity of the genome comes in many flavors. Some changes affect entire chromosomes, others just one letter. There can be changes in the number of chromosomes (most cases of Down syndrome arise from an extra copy of chromosome 21) or re-arrangements of large chunks of chromosomal arms, giving rise to serious birth defects. Segments of chromosomes ranging from a few thousand to millions of letters can be duplicated, inverted, swapped, or deleted, affecting the number, integrity, or expression of potentially hundreds of genes. There are bizarre mutations, such as the striking expansion of repetitive tracts of DNA (such as CAG CAG CAG . . .) that give rise to Huntington's disease and several other neurodegenerative disorders. Sometimes genes are disrupted by the insertion of rogue repetitive elements that jump around the genome, while other genes are disrupted by DNA variations tens of thousands of letters away that affect the gene's expression. Finally, there are far more subtle alterations that switch the identity of just a single letter out of 3 billion in our genetic code.

The impact of these changes is often completely unrelated to their scale. Individuals with Down syndrome can live happy, meaningful lives despite additional copies of hundreds of extra genes, whereas a single typo can have fatal results. Thanks to the heroic efforts of Victor McKusick, the grandfather of medical genetics at Johns Hopkins University, researchers have cataloged more than 5,000 distinct genetic disorders. In more than 1,000 cases, we know the gene(s) and mutation(s) responsible for these diseases.

And yet *mutation* is such a pejorative term. It is supposedly reserved for DNA alterations that have a frequency of less than 2 percent, whereas a variation that is more common is termed a *polymorphism*. This is not entirely satisfactory, however. "Mutations" that may be scarce in the U.S. population may be relatively common in certain ethnic backgrounds but have no bad consequences. On the other hand, some common variations in the human gene pool clearly have detrimental effects. One in twenty-five people of European descent carry the altered gene that causes cystic fibrosis, and an estimated one in ten carry the variant gene responsible for hemochromatosis, an iron overload disorder that is still treated by the medieval technique of bloodletting. Probably arising from a relatively recent Celtic or Viking mutation, hemochromatosis is considered to be the most common genetic disease. However, given that less than 5 percent of the human genome consists of genes, most random alterations to the DNA sequence are thankfully benign. They will probably fall in regions of junk DNA of no significance—or, rather, none that we can currently fathom. Throughout this book, I will use *variations* when referring to harmless sequence alterations and reserve *mutations* for changes that have negative effects.

The daunting task of faithfully copying 3 billion letters of DNA each time a human cell divides takes an army of replication enzymes about seven hours, and it is hardly surprising that the occasional mistake creeps in. The molecular machine that synthesizes new strands of DNA does a remarkable job placing the letter in the new strand that is the complementary partner to its corresponding letter in the existing template, C to G, A to T, and so on. Once every 10,000 letters or so, it inserts the wrong letter. However, the same enzyme quickly proofreads the new copy, correcting obvious mistakes and reducing the net error rate to about 1 mistake in every 10 million bases.

The more times DNA is replicated, the more likely it is that errors will be introduced. This is evident in the link between mutations and parental age in several genetic disorders. The best-known example is Down syndrome, where a fault in separation of chromosomes during the formation of the egg cell results in an extra copy of chromosome 21. But males are by far the worse offender, because sperm cells divide many more times than egg cells. Clear links have been documented between the incidence of hemophilia and other genetic disorders and advanced paternal age. (Perhaps the in vitro fertilization service that recently offered eggs from young female models over the Internet, rather than decrepit Nobel laureates, was onto something after all.)

Until recently, it was generally believed that the mutation rate in males was several times greater than in females. But new studies by David Page's group at the Whitehead Institute have forced a reappraisal. Page compared the sequences of two very similar noncoding stretches of DNA on the X and Y chromosomes with that of chimpanzees, our closest evolutionary neighbor. (By choosing a stretch of DNA that does not contain genes, Page ruled out the possibility that selection for particular variants might influence the mutation rate.) The results were surprising: there were indeed more mutations in the Y sequence compared to the X sequence, but the ratio was less than 2 to 1. The assumption that males harbor much higher mutation rates than females needs to be rethought.

There are other serious sources of damage to our DNA. The double helix is under constant bombardment from cosmic rays, gamma waves, ultraviolet radiation, and radioactive decay, which, if they do not cause direct harm, leave a trail of dangerously reactive chemical species called free radicals in their wake. These chemicals react with everything they come into contact with, including DNA. Every day, the integrity of the genome in every cell is threatened as thousands of bases in the genome are purged, chemically modified, or bound unnaturally to each other. To correct these potentially catastrophic mutations, cells employ an arsenal of

quality-control enzymes that keep the double helix under constant sur-
veillance. In one strategy, a group of three enzymes combine to snip out
the errant base, so that the gap can be correctly filled in with the correct
letter. For more serious errors, another process cuts out a stretch of about
thirty consecutive letters on one strand of the helix, which is then rapidly
filled in. Another repair mechanism, called mismatch repair, which has
been carefully preserved over hundreds of millions of years of evolution,
recognizes incorrect base pairs.

When one or another of these repair mechanisms is faulty, the results
can be devastating. The best example is the inherited disease xeroderma
pigmentosum, actually a group of about a dozen closely related disorders
caused by mutations in genes that code for repair mechanism components. Patients with this disease are extraordinarily susceptible to ultraviolet radiation in sunlight, resulting in skin cancer. They seldom live beyond
thirty years of age. Similarly, mutations in members of the mismatch repair
protein complex give rise to a form of colon cancer. And an inherited
flaw in an enzyme called DNA helicase gives rise to Werner's syndrome,
a maleficent form of premature aging in which young adults have the sad
physical appearance of old age.

THERE ARE MANY WAYS THAT THE DNA SEQUENCE can be damaged to
give rise to genetic diseases and cancer, but often all it takes is the slightest
misspelling, affecting just one of the 3 billion letters of the genome, to
have profound, often distressing, consequences.

In *The Odyssey*, Homer described one-eyed giants called the Cy-
clops who lived as shepherds on the Sicilian coast. While these were
clearly imaginary creatures, it is conceivable that Homer encountered infants with a condition called holoprosencephaly, the most common brain
malformation in humans. The disorder affects about 1 in 250 embryos
during early gestation, although the frequency of live births is much lower
(about 1 in 16,000). In its mildest form, patients have narrowly spaced
eyes, but the most extreme cases suffer cyclopia—a single eye—with a
nose-like structure, or proboscis, above it. Holoprosencephaly can arise
from various genetic aberrations, the first of which was identified by
Maximillian Muenke and Lap-Chee Tsui a few years ago. They discovered a single-letter change in a gene called *sonic hedgehog,* one of the key
signals in early embryonic development. Cyclopia can also be attributed
to environmental causes, notably in sheep that eat plants containing cy-

clopamine. This has led to the intriguing suggestion of using this compound to treat basal cell carcinoma, which is caused by the upregulation of the same pathway affected in holoprosencephaly.

About 2.5 million Americans suffer from alopecia, the loss of clumps of hair from the scalp. In rare cases, the loss of hair can be total. Angela Christiano, a dermatologist at Columbia University in New York, studies rare families with alopecia universalis, a recessively inherited disease in which affected individuals have virtually no hair follicles and no scalp or body hair. Christiano's team traced the wayward gene to the short arm of chromosome 8, in a region containing the human counterpart of a mouse gene called *hairless*. The alteration of an A to a G in the corresponding human gene introduces an incorrect amino acid in a regulatory protein that plays a key role in hair development. Countless other examples of the profound effect of single-letter mutations could be provided, few more dramatic than the changes in sex due to mutations in the male sex-determining gene.

Some mutations are so harmful that their high frequency poses a paradox. The solution to this enigma was offered by British geneticist J. B. S. Haldane, who in 1948 proposed that individuals who carry one copy of a faulty recessive gene (and are consequently symptomless) may be protected against some other disease. This phenomenon is known as heterozygous advantage, and the best-known examples are sickle cell anemia and thalassemia.

The World Health Organization estimates that a staggering 7 percent of the world's population are carriers of an inherited disorder of hemoglobin, the oxygen-carrying protein in red blood cells. These diseases are known collectively as the thalassemias, which literally means "anemia of the seas." The best-known hemoglobin abnormality is sickle cell anemia, which is caused by the swap of a single amino acid in one of the chains of the protein. Carriers of the sickle cell trait are significantly more resistant to infection by *Plasmodium,* the parasite that causes malaria. The sickle cell version of hemoglobin forms fibers that lead to the collapse of the red cell, hindering the growth of *Plasmodium.*

As Haldane predicted, the selective advantage of the sickle cell carriers against malaria has led to the rapid spread of the errant globin gene in the population, condemning millions of children to suffer the painful agony of sickle cell disease. Evolution has selected many other genetic variants in globins and other red cell components because they too offer an advantage against the malarial parasite. Inherited defects in the energy-producing enzyme glucose-6-phosphate dehydrogenase cause favism, but carriers are more resistant to malaria, as are West Africans who lack the Duffy blood group, a protein that normally decorates the red cell surface and unwittingly provides the portal for *Plasmodium*.

Cystic fibrosis is the most common genetic disease that affects people of northern European descent, affecting about 1 in 2,500 live births. Roughly 1 in 25 people carry the faulty gene—an astonishing frequency given that, until recently, patients seldom lived to reproductive age, and even then were likely to be sterile. (New therapies using antibiotics and other drugs to break down the mucus that clogs the lungs of affected patients have led to a welcome rise in life expectancy over the past few decades.)

So why is the disease so common? Some believe that cystic fibrosis carriers hold an advantage against cholera, but this disease reached Europe only in the past two hundred years, not enough time to have had an appreciable effect. By contrast, there is good experimental evidence that cells carrying a single copy of the faulty CF gene, which codes for a membrane protein that transports chloride ions, possess a distinct advantage in warding off the *Salmonella typhi* bacteria that cause typhoid. The most common cystic fibrosis mutation removes just one amino acid in a protein chain containing more than 1,500. But this subtle flaw has devastating consequences, rendering the nascent protein chain incapable of folding into its correct three-dimensional shape. Cells with one faulty cystic fibrosis gene take up 86 percent fewer *S. typhi* bacteria than normal cells. Many more tales of heterozygous advantage against infectious diseases will become apparent as scientists are able to survey the full contents of the human genome.

WITH SO MUCH ATTENTION GIVEN IN THE MEDIA to the battle against harmful mutations that cause literally thousands of cancers and genetic diseases, it is easy to overlook that some spontaneous changes to the genetic sequence are clearly beneficial to humans. For years, athletes have illegally boosted their body's oxygen capacity by doping with erythropoietin (Epo), a natural hormone produced by the kidneys that stimulates the proliferation and differentiation of red blood cells. But Finnish cross-country skier Eero Mäntyranta had no need for such duplicity. Winner of two gold medals at the 1964 Winter Olympics in Innsbruck, Mäntyranta carried a mutation in the gene for the Epo receptor, which receives and

transmits the Epo signal. As a result, Mäntyranta's body continued producing red cells until his count was 25 to 50 percent higher than normal,
boosting the skier's oxygen capacity and endurance. Mäntyranta's mutation (known in the trade as autosomal dominant benign erythrocytosis)
traces back to one of his great grandparents and affects about thirty people in his extended family but is virtually unknown elsewhere.

Gold medals are nice, but immunity against HIV is truly precious. A
naturally occurring mutation in the CCR5 receptor, which is found on
the surface of white blood cells, robs the AIDS virus of a convenient
foothold that allows it to breach the surface of the host cell. However, a
small percentage of the population possesses a corrupt version of the
CCR5 gene that lacks a stretch of thirty-two letters, preventing production of the receptor. Even when repeatedly exposed to the AIDS virus,
these individuals do not become infected, because the virus is unable to
gain entry into the cells where it replicates.

The picturesque Italian village of Limone Sul Garda, seemingly un-
touched by the ravages of modern life, holds a secret. About forty residents have lived to be centenarians. While much of their vigor can be
attributed to a stress-free lifestyle and a healthy Mediterranean diet (no
doubt supplemented with lashings of chianti), about one in twenty villagers have the added advantage of a genetic glitch. They possess a variant
form of apolipoprotein A1 called A1 Milano (named after the university
where it was discovered), which scavenges cholesterol in the blood and
escorts it to the liver, where it is broken down. The mutation—the simple
substitution of a cysteine for an arginine in the A1 protein—probably
arose more than two hundred years ago, remaining confined in this isolated community by a lake. With some 600,000 angioplasties performed
in the United States alone each year, hopes are high that the recombinant
form of Apo A1 Milano will provide a new and effective therapy for treating coronary artery disease by boosting the levels of "good" cholesterol.

Another common advantage that most people take for granted has
its origins ten thousand years ago with the advent of livestock farming.
Since then, many humans have been able to consume milk into adulthood thanks to the sustained production (unlike any other mammal) of
lactase, the enzyme that breaks down the milk sugar lactose. Some people
remain lactose intolerant: in the absence of lactase, drinking milk produces rather unpleasant gastrointestinal problems. The incidence of milk
drinking in Europe declines markedly from the north, where the rate in
Scandinavia is about 90 percent among adults, to the south, where in Italy

it is just 50 percent. Interestingly, this gradient mimics the incidence of another apolipoprotein, Apo AIV-2, which probably originated in Scandinavia, and may play a role in aiding milk metabolism.

THERE IS ANOTHER CLASS OF MUTATION, poorly understood at present, which will become more apparent with the aid of the complete human sequence. An atavism, or "throwback" mutation, is the sudden reappearance of a trait long since consigned to the evolutionary recycle bin (the term comes from the Latin *atavus,* meaning a great-grandfather's grandfather). Some humans are born with three nipples or a trace of webbing beneath their digits. The most dramatic is a vanishingly rare disorder known as congenital generalized hypertrichosis.

In the small town of Zacadecas in Mexico lives a large family in which affected persons live with a thick matte of hair over their face and upper bodies. Especially in affected males, every pore of their face is covered in hair, lending them the sad but incontrovertible appearance of a werewolf. Two men in this family work as sideshows in the circus, parading as "people of the forest" and "dog men." Another male relative works as a night watchman so he will not have to see anyone. A Mexican-American team of scientists led by Luis Figuera of the University of Guadalajara and Pragna Patel at Baylor College of Medicine has traced the responsible gene to a region of the X chromosome. This explains why the women are less severely affected than their male relatives—the second X chromosome partially masks the effect of the atavism—but they refuse to shave their faces for fear that the hair will grow back thicker and darker.

Atavisms can give rise to new patterns or new structures. Examples of the latter include extra toes in horses—the horses belonging to Julius Caesar and Alexander the Great were prized for their extra digits—and hind limbs in whales. The expectation is that during the course of human evolution, this hirsutism gene fell into neglect and has sat mute for hundreds of thousands of years. In this poor Mexican family, some schism on the X chromosome sparked this gene back into life. Tracing the precise genetic flaw will be difficult, but once the sequence of the X chromosome is completed, the "werewolf" gene will eventually be found.

How many mutations accumulate in the human genome? While we wait for a comprehensive study of the full human genome sequence, we can consider a recent comparison study of DNA sequences of forty-six genes from humans and chimpanzees. Researchers observed 143 base substitutions that would alter the sequence of the corresponding protein, whereas the expected number was 231. The difference—88 mutations—may have been so deleterious that they were discarded by natural selection. Extrapolated to the entire human genome (say 60,000 genes) and allowing for the 5 million years since humans and chimpanzees last shared a common ancestor, the results suggest that there are about four new mutations that change the sequence of a protein every generation. Of these variants, two or three could be severely deleterious.

For some evolutionary geneticists, the accumulation of mutations in the human genome is a cause for concern. Geneticist James Crow has speculated that headaches, stomach upsets, poor vision, and other ailments may be the consequence of mutation accumulation.

In fact, geneticists believe that every person has about five to ten recessive mutations lurking in their genome. In most cases, we would learn about them only if we were unlucky enough to have a child with a partner carrying the same mutation. Within the next decade or two, when personalized DNA sequence surveys become available, doctors will be able to offer an instant diagnosis of the phantom menace lurking in our genomes. No genome is immune to the ravages of mutation. Everybody harbors flawed genes.

VARIATIONS IN THE HUMAN GENOME are the key to understanding human health and evolution. But many occur in regions of junk or repetitive DNA and have no bearing on the function of a gene. As Stanford University geneticist David Botstein realized in 1978, such variations hold the key to producing a complete map of the human genome.

Botstein was a guest at a University of Utah retreat in the Wasatch Mountains, listening to a student explain the difficulties he was having trying to locate the mutant gene that causes the common iron-overload disorder hemochromatosis. Botstein suddenly realized that this problem could be solved with the availability of a series of DNA markers strung along each chromosome like mileposts along a highway. By identifying random segments of DNA that vary slightly in sequence between different people, the inheritance of these markers could be followed in families with a genetic disease. If the inheritance pattern of the marker corresponded to that of the disease (in other words, the mutant gene) and the chromosomal location of the marker was known, then by inference, the map position of the disease gene was also revealed.

The same principle had been shown to work for the prenatal diagnosis of sickle cell anemia in 1978 by Y. W. Kan and Andres Dozy at the University of California, San Francisco. But Botstein and three colleagues showed how a few hundred DNA markers could lead to a map of the human genome, allowing any disease gene, in principle, to be localized. Their paper, published in 1980, proved a major turning point in genome research, particularly when the gene for Huntington's disease was mapped three years later. As more and more polymorphic DNA markers were mapped to different chromosomes, the localization of genes for major diseases such as cystic fibrosis and muscular dystrophy soon followed. But trying to map genes for mental illness such as schizophrenia and bipolar disorder proved more complicated. A rash of high-profile papers near the end of the 1980s nearly all ended in disappointment, underlining how difficult it would be to map and identify genes for complex traits, where deciding where genetic influences leave off and environmental factors begin is virtually impossible.

Moreover, many researchers were not interested in studying genetic disease, but were anxiously trying to identify genes based on their functional roles. These searches could take years and consume millions of dollars. The widespread frustration with the difficulty of tracking human genes, whether they be those implicated in diseases or simply because of their importance to biology, convinced some scientists that there had to be more efficient solutions. One, of course, was the creation of the Human Genome Project. Another was percolating in the mind of an NIH researcher who had learned in Vietnam that life is cheap, time is short, and speed matters.

CHAPTER THREE

The Eye of the TIGR

Rutherford said there were two kinds of science—physics and stamp collecting. But what he forgot is that there are some stamps worth collecting.

—SYDNEY BRENNER

*T*IME MAGAZINE'S LIST OF THE TOP 100 GREATEST MINDS of the twentieth century finds room for just four biologists: two giants of the microbial world, Alexander Fleming and Jonas Salk, and the co-discoverers of the double helix, James Watson and Francis Crick. But if there were to be an award for the most influential scientist of the 1990s, the odds-on favorite would have to be John Craig Venter.

Over the past decade, it is probably fair to say that no other researcher has approached the impact that this controversial Californian has had on biomedical research. As a neuroscientist at NIH, Venter launched a program to identify tens of thousands of novel human genes. As president of a nonprofit research institute, he pioneered the sequencing of the complete genomes of a number of deadly pathogenic microorganisms. And as the founder of Celera Genomics, he almost single-handedly transformed the odyssey to obtain the sequence of the human genome. In the process, he became one of the most powerful, controversial, not to mention wealthiest, scientists in the world. If he had not stirred up so much controversy and resentment over the past decade, he would be a certainty for Stockholm's annual Nobel party. He may still get there.

Craig Venter was born on October 14, 1946, in Salt Lake City and grew up in Millbrae, just south of San Francisco. His father was a tax accountant, his mother an artist. A rebellious child, he frequently skipped classes at Mills High School in favor of surfing at Half Moon Bay, before barely scraping his way to a diploma in 1964. Instead of going to college, he traveled south to hang out at Newport Beach, surfing during the days and working at Sears at night. He was the epitome of southern California health and vitality: sun-bleached hair, strong tanned body, living the good life. But the fun times were cut short by the looming presence of the Vietnam War. In 1965, Venter enlisted in the navy in the hope of joining the swim team and possibly making the Olympic swim team. His plans were dashed, however, when President Johnson shut down military sports teams. Venter jokes now, "First Lyndon Johnson messed up my surfing career, then he messed up my swimming career!" But after scoring top marks in an intelligence test given to more than 30,000 recruits, Venter was sent to San Diego to train as a hospital corpsman, because he thought it would be relatively safe and would ensure that he could leave after three years of service.

In 1967, not even twenty-one years old, Venter found himself in Da Nang, one of the largest American bases during the Vietnam War, confronted with the hideous task of triaging troops who had been severely wounded during the Tet offensive in the hope of identifying a few salvageable lives. He witnessed the deaths of hundreds of men, but the fate of two young soldiers had a particularly profound effect on the corpsman. One soldier, who looked unmarked save for a .22 bullet wound in his head, died shortly after his arrival. By contrast, an eighteen-year-old boy was carried in with horrendous injuries to his chest and abdomen, yet somehow managed to cling to life for several more weeks, dreaming of going home.

Not surprisingly, Venter was severely shaken by Vietnam, and to this day, he becomes uncharacteristically emotional when asked to relive his experiences there. "You can't live through a situation like [Vietnam] and come out of it with the same laissez-faire attitude that you might approach surfing with before that," Venter told CNN. "I dealt with the death of thousands of young men, my age and younger, dealt with the failure of medical tools to do anything functional to help most of these people."

The man who returned to the United States was very different from the free-spirited boy who had left. "I felt very fortunate to survive the year and didn't want to waste any of my life going forward," Venter recalls. He immersed himself in studying instead of surfing. After a brief stint at junior college, he enrolled at the University of California, San Diego, initially to study medicine. His interests shifted to basic science, however, and within just six years, he had graduated with a bachelor's degree in biochemistry and a Ph.D in physiology and pharmacology. Although he doubtless would have made a good doctor and saved lives, Venter felt that he might be able to have a bigger impact by striving for real scientific breakthroughs in the research laboratory. In 1976, he moved to Buffalo, working at the State University of New York and the Roswell Park Cancer Institute. Here he met a charming microbiology graduate student, Claire Fraser, who would become his wife and steadfast scientific partner. They even wrote a grant proposal on their honeymoon.

In 1984, Venter moved to the National Institute of Neurological Disorders and Stroke at the NIH, while Fraser took a post in another branch of the NIH. Venter's laboratory, part of the Section of Receptor Biochemistry, was unremarkable but well funded (receiving between $1 million and $2 million a year for salaries and research costs). Venter's chief goal was to identify a solitary protein: the protein on the surface of heart cells that senses adrenaline—the "fight-or-flight" hormone. Although ultimately successful, the search took years of tedious purification of proteins, at a cost that Venter estimates to be about $10 million. Despite his success, Venter was forced to ask whether the ends justified the means. If the cost of the hunt for the adrenaline sensor was extrapolated for 100,000 genes, the pace of progress in biomedical research would be insufferably slow.

Venter's solution to this problem was characteristically spontaneous and daring. In 1986, he flew to California to meet with Michael Hunkapiller, one of the designers of an exciting new automated DNA sequencing machine, made by the Applied Biosystems (ABI) division of Perkin Elmer Corporation (PE). In conjunction with Leroy Hood, then professor of biology at Caltech, ABI had developed an advanced method of automated DNA sequencing. Although relying on the same chemistry that Fred Sanger's team had developed, the new system employed fluorescent dyes rather than radioactive compounds to tag each letter a different color; for example, A might be green, T red, and so on. Because the four dyes could be distinguished easily, the reactions could be pooled and run together on one lane of the sequencer, with the signature fluorescence of each dye read by a laser scanning the bottom of the gel. The ABI 373A could analyze twenty-four samples simultaneously, yielding about twelve thousand letters of DNA per day. The only downside was the cost: about $100,000 per machine.

In February 1987, Venter's NIH laboratory became one of the first alpha test sites for the ABI sequencer and the prototype ABI 800 Catalyst robotic workstation, which mixed the DNA, chemicals, and enzymes for the sequencing reactions. Venter immediately tested his new toys by sequencing a pair of rat genes related to the adrenaline receptor. The ABI machine was as good as advertised, proving much faster and cheaper than the tedious manual methods of DNA sequencing.

But what Venter really wanted to do was to sequence interesting regions of the human genome and examine the chromosomal DNA sequence to discover interesting genes. He was anxious to try a completely different tack to the slow, single-minded pursuit of one gene out of tens of thousands. His dream was to tackle one of the most fertile regions of the human genome: a segment of the long arm of the X chromosome called Xq28, to which dozens of genetic diseases had been traced. To do this, Venter requested funding from Watson's genome center, only to be repeatedly frustrated by government bureaucracy and skepticism toward automated DNA sequencing. While the National Center for Human Genome Research was debating the Xq28 project, Venter launched two smaller sequencing projects of prime interest to human geneticists. These were regions likely to contain the genes mutated in two important hereditary disorders that were on the geneticists' top ten most wanted list: Huntington's disease on chromosome 4 and a muscle-wasting disorder called myotonic dystrophy on chromosome 19.

Using the ABI 373A, Venter's team sequenced approximately 60,000 bases of DNA in the general vicinity of the Huntington's gene and 106,000 from the myotonic dystrophy candidate region. By analyzing the chromosome 19 region, Venter's team predicted the location of five genes, but nothing relevant to muscular dystrophy. And although Venter did not find the Huntington's gene in his foray on chromosome 4, the results were interesting in their own right. Half of the interval was sequenced twice from two different DNA donors, allowing Venter to compare systematically the rate of variation in the genetic code between two unrelated individuals. Venter logged seventy-two differences in all, corresponding to a surprisingly high frequency of 1 polymorphism per 450 bases, and he immediately recognized their intrinsic value. Venter predicted that "informative polymorphisms will be present near [any] gene or regulatory region, allowing abnormalities in these sequences to be linked to phenotypes at high resolution." Five years later, these single nucleotide polymorphisms—as these variations are now known—would become the hot trend in human genetics. The chromosome 4 report was also notable in one other respect: among Venter's coauthors was one Francis Collins.

Although delighted with the performance of the automated sequencing machines, overall Venter was disappointed with his genome sequencing efforts and the problems in identifying genes in uncharted genomic DNA sequence using computational methods alone. The problem was not that the software had difficulty in recognizing genes, but rather that it predicted too many false positives, requiring Venter's group to go back and confirm their existence experimentally—a laborious exercise at the best of times.

Venter started to wonder if there wasn't an easier way to identify human genes.

VENTER'S EPIPHANY came some 35,000 feet above the Pacific Ocean while returning from a conference in Japan. One of Venter's virtues is that he is an impatient man, and everything he had tried to that point simply took too long. His obsessive quest for the adrenaline receptor had taken years to pull off, and even the automated sequencing of two chromosomal regions had limitations in finding interesting genes. Venter needed a system that would enrich genes so that time would not be wasted sequencing long stretches of useless junk DNA.

The answer, at least in hindsight, was obvious: the most efficient gene enrichment process is one that takes place naturally in almost all of the trillions of cells in the body, as gene instructions are read and translated into proteins. Venter would adopt the natural biological process of gene transcription as a biological computer to sort through billions of letters of DNA in search of the small percentage that encode genes. What could be simpler?

The process of converting the instructions carried by a gene into the corresponding protein relies on an intermediary called RNA. In any given cell, only a small proportion of genes are turned on at any one time. An enzyme called RNA polymerase reads along the sequence of the gene, producing a complementary strand of RNA that is escorted out of the nucleus and into the body of the cell. Molecular machines called ribosomes clasp the RNA strand and read the base code in triplets, or codons. At each codon, the appropriate amino acid is carried to the ribosome by an adaptor RNA molecule that specifically recognizes each codon. One by one, the string of amino acids that constitute the protein are linked until the complete protein is assembled.

Since the beginning of the genetic engineering era in the early 1970s, the standard way to produce a pure protein was first to isolate the relevant messenger RNA molecule. A liver cell and a pancreatic cell both contain two copies of the factor IX gene and two copies of the insulin gene, because each cell contains the complete human genome. But factor IX is manufactured in the liver; hence this organ is a good source for the factor IX message. By contrast, insulin is produced in the pancreas; therefore, efforts to isolate the insulin gene start logically with the most abundant source of insulin RNA.

The first genes were isolated by purifying RNA molecules from a given cell type. Because of its unstable nature, the purified RNA is first converted back into DNA; these DNA copies of RNA molecules are known as "complementary DNA" (cDNA). To store and propagate these genes, the cDNA molecules are stitched into an appropriate vehicle, such as a small bacterial chromosome. This collection, known as a cDNA library, is stored in bacteria and represents an accurate snapshot of the active genes in a given cell type. Any cDNA can be isolated by growing the bacteria on a petri dish and selecting colonies.

Venter's idea was shockingly simple. cDNA libraries are a standard resource in molecular biology labs, but researchers were usually interested only in identifying the one bacterial colony that contained their gene, or cDNA, of interest. The other clones, representing thousands of genes, were essentially irrelevant. So what was to stop Venter from making use of all of these clones by analyzing hundreds of randomly selected genes at a time?

Together with Mark Adams, a research associate in his lab, Venter chose a brain cDNA library containing copies of potentially tens of thousands of genes that are active in the brain. It was a routine process to pick a few dozen bacterial colonies, each containing the cDNA of a mystery gene expressed in the brain, purify the DNA, perform the sequencing reactions, and determine the sequence on the ABI machine. Finally, Venter would compare the DNA sequence of the two hundred to three hundred bases that he typically obtained from each cDNA with previously identified genes from a variety of species whose sequences were already in the public gene database. Each snippet of cDNA sequence was dubbed an "expressed sequence tag" (EST)—a phrase destined to enter the lexicon of the molecular biologist.

It was such an elegant, simple plan, Venter could not believe he was the first to dream of it. Sure enough, he was not. In 1983, Paul Schimmel and two colleagues at MIT had published a paper showing that they could identify most of the then known muscle protein genes simply by sequencing DNA from two hundred randomly selected muscle cDNA clones. Schimmel suggested that the strategy could be used to identify novel variants of muscle proteins (and in fact he documented one such example). Schimmel's proof-of-principle of the surprising power of the almost effortless process of sorting through a random collection of cDNA clones appeared in *Nature* and thus was widely read, but nobody quite grasped the conceptual implications. Even Schimmel moved onto other projects.

The man generally credited with putting the cDNA concept on the map is the renowned geneticist Sydney Brenner, the man who had helped to discover messenger RNA and decipher the genetic code in the early 1960s. During one of the early meetings in Santa Fe in 1986 held to discuss the genome project, Brenner publicly argued in favor of a cDNA-sequencing program. His view was that since 97 percent of the human genome was considered "junk," priority should be given to sequencing the remaining 3 percent of the genome that contains genes. The sequencing of junk DNA, said Brenner, was like income tax: "It cannot be evaded but there are ways of avoiding it, and, anyway, it is one of those problems we can and should leave to our successors."

Automated DNA sequencing was still in its infancy in the late 1980s, so Brenner believed it would be more efficient to sequence genes and then map them, rather than map and sequence chromosomes in a blinded fashion. Moreover, focusing on cDNAs would yield immediate benefits for any organism, regardless of how well it had been studied. "The development of methods of cloning and sequencing DNA," Brenner wrote in 1990, "has liberated genetics from the constraints of breeding experiments, and all organisms are now accessible to genetical analysis by the new methods."

By 1988, Brenner's group was sampling cDNAs from various tissues, beginning with the placenta because of its easy availability. Brenner had his own funding from a couple of prizes he had been awarded. Meanwhile, at the University of Colorado, James Sikela was starting a small effort of his own to sequence brain cDNAs, with a particular emphasis on identifying genes potentially involved in psychiatric disorders, and mapping the chromosomal location of the gene fragment. But none of these efforts could match the sequencing efficiency that Venter's laboratory had honed to a fine art.

In April 1991, Venter withdrew his stalled proposal to sequence part of the long arm of the X chromosome. If it had been funded when first proposed more than two years earlier, he complained to Watson, 2 million letters of DNA might have been sequenced by now. Venter felt betrayed that Watson had failed to make good on his initial pledge of millions of dollars for his sequencing project. But Venter put his hard feelings aside. He was about to change the course of biomedical research.

ON JUNE 21, 1991, Venter presented his new EST strategy in a landmark article in *Science,* with the prepositionally challenged title "Complementary DNA Sequencing: Expressed Sequence Tags and Human Genome Project." (The article was dedicated to his late father, John, who had died of a heart attack at age fifty-nine.) Whereas most other research reports typically dealt with a single gene or two, Venter's paper revealed the identity of more than 330 novel genes active in the human brain, deduced from the sequence analysis of more than 600 randomly selected brain cDNA clones. The choice of the brain for the pilot EST project was dictated by two things: first, more genes are active in brain than any other tissue—as many as   one-half of all genes. Second, approximately one-quarter of the more than 5,000 known genetic diseases affect the nervous system.

In addition to a few dozen matches to previously known human genes, Venter found novel human genes that showed strong sequence similarities to interesting genes from other species. For example, one strong match occurred with the sequence of a *Drosophila* gene called *Notch,* which codes for an important intercellular signaling molecule. The striking conservation of genes in species that are separated by hundreds of millions of years of evolution is strongly suggestive that they perform important roles. That prediction was borne out six years later, when a French team showed that mutations in the human *Notch3* gene on chromosome 19 cause an inherited form of stroke called CADASIL (cerebral autosomal dominant arteriopathy with subcortical infarcts and leukoencephalopathy).

But the overall impact of Venter's paper was far greater than the sum of its parts; it sent shock waves through the scientific community. What he had done was conceptually redundant and technically mundane—and yet in one stroke he had uncovered more new genes than any other report before then. At the time of publication, the public gene database, Gen-Bank, contained the sequences of fewer than three thousand human genes. Venter had almost single-handedly identified more than 10 percent of the world total in barely a few months.

In contrast to the usual impenetrable style of scientific prose, Venter was already rehearsing a brand of straight talk that did little to endear him to his peers. He drew a sharp contrast between the dreary timetable for the full genome sequence, which he said might take the better part of two decades to finish, with his own lab's productivity, which could generate 10,000 ESTs a year at a cost of just twelve cents per base. "In our own laboratory," Venter concluded, "the EST approach should result in the partial sequencing of most human brain cDNAs in a few years." He also predicted that "improvements in DNA sequencing technologies have now made feasible essentially complete screening of the expressed gene complement of an organism."

Not surprisingly, Venter's pronouncements upset several senior figures in the Human Genome Project. John Sulston, the British director of the nematode genome project, dismissed Venter's prediction that he could identify the majority of genes by sampling cDNAs, especially given the poor quality of some commercial products. But the biggest scientific concern was that the EST approach would skip crucial genetic information contained in the promoters, the DNA sequences located ahead of the start of a gene sequence that control where and when a gene should be switched on. James Watson, director of the NIH genome project, was adamant that cDNAs were not a replacement for genomic analysis. Venter acknowledged this shortcoming, maintaining that his EST approach should complement, rather than replace, the Human Genome Project. But in his own lab, all resources were now directed to generating ESTs.

Eight months later, in February 1992, Venter was in the news again. In a report in *Nature,* he described partial DNA sequences of another 2,375 human genes expressed in the brain. Although this report lacked the element of surprise of his 1991 *Science* article, Venter's group of fifteen researchers was producing a staggering volume of data. In less than a year, Venter's group had sequenced part or all of more than 2,500 human genes—double the total number of genes sequenced by the rest of the world at that time. Moreover, he was adding to the tally at a rate of more than 100 per day. Criticism flared briefly that some of the ESTs contained intron DNA sequences, but this had negligible impact on the utility of the sequences.

While scientists pondered the merits of EST sequencing, *Nature's* editor, Sir John Maddox, interjected a word of caution. A few weeks after publishing Venter's second opus, he wrote: "In the wake of James Watson's departure from the US Human Genome Project, there is a danger that the cDNA approach will be offered as a cheaper alternative to complete [genome] sequencing, which it is not."

EVEN AS VENTER'S NOTORIETY as a leader in the DNA sequencing field was soaring, he was becoming embroiled in a bitter controversy over the issue of gene patents. The issue of patent protection for DNA sequences is an emotionally charged and legally intricate issue. To qualify for a patent, an invention must be novel, nonobvious, and useful. Before the EST landslide, only a handful of patented genes had been used successfully by biotech companies to make therapeutic products, including insulin (for diabetes), clotting factor 8 (for hemophilia), and erythropoietin (for kidney failure).

In order to claim precedence for international patent rights, the patent application must be filed before the material is published. When Reid Adler, head of the NIH Office of Technology Transfer, received a tip in 1991 alerting him to the pending publication of Venter's first EST paper, he quickly filed a patent application on the first batch of 347 ESTs before the report was published in *Science.* "I didn't want to miss the boat," admitted Adler, who was nervous that other countries might seek patents on some ESTs if the NIH did not preempt them. He defended the NIH's policy, saying, "Our primary goal is to get products developed. Having patent protection will enhance our ability to transfer this technology to companies. Without patent protection, companies won't spend the money to develop it." Just before the publication of Venter's landmark *Science* article, Adler delivered a 400-page application to the U.S. Patent and Trademark Office detailing information on Venter's first batch of 337 ESTs. Early the next year, one day before Venter's *Nature* publication, Adler modified the application by adding a further 2,421 ESTs. The application now covered more than 2,500 ESTs, potentially as much as 5 percent of the total number of human genes.

The NIH patent application did not become widely known until

Venter disclosed it during a Senate meeting held shortly after the *Science* publication. The reaction was one of incredulity and anger. Scientists charged that Venter's actions—akin to trying to patent the periodic table of chemical elements—would ignite an international arms race. As director of the genome center, James Watson had been forewarned of the patent application, but this was his first opportunity to comment on the record. He challenged the "non-obviousness" of Venter's sequencing operation, branding it as "sheer lunacy" and "brainless work . . . this is a perfect case of a brainless robot." In his most incendiary remark, he suggested that the EST program "could be run by monkeys." Some members of Venter's laboratory responded by buying gorilla masks and posing sarcastically for photographs in front of the sequencing machines.

An unwitting victim of the NIH proposal was Brenner, who had spent several years on his own cDNA sequencing program. Brenner was instructed by Britain's Medical Research Council not to publish his data, because if there was a chance that the NIH group could obtain a patent, then the British wanted to follow suit. Prior publication would invalidate any subsequent patent application.

More germane objections centered on the dubious utility of Venter's sequences. Most of Venter's ESTs were partial gene sequences, with only a few corresponding to full-length genes. Moreover, aside from computer comparisons of the EST sequences to identify potential relationships with known gene families, Venter had no direct experimental data on the function of the genes with which to claim a new utility. And yet in its first application, the NIH was seeking patents not only on the ESTs but also by extrapolation on the full-length genes and the corresponding proteins. How could the NIH justify such premature applications on fragments of genes, most of which had no known function?

During a meeting at the National Academy of Sciences to discuss gene patenting, Axel Kahn, a leading French geneticist, sardonically declared, "I compare this information to the discovery of celestial galaxies. I would patent the moon!" Even with patent protection, the information that could be gleaned from incomplete gene sequences, many with no identity to any known genes, would be virtually useless for drug barons. Sydney Brenner, who invented the cDNA approach, compared the pharmaceutical purchasers of ESTs to the buyers of uncut diamonds who must patiently sift through flawed stones in the hope of uncovering the occasional gem. Ironically, a few weeks after Venter's second round of ESTs were published in *Nature,* Britain's Medical Research Council filed applications with the U.S. Patent Office for its own ESTs, based on the unpublished work initiated by Brenner.

A final fear was the unprecedented scale of the applications, raising the specter of a patent gold rush, with companies and perhaps even countries racing to lock up the rights to thousands of genes. To many people, the human genome is the sacred birthright of humanity, a priceless heritage that should not be desecrated by an ugly stampede to claim exclusive rights to vast collections of genes in the form of ESTs. "It makes a mockery of what most people feel is the right way to do the genome project," declared Stanford Nobel laureate Paul Berg. Professional genetics societies and other organizations lined up to issue official statements deploring the NIH's action. Even politicians weighed in. Vice President Al Gore charged that the patent applications were "a preemptive strike [that] is universally viewed as an attempt to corner the market on human genetic information." The French research minister, Hubert Curien, argued that "a patent should not be granted for something that is part of our universal heritage."

But NIH director Bernadine Healy steadfastly believed that the applications should stand and dismissed the controversy as "a little silly . . . a tempest in a teapot." She argued that the applications were simply a defensive maneuver to bide time until the debate over the issue of gene patenting was resolved. "The rationale," Healy explained later, "is not to make money, but rather to promote and encourage the development and commercialization of products to benefit the public and to do so in a socially responsible way."

On August 20, 1992, to no one's great surprise, the patent office rejected the first round of applications. It categorically ruled that the claims failed to meet any of the three major criteria—nonobviousness, novelty, and utility—because they were (among other things) "vague, indefinite, misdescriptive [sic], incomplete, inaccurate, and incomprehensible." Particularly damaging to the NIH case was the identification of fifteen-letter segments in some of the ESTs in other genes in the database. Thus, it could be argued, a researcher could have theoretically isolated the EST on the basis of the presence of the fifteen-base sequence, invalidating the nonobviousness of the invention.

The NIH appealed the decision, and in September also filed applications for a further 4,448 ESTs. But in early 1994, Healy's successor, Nobel laureate Harold Varmus, decided not to appeal a further rejection of the original application and withdrew the subsequent application. Varmus believed that patents should be granted only to genes that were full-length and of known function.

By this time, however, the three principals in the patenting dispute—Watson, Healy, and Venter—had all left the employ of the federal government.

BY THE TIME THAT JAMES WATSON faxed his resignation letter to Bernadine Healy in April 1992, Venter was contemplating his own departure from the NIH. His ambitious application for $10 million to expand his sequencing program had been firmly rejected by the NIH. "Craig wasn't just turned down," says Sam Broder, former chief of the National Cancer Institute, "he was vilified." Venter believed that the NIH did not want a major intramural genome program being run from a neuroscience institute.

But even as his relations with the NIH were souring over the issues of patenting and funding, Venter had no shortage of suitors from the biotechnology industry. Reluctant to jump directly into the private sector, Venter spurned a $70 million offer from the biotech giant Amgen in favor of an unusual approach from the late Wallace Steinberg, chairman of the venture fund HealthCare Investment Corporation, and inventor of the Reach toothbrush. Steinberg had helped to set up more than three dozen biotech companies since 1986, and after hearing rumors that Venter was interested in leaving the NIH, he began putting together a proposal to expand and commercialize Venter's operation.

Steinberg's plan, outlined in a fifteen-minute meeting with Venter, was to set up a nonprofit research center: The Institute for Genomic Research (TIGR). Venter could conduct his research without interference from the venture fund, exert total control over the direction of his research, and enjoy almost complete freedom to publish his findings. Even so, Venter rejected Steinberg's first offer—a paltry $20 million—finally accepting a deal for a $70 million grant over ten years (and that sum was later increased to $85 million). Venter was ecstatic. "It's really remarkable," he exclaimed. "Not a single document was exchanged. . . . It's every scientist's dream to have a benefactor invest in their ideas, dreams and capabilities."

Steinberg's mission was simple: "By the year 2000, [all] drug companies in the world will use genomic data as their Rosetta stone for the development of new drugs and diagnostic procedures. No science will be

THE EYE OF THE TIGR 65

more important to the future of medicine than genomic research." To recoup his investment in TIGR, Steinberg set up a sister company, called Human Genome Sciences (HGS), to market the discoveries made by TIGR sequence data. Moreover, Steinberg knew exactly who Venter's new business partner should be.

William Haseltine was a highly respected Harvard professor, known (in no particular order) for his pioneering work on the AIDS virus; a peerless group of mentors that includes three Nobel laureates (James Watson, Walter Gilbert, and David Baltimore); a supercilious personality; and his marriage to Gale Hayman, the multimillionaire socialite who created Giorgio, one of the best-selling perfumes in history. While at Harvard, Haseltine had helped Steinberg launch several successful biotech companies—in one interview he indignantly reminded a reporter that "when we got married, I had made millions too!"—but remained tied (albeit tenuously) to academia.

That ended when Steinberg invited Haseltine down to Washington to meet Venter. As they feasted on fast food above the Bethesda Metro station, Haseltine heard about the scientific and commercial potential of Venter's EST operation. "It took me about 10 minutes to see the future," says Haseltine. That future was one where genes would be turned into drugs—and lots of money. Venter would be absolved from any direct business responsibilities and have free rein to publish his gene discoveries. The only serious stipulation was that HGS would have six months to review the data before publication, twelve months in the case of potential drug candidates.

On July 10, 1992, Venter resigned from the NIH and began preparing to move his group a few miles north to TIGR's new facility in Rockville, Maryland. In what was once a ceramics factory, Venter set about creating the world's largest DNA sequencing institute, housing thirty ABI 373A automated DNA sequencers, seventeen ABI Catalyst workstations, and a custom relational database installed on a Sun SPARCenter 2000 computer. His immediate goals were to increase EST production tenfold, establish the location of each gene on one of the twenty-three pairs of chromosomes, and begin making plans to sequence genes from model organisms as well. His long-term interest was in evolution; his dream, he told me once, was to sail to South America and retrace Darwin's voyage to the Galapagos Islands, sampling DNA en route.

Venter wasted little time in establishing TIGR as one of the, if not the most, prodigious DNA sequencing centers in the world. His first priority was to continue to build a comprehensive repository of human ESTs. Genes vary greatly in their patterns of expression. Some genes are termed "housekeeping" genes, because they encode proteins that perform essential roles in virtually all cell types in the body. By contrast, others may be expressed only fleetingly in one or two cell types, perhaps for just a matter of hours during some crucial phase of early development. In order to identify ESTs for as many genes as possible, Venter had to sequence cDNA clones obtained from as many different sources as possible: a total of three hundred different libraries running the gamut—fetal and adult tissues, healthy and diseased, male and female. In less than three years, what had originally looked like a convenient shortcut to isolating genes was now a full-scale assault on the treasures of the human genome.

Haseltine became chief executive officer of HGS in May 1993. His aspiration was to build HGS into a fully fledged pharmaceutical company but knew this would take years to achieve. The short-term solution was to sell access to the EST data to forward-looking pharmaceutical partners. Several companies, including Glaxo and Rhone-Poulenc Rorer, shied away, nervous about the task of managing and interpreting hundreds of thousands of DNA sequences. But Steinberg found a willing buyer in SmithKline-Beecham. On May 20, 1993, the British pharmaceutical company paid an unprecedented $125 million for 7 percent of HGS and exclusive commercial rights to Venter's gene portfolio. The signing triggered a wave of similar liaisons between start-up genomics companies with fanciful names like Mercator Genetics and Sequana Therapeutics, and established "big pharma," notably Hoffman-LaRoche's $70-million investment in Millennium Pharmaceuticals in Cambridge, Massachusetts.

VENTER'S METEORIC RISE TO FAME AND FORTUNE—and the furor over the NIH patent applications—made him an easy target for scientists around the world. Scientists were incredulous at the mass of data garnered by such an elementary sequencing strategy, even if no one echoed Watson's crude remarks. "He has never invented anything," complained Leroy Hood, the chief developer of automated DNA sequencing. "The only thing he deserves credit for is scaling up the process." The name-calling became increasingly personal, with "Darth Venter" being one of the least offensive suggestions. One NIH scientist contemptuously insisted on pronouncing Venter's new institute "tigger" (rather than "tiger").

Many scientists were plainly jealous of Venter's high media profile, epitomized when *Business Week* dubbed Venter and HGS chairman Bill Haseltine "The Gene Kings" on its cover. Even before that, Venter was becoming the butt of scientists' jokes. The authors of a spoof of *Nature* magazine called *Denatured* took pleasure in reports that some ESTs were contaminated with bacterial DNA sequences. The mock quote from Venter read: "You mean, I should have cloned in the cDNA before doing the sequencing?"

Venter had no doubt as to why he was being subjected to such harsh criticism. "I had a radical idea, it worked, and I was an outsider," he says. Of course, being shot at for a year in Vietnam puts things in perspective, but Venter was not the sort of person to take the jibes lying down. The worse the criticism, the more outspoken he became in defending the significance of his gene-hunting strategy and denigrating that of others.

But what upset Venter's peers the most was that in just two years, Venter had been transformed from a solid, respectable scientist working for the federal government into a stunningly successful, fabulously wealthy entrepreneur. Venter said he founded TIGR not to become rich but to perform sequencing at a level he could not achieve at the NIH. However, it was Steinberg's practice to give company founders a 10 percent stake in their new venture—in Venter's case, more than 750,000 shares in HGS. In 1994, an article in the *New York Times* revealed that Venter's financial stake in HGS was worth a cool $13.4 million. Venter was angry that his newfound wealth had just been broadcast across the country.

Still, for a man who says he had just $2,000 in savings when he started TIGR, his sudden windfall was welcome. He wasted no time in buying a $2 million, five-acre mansion in a plush Washington suburb and, as if to prove he was not beholden to HGS, sold his shares in the spring of 1994 for about $9 million. Money may not have been the prime motive for his leaving the NIH, but there was an element of truth when Venter jokingly told a *Washington Post* reporter that he left because he "wanted a bigger sailboat."

Venter first caught the sailing bug off the coast of Vietnam by agreeing to remove a soldier's unwanted tattoo in return for a modest nineteen-foot boat. Following his windfall, Venter could finally afford the boat of his dreams: a lavish $4-million, 100,000-pound, 82-foot aluminum racing sloop previously belonging to the owner of the Lands End clothing company. Venter changed the name from *Turmoil* to *Sorcerer* (the outboard was called *Apprentice),* and spent liberally on new sails, including a spinnaker that unashamedly bears a 20-foot-high caricature of its new owner sporting a peaked wizard's hat. Venter's passion for the water was demonstrated a few years later in May 1997, when he proudly captained *Sorcerer* to victory in the Atlantic Cup Challenge race from New York to Falmouth, competing against schooners twice the size of *Sorcerer*. It would not be the last time his competitive instincts rose to the fore.

CHAPTER FOUR

Loading the Bases

> THE DNA MOLECULE
> THE DNA MOLECULE
> THE DNA MOLECULE
> is *The Nude Descending a Staircase*
> *a circular one.*
> *See the undersurfaces*
> *of the spiral*
> *treads and*
> *the spaces*
> *in between.*
> —From "The DNA Molecule," MAY SWENSON

"MY WHOLE CAREER HAS BEEN SPENT training for this job—this is more important than putting a man on the moon or splitting the atom." With those giddy words at an April 1993 press conference in Washington, D.C., Francis Collins finally confirmed his appointment as the second director of the National Center for Human Genome Research and the de facto leader of biology's version of the Manhattan Project.

Bernadine Healy's appointment of the University of Michigan medical geneticist after a protracted year-long courtship was greeted rapturously by the genome community, which regarded Collins as the logical choice to succeed James Watson. Nobody else had played a more successful or visible part in the spectacular successes during the early years of the Human Genome Project than Collins had. International teams of scientists had made a series of spectacular gene discoveries and identified mutations responsible for many of the most common and life-threatening genetic disorders. This gene bonanza made minor celebrities of many genetics researchers, but none more so than Collins. It was a remarkable rise for a gifted geneticist who just five years earlier had been a virtual unknown outside the genetics community.

Born in April 1950, Collins was the fourth son of a teacher father and playwright mother who lived on a small farm in the Shenandoah valley of Virginia. He was schooled at home by his mother, who nurtured his early interests in chemistry and mathematics. His interest in medicine did not materialize until after he obtained his Ph.D. in physical chemistry from Yale University at the age of twenty-four, sparked by some biochemistry courses where he learned about nucleic acids. After completing his residency in 1981, Collins returned to Yale to do research with geneticist Sherman Weissman, whom Collins calls "the smartest guy" he has ever met. Collins wasted no time in making his mark. In 1985, he discovered a mutation in one of the globin genes that causes a fetal form of thalassemia. But his attention was really caught later that year by a triad of reports that traced a deadly disease gene to a spot on chromosome 7.

The most common recessively inherited disease among people of European descent is cystic fibrosis, a fatal lung disease in which bacteria thrive in the airways due to an altered salt balance and reduced efficiency of the host's defenses. In 1985, Canadian geneticist Lap-Chee Tsui, working with a Massachusetts biotech company, Collaborative Research, had mapped the cystic fibrosis gene to the long arm of chromosome 7, with two other groups snapping at his heels. The scene was set for an all-out race to snare the errant gene.

By contrast to Collins's routine studies on globin gene mutations, the search for the gene mutated in cystic fibrosis patients was altogether more dynamic and exciting. Finding the rough location of the gene was a tremendous breakthrough, but now came the hard part: navigating from the linked marker to the gene itself, across a region of possibly 10 million to 20 million bases of unmapped DNA containing hundreds of genes that could last years. Moreover, there were no reports of any chromosomal rearrangements that might point researchers to the location of the faulty gene. Such clues were priceless in the hunt for disease genes, as researchers in Boston soon showed.

In 1966, a boy named Bruce Bryer was born in Washington State. At the age of three, Bruce was diagnosed with chronic granulomatous disease, an inherited abnormality of the immune system that left him prone to serious infections. By age twelve, Bruce was confined to a wheelchair suffering from Duchenne muscular dystrophy, and a few years later, he was diagnosed with retinitis pigmentosa, a progressive form of blindness. Bruce died in 1983, but his DNA proved to be a blessing for genetics research. Careful analysis of Bruce's chromosomes revealed that his X chromosome was slightly smaller than normal; there was a missing segment in the middle of the short arm. The logical conclusion was that the absence of three critical genes in this region led to Bruce's ilnesses. In 1986, Stuart Orkin's group at Boston's Children's Hospital isolated the gene for chronic granulomatous disease, marking the first isolation of a mutated gene purely by its chromosome position, without any knowledge of its function. The following year, Lou Kunkel's team identified mutations in Duchenne patients in a massive nearby gene called dystrophin.

Unluckily for Collins, there were no obvious DNA deletions to guide their journey along chromosome 7. What was needed was an expedited method to jump from the marker to the disease gene, and Collins thought he had the solution. Together with Weissman, Collins conceived of a clever scheme called "positional cloning," which Collins later refined in his own lab at the University of Michigan. Instead of "walking" along the chromosome by using one DNA fragment to fish out an overlapping fragment and then repeating the process with this new fragment, Collins's new "jumping" method bypassed many of these steps. He took a much larger piece of the chromosomal DNA, about 100,000 letters in length, and joined the ends to form a circle. Then by isolating a smaller segment that contained the two fused ends, Collins had effectively jumped 100,000 bases, saving weeks of work. Now Collins could systematically work his way from the linked marker toward the gene itself. Even when it briefly looked as if Collins's chief rivals, Bob Williamson's group in London, had found the gene, Tsui and Collins kept going and decided to pool their resources. In the summer of 1989, just weeks from potentially finding the gene, Collins stunned his Canadian colleagues by flying to Nigeria with one of his daughters to work in a small missionary hospital. For those who know Collins, the trip was not a complete surprise. He visits every few years if possible; on one occasion, he performed a risky surgery to save the life of a man suffering from heart failure. "It helps me to remember what matters," he explains.

Shortly after his return, Collins and Tsui found the definitive piece of evidence that they had snared the cystic fibrosis gene: three missing letters of DNA on chromosome 7 in 70 percent of cystic fibrosis patients. Collins drove the short distance from Ann Arbor to the Hospital for Sick

Children in Toronto, where he celebrated with his partners by sharing a bottle of Canadian whiskey that Tsui had given him two years earlier to seal their collaboration.

While members of his group began studying the function of the cystic fibrosis gene, Collins was eager to apply his cloning skills to a new challenge. He chose neurofibromatosis, another common inherited disease, the gene for which is on chromosome 17. Sometimes mistaken for Proteus syndrome, the disorder that afflicted Joseph Merrick, the Elephant Man, neurofibromatosis patients develop a multitude of frequently disfiguring benign and malignant tumors on their skin and so-called café-au-lait spots. Once again, Collins's competitive spirit came to the fore. A once amicable collaboration with University of Utah geneticist Ray White suddenly fell apart, turning former allies into bitter competitors. Collins reached the neurofibromatosis gene first and fired off a report to *Science*. White countered by calling the editor of *Cell*, a leading molecular biology journal, to arrange expedited publication of his article—though he had yet to write the manuscript. Some judicious lobbying by Max Cowan, vice president of the Howard Hughes Medical Institute, the financial benefactor for both investigators, ensured that both articles were published on the same day and credit was shared.

But squabbles among scientists over collaborations and credit are nothing new, and the whiff of controversy soon disappeared. There was no doubt by now that Collins's career was on the fast track. Using tools that he helped develop, Collins had identified two precious disease genes that had eluded researchers for years, catapulting him into the public eye, where he excels with a natural laid-back charm and a sixth sense for knowing how to pitch his talks to his audience. This is on display whether he is entertaining an auditorium of medical students by strumming his acoustic guitar, delivering the keynote lecture before thousands of people at a medical convention, testifying on Capitol Hill, counseling an anxious family, or being interviewed live on television.

The media, much to its delight, also found a surprisingly engaging personality, who was happy to share his views on his Nighthawk 750 for an Italian motorcycling magazine and his religious beliefs for anyone else. Collins, who became a born-again Christian about the time he earned his M.D., says his passion for tracing disease genes is like "appreciating something that up until then, no human had known, but God knew it. . . . In a way, perhaps, those moments of discovery also become moments of worship."

• • •

IN MARCH 1993, just a couple of weeks before his induction as the head of the NIH genome project, Collins celebrated yet another coup—his third major gene discovery in four years—although in a less prominent role. After an arduous search lasting ten years, an international research consortium of sixty researchers finally ended what some people were calling the greatest gene hunt in history by identifying the gene for Huntington's disease (or chorea), an incurable neurodegenerative disorder. Huntington's disease is named after George Huntington, a physician's son who made a surprising discovery as a boy while traveling in Long Island, New York: "Driving with my father through a wooded road . . . we suddenly came upon two women, mother and daughter, both tall, thin, almost cadaverous, both bowing, twisting, grimacing. I stared in wonderment, almost in fear. What could it mean?"

Huntington went on to study a number of affected families, mostly descendants of English colonists who had settled on the east coast of the United States in about 1630. He noted that the disease "has been transmitted to them, an heirloom from generations away back in the dim past." In addition to the inexorable physical decline, the young physician noted something else: a gruesome cognitive decline, frequently leading to insanity and suicide. It is, as one neurologist wrote, "the complete ruin of a human being." Huntington's disease is a dominant disorder: just a single mutant gene is sufficient to trigger the development of the disease. But Huntington's disease plays a cruel trick: there are no traces of illness at birth or during childhood. It is only as an adult, typically in the forties or fifties, when the first distressing signs of the disease begin to occur.

One year after Huntington's disease claimed its most famous victim, the American folk singer Woody Guthrie in 1967, a fifty-three-year-old woman was diagnosed with the same disease. Her name was Leonore Wexler, and her family members made the search for the killer gene their life mission. Her husband, Milton, created the Hereditary Diseases Foundation to sponsor research into the causes of Huntington's disease, while her daughter, Nancy, a psychologist at Columbia University, spearheaded the effort to find a genetic breakthrough. Shortly after her mother's death in 1978, Nancy traveled to Lake Maracaibo in Venezuela, where scores of villagers are afflicted with *El mal de San Vito,* as the disease is known. As she recorded the family histories of the villagers, she realized that the vast pedigree, numbering more than three thousand people, might provide the means to map the Huntington's gene. Many of Wexler's confidantes were enthused about the prospects of using DNA markers to map disease genes, in the manner outlined by Botstein and colleagues in their seminal 1980 article. Even so, the chances of finding a marker adjacent to the Huntington's gene would be like pulling the proverbial needle out of the genome haystack.

Nevertheless, a brave young Canadian geneticist named Jim Gusella took up the challenge. He began testing polymorphic DNA markers one at a time, hoping one would reveal a sequence variation that would track with the Huntington's gene. The twelfth marker he tried was called G8; this was not a gene, simply a random fragment of DNA with a common sequence variation in different people. Incredibly, all of the Huntington's patients tested had inherited one form of G8 as assuredly as they had inherited the Huntington's gene. The results produced odds of more than 1 million to 1 that G8, known to reside on chromosome 4, must lie nearby the disease gene. Gusella's serendipitous discovery—the first time the location of a disease gene had been traced without any prior knowledge of its location—was published in 1983 to worldwide acclaim.

The surprising ease with which Gusella traced the Huntington's gene fueled expectations that the gene would be isolated shortly after. Collins's exciting positional cloning techniques made him a natural fit for the international alliance of American and European groups searching for the gene. But those hopes soon faded. Without any DNA sequence of the key region of chromosome 4, the researchers had first to isolate and map millions of letters of DNA before they could contemplate finding the gene. It was like conducting a house-to-house search for a criminal in New York City without a street guide. The researchers played hunches as to which spot they should concentrate their energies; at times, Collins said, the frenetic confusion was worthy of the Keystone Kops.

As the tenth anniversary of the mapping of the Huntington's disease gene approached, the dozens of dedicated researchers obsessed with the quest had little to show. The only thing that kept them going was to dream of the moment when the gene was finally found. One of the consortium leaders, John Wasmuth of the University of California, Irvine, summed up his emotions like this:

> It's just like reading a great mystery novel, you're up to the last chapter and it's been so involving and exciting you want it to continue and yet you can't wait to find out who did it. At the

same time, I'll be a little bit sad [when it's over] because it's the
end of the most exciting project that's ever gone on in my lab.
. . . Like Buzz Aldrin, the second guy to walk on the moon,
says, "Well, what the hell do I do now?"

Finally, in March 1993, the rogue gene was finally caught. Marcy
MacDonald, a senior member of Gusella's laboratory, detected a telltale
difference in the DNA from chromosome 4 of Huntington's patients versus healthy individuals. A small stuttering stretch of three letters—CAG—
normally repeated about six times, had expanded like a concertina to
more than forty in patients. The significance of this stuttering segment of
DNA was plain. Two years earlier, researchers had shown that fragile X
syndrome, the most common inherited form of mental retardation, was
caused by a similar expansion in a three-letter riff in the DNA sequence.
In general, the longer the repetitive tract of DNA is, the earlier the symptoms appear. The protein that is ravaged by this bizarre expansion of repeats is named Huntingtin, in honor of the disease discoverer.

WHILE THE DISCOVERY OF THE GENES for major genetic diseases captured most of the headlines, other groups were working to find more navigational aids to help the gene mapping community. To many people's
surprise, the most important contributions emerged not from the United
States but from a remarkable organization based just south of Paris, the
Centre d'Etude du Polymorphisme Humain (CEPH).

In 1982, Sotheby's auctioned the art collection of Helene Anavi,
raising 50 million francs (about $8 million). The proceeds were left to Jean
Dausset, who had won the Nobel Prize in 1980 for his studies on the immune system and was perhaps the most famous French scientist of his
generation. Dausset used his bequest to create CEPH. His chief goal was
to disseminate DNA from members of forty multigeneration families (including twenty-seven Mormon families) to more than one hundred research groups around the world, laying the international foundation for
the construction of the genetic map. In return for distributing CEPH
family DNA, researchers agreed to deposit their genotyping results in a
central database. Researchers in San Francisco, for example, could compare the inheritance pattern of a DNA marker with results on a second
probe produced by a group in Strasbourg. The blood cells from these
family members were treated with a tumor virus to become immortal-

ized, providing a perpetual source of DNA. By 1987, teams led by Ray White and Helen Donnis-Keller at Collaborative Research in Massachusetts had assigned markers to every human chromosome and constructed the first patchy framework maps of the human genome, guiding researchers toward the location of disease genes.

Dausset quickly found an unusual ally in the form of the French Muscular Dystrophy Association, and in particular Bernard Barataud, whose son suffered from the muscle-wasting disease. In 1986, Barataud listened in awe to a lecture given by a young American geneticist at his association's annual conference. He was Tony Monaco, a member of Lou Kunkel's laboratory, and he was describing the identification of the gene on the X chromosome responsible for Duchenne muscular dystrophy. "Good God," thought Barataud, "this was no old distinguished scientist who was announcing this to us with his Legion of Honor ribbon in his buttonhole. It is [an] adolescent in sneakers and jeans who looks like he just left a party."

Barataud's emotions veered from unbridled joy to intense anger—anger directed not so much at the genetic curse in his family but more at the indifference of the French government to support the genetics revolution. The Duchenne discovery galvanized the French muscular dystrophy association, which followed the successful model of Jerry Lewis's annual muscular dystrophy telethons in the United States. The first French *téléthon* in 1987 brought in donations of roughly $35 million. The following year, Barataud met with CEPH director Daniel Cohen, a talented and ambitious physician-turned-geneticist who hailed from Tunisia. Cohen believed that the best use of *téléthon* funds lay in bolstering the genetic infrastructure, the chromosome maps, before his team could make strides in identifying other genes that cause muscular dystrophy. Cohen's specific idea was to produce an exhaustive collection, a library in effect, of purified human DNA fragments spanning every chromosome. These fragments would be housed in special cloning vehicles called yeast artificial chromosomes (YACs), devised by Washington University's Maynard Olson, which could house much larger DNA fragments than conventional viral and bacterial vectors.

Barataud's success in funding such an abstruse project was impressive, but he did not want to stop there. He asked Cohen whether it would be feasible to build an industrial-scale factory for the discovery of human disease genes. It was, Cohen says modestly, "a turning point in the history of humanity." In just four months, CEPH and the French Muscular Dystrophy Association built Généthon—Europe's largest genome institute, a veritable assembly plant for the mass production of genetic resources and information. Cohen installed an army of machines to do the rote work of DNA preparation and analysis to allow researchers more time for constructive research. The robots were the prime attraction for visitors and helped establish Cohen as the Henry Ford of the new era of industrial genetics.

By 1992, Cohen and team leader Ilya Chumakov had assembled a set of YACs that spanned the entire length of chromosome 21 and went on to produce a preliminary physical map of the human genome by assembling overlapping YACs for each chromosome. It was admittedly rough in places, but it signaled to researchers that the genome could be subdivided into ordered fragments suitable for DNA sequencing. While Chumakov headed production of the physical map, Jean Weissenbach was making equally rapid progress in constructing a different kind of map—the "genetic map."

Building on the genetic maps created by White and Collaborative in the late 1980s, Weissenbach was intent on charting the locations of thousands of DNA markers along each chromosome to expedite the search for disease genes. Weissenbach concentrated on a new breed of marker called microsatellite repeats—stretches of DNA containing variable lengths of two alternating letters (such as CACACACACA . . .). These microsatellite markers rapidly gained favor because of their abundance, extensive variability in families, and suitability for automated analysis. By the end of 1992, Weissenbach had produced a genetic map containing 800 markers, twice the number Collaborative Research had reported five years earlier. Eighteen months later, in June 1994, that number had surpassed 2,000, with an average spacing between markers of about 2.5 million bases, essentially fulfilling (three years early) one of the main targets of the genome project's five-year plan.

Cohen's ambitions did not stop with CEPH. He established the Science for Peace Foundation to create a state-of-the-art facility in his native Tunisia for research into Third World diseases (Craig Venter volunteered to help raise funds). He authored a provocative book, *Les Gènes de l'Espoir (The Genes of Hope),* in which he anticipated the future "amelioration of our genetic patrimony." This, he says, "will be a form of eugenics . . . that doesn't frighten me more than the discovery of vaccinations, antibiotics, or improvements in childbirth." Cohen also predicted the creation of "a human that is more complex, more refined, more subtle, farther from animals, than the ones we have today."

Together with Eric Lander, who was rapidly building mouse maps analogous to Généthon's human maps, Cohen became cofounder of a new biotechnology company called Millennium Pharmaceuticals, based in Cambridge, Massachusetts. Millennium sought to find the genes for a variety of complex disorders, including diabetes and obesity, and Cohen just happened to be sitting on a priceless resource. Phillipe Froguel had recruited more than five thousand subjects from eight hundred families to give blood for the study of genetic factors in diabetes. In early 1994, Cohen and Froguel met with Millennium executives and lawyers to discuss terms of a possible collaboration, but Froguel was unwilling to cede control of his patient samples to an American company, especially one in which Cohen owned a 2 percent stake. On March 7, 1994, Froguel appealed in writing to the French prime minister seeking his "urgent intervention to protect the national 'patrimony,' " to save the exploitation of "AND français"—French DNA.

The scandal became front-page news in France, as the French press depicted American biotech companies capitalizing on French donations of DNA. While Cohen claimed no French company could commercialize Froguel's diabetes research, CEPH's deal with Millennium floundered. In February 1996, Cohen resigned from CEPH to take a senior post with GenSet, a French genomics company that seeks to identify genes for common disorders including cancer, Alzheimer's disease, and schizophrenia. But Cohen's assault on the French national patrimony was eventually forgiven. He was awarded the Legion of Honor by the French government in 1997.

As COLLINS SETTLED IN at the helm of the genome project, the explosion in human disease gene discoveries that he had helped foster continued apace. Fittingly, in the Decade of the Brain, many advances centered on disorders of the nervous system. Three genes were found to be mutated in rare early-onset cases of Alzheimer's disease, and variations in the apolipoprotein E gene on chromosome 19 were associated with later-onset cases of Alzheimer's. Another tense search turned up the gene for a familial form of amyotrophic lateral sclerosis (ALS, also known as Lou Gehrig's disease), the relentless loss of motor skills that affects British physicist Stephen Hawking. Surprisingly, the mutated gene on chromosome 21 turned out to be one of the best-known genes on the chromosome, coding for an enzyme called superoxide dismutase, which mops up highly reactive oxygen radicals. A group at Collins's own research center at the NIH, studying rare Mediterranean families with a hereditary form of Parkinson's disease, discovered mutations in a gene called alpha-synuclein, a protein deposited in the brains of Parkinson's patients. More neurological diseases, including spinocerebellar ataxia and Friedreich ataxia, were added to the family of disorders such as Huntington's caused by expanded triplet repeat sequences.

In 1993, a French team led by Jean-Louis Mandel discovered the gene for adrenoleukodystrophy, the disease made famous by the film *Lorenzo's Oil,* starring Susan Sarandon and Nick Nolte as Michaela and Augusto Odone, the parents of an affected boy. The Odones lobbied for the use of polyunsaturated oils to treat the disease, which experiments have shown to be beneficial in some cases if administered early in the course of the disease. Ironically, Lorenzo outlived his mother, who died of lung cancer in June 2000.

The list continued to grow. In 1994, John Wasmuth discovered the gene for the dominant disorder achondroplasia, the most common genetic form of dwarfism. The discovery was a welcome bonus after the decade of despair working on Huntington's disease, because the dwarfism gene, near the tip of chromosome 4, had been bypassed by Wasmuth during his earlier search. The identical nucleotide was mutated in every patient, resulting in the substitution of a single amino acid. Wasmuth was concerned that his discovery not be misused to abort fetuses with what he considered a benign trait. Many couples with achondroplasia would rather raise a similarly affected child than a "healthy" unaffected individual. "I don't want to be alive when the last dwarf is born," says John Brookfield of the Little People of America.

Another French team discovered the gene for spinal muscular atrophy, considered to be the second most common recessive disorder among Europeans (after cystic fibrosis). Another large collaboration pulled out the gene for polycystic kidney disease, a common dominantly inherited disorder that, like Huntington's disease, manifests itself only in adulthood. Several genes for hereditary blindness, including retinitis pigmentosa, and hereditary deafness were found. Mary-Claire King made headlines by finding the first deafness gene, in a large family from Costa Rica. The hearing loss grew progressively, taking away the lower tones first and higher pitches later. Grandparents forced their grandchildren to learn to lip-read in case they had inherited the deafness mutation.

By 1995, researchers had unearthed more than fifty disease genes by the technique of positional cloning (or some variant of it). With understandable paternal satisfaction, Collins said the technique had moved "from perditional to traditional," and there were occasions when it looked almost too easy. To Oxford University geneticist Brian Sykes, the quest for an elusive disease gene had all the trappings of an addictive soap opera. In some episodes, however, the courtship of a closely linked DNA segment quickly turned into a monogamous relationship. "Gone were the opportunities for brief flirtations with different markers, each in turn discarded for a better-looking rival in an atmosphere of secrecy, treachery and betrayal. Lost were the chances of leading the audience step by panting step to the ultimate climax—the isolation of the gene itself."

As the list of identified genes responsible for Mendelian, or single-gene, diseases increased, interest naturally turned to much more common multigene disorders, where variations in several genes likely conspire to cause disease, frequently in conjunction with environmental factors. Researchers have used two complementary methods to begin finding genes responsible for these diseases. One involved studying families with inherited versions of more complex diseases, a good example being elevated blood pressure, or hypertension. At Yale University, Richard Lifton's group has collected families with an assortment of syndromes characterized by increases in blood pressure and documented mutations in almost a dozen genes. Many of these genes code for proteins in the kidney that regulate the passage of salt and other metabolites. He can now assess how variants in these genes contribute to common hypertension or if other factors are involved.

Another strategy is to study mutations in mice that, because of their genetic relationship with humans, can provide valuable clues to understanding genetic disease. In 1994, Rockefeller University scientist Jeff Friedman isolated the mouse "obesity" gene, a gene that is mutated in profoundly obese animals. The gene turns out to code for a hormone-like molecule called leptin (after the Greek *leptos* for thin). In humans, as in mice, leptin acts as a satiety signal, secreted from fat cells in the body and detected in the brain, which tells the animal to stop eating. Loss of leptin production (or recognition in the brain) causes massive obesity. The commercial rights to leptin were auctioned off by the Rockefeller group to Amgen for a staggering $20 million. Friedman's gene sparked an extraordinary upswing in obesity research, leading to the identification of many more appetite-regulatory factors and new leads for treating the disorder.

As the explosion in gene discoveries associated with inherited diseases reached its peak in the mid-1990s, with researchers triumphantly announcing major genetic breakthroughs almost weekly, a "gene of the week" trend became evident, which exasperated many commentators, who tired of the importance being ascribed to genes. This was typified by this satirical extract from Horace Judson:

> It is announced that molecular biologists have discovered the gene for schizophrenia. The gene for diabetes. The gene for alcoholism. The gene for Huntington's chorea. The gene for homosexuality. The gene for Alzheimer's disease. Another gene for schizophrenia. The gene for obesity. Another gene for pancreatic cancer (did you miss the first one?) The gene for vivacity . . .

Judson's complaint was that with a handful of exceptions, scientists were failing to explain the nuances of their discoveries and could not blame the public's perceived ignorance on the sensationalizing media. It is easy to say that someone "has the muscular dystrophy gene" or "the Huntington's gene," but in fact the opposite is true: such individuals *lack* a functional copy of the gene that makes the dystrophin or Huntingtin protein. These proteins actually perform roles in maintaining the structural framework of muscle fibers or nerve cells. But for most people, the normal function of these gene products is irrelevant. The genes in these two examples were isolated solely because of their ties to two fatal genetic diseases.

PERHAPS THE MOST CELEBRATED PIECE of genetic real estate claimed in the early 1990s was *BRCA1,* popularly known as the breast cancer gene. The sharply rising incidence of breast cancer in the United States over the past few decades triggered an explosion of feminist anger, congressional sympathy, and research funding. The lifetime risk of breast cancer in the United States is one in nine, yet with no definitive evidence as to the environmental causes, hope turned to genetics. For years, nobody believed that the risk of breast cancer could run in families. It was only through the tireless research of Berkeley geneticist Mary-Claire King, beginning in the mid-1970s, that this began to change. Nancy Reagan's disclosure in

1987 that she had been diagnosed with breast cancer led to thousands of women contacting King's group at Berkeley. Many had several relatives with breast or ovarian cancer, providing tantalizing evidence for a susceptibility gene. King's group trawled through over two hundred DNA markers looking for one that tracked in the same pattern as the incidence of cancer in the pedigrees.

The crucial breakthrough came in 1990 when one of King's students suggested sorting the families based on the age of diagnosis of the cancer. When King confined her analysis to just those families with a relatively early age of onset (under fifty years of age), there was convincing linkage with a marker on chromosome 17. King presented her stunning results at the American Society of Human Genetics conference in Cincinnati in October 1990. Inheritance of this faulty *BRCA1* gene increased a woman's chances of developing breast cancer to 85 percent and ovarian cancer 50 percent. Among the packed crowd hanging on King's every word that evening was Francis Collins, who was already marshaling resources on chromosome 17 to find the neurofibromatosis gene. Collins immediately set about persuading King to collaborate with him. She was flattered but hesitated until James Watson counseled her: "I needed a Francis, and so do you." The alliance looked formidable on paper, combining King's flair in epidemiology and statistical analysis with Collins's forte of snaring disease genes.

Once King had announced the rough location of *BRCA1*, the prize was up for grabs. The stiffest competition came in the shape of a young biotechnology company in Salt Lake City called Myriad Genetics. Myriad was cofounded by Nobel laureate Walter Gilbert (the man who compared the genome project to the holy grail) and Mark Skolnick, a genetic epidemiologist from the University of Utah who had helped David Botstein conceptualize the human genetic map in 1980. Both teams zeroed in on the whereabouts of *BRCA1*, trawling for telltale mutations in the DNA of women with familial breast cancer.

Like other groups, Collins and his associate, Barbara Weber, sought to identify large families with breast cancer. One large family interested them in particular because of the devastation that had been caused by breast cancer. One young woman (known simply as Susan M. to protect her anonymity) decided to have her breasts prophylactically removed rather than endure the near certainty (in her mind) of developing cancer, which had struck her two sisters and several cousins and had killed her mother. Unaware that she was preparing for surgery, Collins and Weber found strong evidence that the cancer in Susan's family was probably due to a mutation in *BRCA1*. It was only when Weber made a timely call to Susan's sister informing her that the genetic tests in her family would be informative that the surgery was postponed—for good. Susan tested negative for a *BRCA1* mutation, but suffered from survivor's guilt for being the only sibling not to inherit the faulty *BRCA1* gene.

Collins continued the quest for *BRCA1* after moving his laboratory to the NIH in 1993, but Myriad was making the faster progress, thanks in part to the decision to build a physical map of the *BRCA1* region with bacterial artificial chromosomes (BACs). These vectors could not house fragments as large as YACs, but were much more stable. Finally, in September 1994, Collins received a phone call from Robert Bazell, the chief science correspondent for NBC News, confirming his worst fears: Myriad had won the race for one of the most prized genes in biomedical research. Bazell broke the news that night, even though Myriad's formal report was still weeks from being published. The discovery made headlines around the world. Natalie Angier of the *New York Times* hailed the quest for the breast cancer gene as "a genetic trophy so ferociously coveted and loudly heralded that it had taken on a near-mythic aura."

King was also devastated by the loss of the *BRCA1* race and the abrupt termination of her collaboration with Collins. But she recovered her poise to deliver a magnanimous talk in front of thousands of researchers at the American Society of Human Genetics convention in Montreal, just weeks after Myriad released the DNA sequence of *BRCA1*. What many expected to be a concession speech turned into a classic demonstration of mutational analysis in a series of breast cancer families. Credit for the discovery was of limited importance, King said, because she would have been searching for breast cancer mutations whether her lab had found the gene first or not. She closed with an emotional appeal for her colleagues to fight on:

> An important part of the last few weeks has been [to] distinguish reality from fantasy. Fantasy has been the race—*New York Times* profile, *60 Minutes,* guys on motorcycles in *Time* magazine. Reality is having the gene, not knowing what it does, and the realization that in the 20 years since we have been working on this project, more than a million women have died of breast

cancer. We very much hope that something we do in the next 20 years will preclude another million women dying of the disease.

To judge from the standing ovation that greeted these words, anyone would have thought that it was King who had discovered BRCA1, not a biotech company that intended to profit from the commercialization of a breast cancer test. But in a way, she *had* discovered the breast cancer gene, even though she narrowly lost the race to determine its sequence.

Ironically, after all the hype, Myriad's discovery had a relatively subdued impact on the diagnosis of breast cancer. The company offered a comprehensive sequencing test that was both expensive (more than $2,000) and yet equivocal. A negative result (that is, no detected mutations in BRCA1) still carries the risk of breast cancer of the general population, whereas a positive test greatly increases the chances of developing breast cancer but does not make it a certainty. Nevertheless, some women found to carry a BRCA1 mutation do decide to have a prophylactic mastectomy or have their ovaries removed. In one dramatic example, the wife of Joseph Schulman, the director of the Genetics and IVF clinic in Fairfax, Virginia, decided to have her BRCA1 genes tested because of a family history of breast cancer. The result was positive: one copy of her BRCA1 gene was missing two letters, putting her at heightened risk of a breast tumor. She chose to undergo a prophylactic mastectomy, even though she had no signs of cancer.

Other women were not so fortunate. One woman had both her ovaries removed after a commercial genetic test for BRCA1 mutations came back positive. Follow-up studies by a university team revealed that a mistake had been made: she did not have a mutation. The company that initially performed the test, Oncormed, refunded the $350 testing fee "as a small token of our concern."

THE LOSS OF THE RACE for BRCA1 was a bitter blow for Collins; it was the first time he had come out on the losing end of a high-profile gene hunt. But thanks to another far-ranging collaboration, he was able to claim a share of a significant consolation prize. The following summer, a large team headed by geneticist Yosef Shiloh, from Tel Aviv University, identified a gene on chromosome 11 mutated in patients with ataxia telangiectasia. This puzzling recessive disorder affects about 1 in 50,000 people and is characterized by ataxia (disturbance of balance), increased risk of cancer and heightened sensitivity to X rays. But there is also evidence that carriers of the mutated gene (given the catchy title *ATM,* for ataxia telangiectasia, mutated) are also at increased risk for cancer, with perhaps a fivefold increased risk of breast cancer in women. Given that 1 in 100 people carry a defective copy of *ATM,* these findings are a cause for concern, especially given the suggestion that the dose of X rays used in mammography might contribute to cancer in *ATM* carriers. Collins was also involved in a large project to map the location of a gene for familial prostate cancer, which paid off in 1995 when such a gene was located on chromosome 1.

But amid the celebrations came a major crisis. In 1993, Collins's group had reported an important discovery about the genetic origins of a subset of acute myeloid leukemia, which are initiated when the middle portion of chromosome 16 mysteriously becomes inverted. Collins's group deciphered the DNA sequence at the junctions of the two breakpoints and showed that the inversion produced a genetic chimera—half of one gene fused into a completely different gene. This fusion product was speculated to cause a drastic alteration in the properties of a transcription factor, potentially turning white blood cells cancerous. Following this important discovery, Collins's team set about fleshing out the story. Over the next two years, a steady stream of papers was published from the Collins lab, mostly based on experiments conducted by Amitav Hajra, a brilliant graduate student in the Collins group.

In 1996, Collins submitted another manuscript, detailing some of Hajra's experiments to the British cancer journal *Oncogene.* One unusually conscientious reviewer, who was troubled by some of the data, made a depressing discovery: one of the photographs in the manuscript looked as if it had been doctored.

Hajra soon admitted to the fraud, but the deception went deeper than anybody realized. A subsequent inquiry by the Office of Research Integrity concluded that no fewer than five papers coauthored by Hajra, Collins, and others contained falsified data. Scientifically, the damage was minor: the publications in question were the standard follow-up analyses that follow the identification of an important disease gene and did not invalidate the discovery of the leukemia-causing rearrangement on chromosome 16. But the extent of the fraud—spanning more than two years and wasting tens of thousands of dollars—shocked researchers, none more so than Collins, who felt betrayed by his star student.

Calling it "the most painful experience of my professional career," Collins issued a letter of apology to a hundred close colleagues in which he rhetorically asked the question on many people's minds: "Many will wonder whether I as the research mentor was paying sufficient attention to the individual, if such deliberate and systematic assaults on scientific truth were occurring." Collins drew praise for his handling of the crisis, but the episode illustrated how time-consuming and distracting the bureaucratic burden of managing the genome project was becoming.

NO MATTER HOW EXCITING the progress in identifying disease genes during the 1990s, Collins knew that these successes alone were not sufficient justification for the investment in the genome program. From the moment he took charge of the genome project, Collins knew that two things were required to have any hope of sequencing the human genome by the target date of 2005. The first priority was the construction of a complete physical map for each chromosome, consisting of a series of purified overlapping fragments of DNA that would provide the raw materials for DNA sequencing. The second was for major improvements in the speed and efficiency of DNA sequencing. Neither was progressing as hoped.

In contrast to their spectacular success in producing dense maps of genetic markers, the physical maps compiled by Daniel Cohen and his colleagues at CEPH were less reliable. The French team produced a library of large chromosome fragments in "megaYACs," which could house DNA fragments of more than 1 million bases. However, these clones were notoriously unstable: the DNA inserts would often become rearranged, wreaking havoc with attempts to construct physical maps. The first physical map of any human chromosome was that of the Y chromosome, produced by the Whitehead Institute's David Page, despite the fact that almost two-thirds of the clones had rearrangements. And both Collins and King blamed their failure to isolate *BRCA1* on the disadvantages of working with YACs compared to BACs, created in 1992 by Mel Simon's group at Caltech.

Despite the well-documented problems with the physical map of the genome, many experts were getting impatient to begin full-scale sequencing. University of Washington genome center chief Maynard Olson is a quiet-spoken, intellectual man, so when he voices a strong opinion, people tend to listen. In October 1995, Olson penned a commentary in *Science* entitled "A time to sequence," in which he concluded, "The development of a human being is guided by just 750 megabytes of digital information. In vivo, this information is stored as DNA molecules in an egg or sperm cell. In a biologist's personal computer, it could be stored on a single CD-ROM. The Human Genome Project should get on with producing this disk, on time and under budget."

In February 1996, the Wellcome Trust organized the first International Strategy Meeting on Human Genome Sequencing, in Bermuda, which yielded the Bermuda Accord. This stated: "All human genome sequence information should be freely available and in the public domain in order to encourage research and development and to maximize its benefit to society." All of the major international sequencing laboratories agreed to abide by the terms, including the principal genome sequencing laboratories in the United States (Whitehead Institute, Baylor College of Medicine, Washington University, Joint Genome Institute, Stanford, TIGR, and others), as well as laboratories in England (Sanger Centre), France, Germany, and Japan.

But Collins was facing a second major concern: where was he going to find the improvements in the technology and economy of DNA sequencing necessary to meet the 2005 deadline? He entrusted six of the preeminent sequencing centers in the United States (Stanford, Houston, Seattle, St. Louis, TIGR, and MIT) to develop new technologies and faster methods of DNA sequencing, but although they adopted a variety of strategies and techniques to boost sequencing output, the results were lackluster. The standard sequencing machine, the ABI 377 sequencer, was impressive—with two runs squeezed out of it per day, a machine could average about 50,000 bases—but it was not nearly enough. By early 1998, only two of the pilot centers had reduced costs to less than $1 per base, and one team had barely cracked the $10 mark. The centers were producing between 2 million and 30 million bases of sequence a year, well below their targets, such that only 2 to 3 percent of the DNA in the human genome had been sequenced. In fact, researchers had finished more nematode sequence than human DNA. To complete the genome sequence by 2005, the major centers would have to increase their productivity to at least 75 million bases per year for seven consecutive years. As TIGR's Mark Adams told *Science* in May 1998, "I don't know how we're going to do that."

Despite the productivity problems, there were encouraging signs that international genome sequencing with a loose consortium of laboratories could achieve remarkable results. In April 1996, a coalition of 633 scientists based in one hundred different laboratories in Europe, the United States, Canada, and Japan completed the DNA sequence of the sixteen chromosomes, composed of 12 million bases, that compose the genome of brewer's yeast, *Saccharomyces cerevisiae*. Yeast is a charter member of what one geneticist calls "the security council of model genetic organisms." Although a single-celled organism, yeast is far more sophisticated than a bacterium, containing a nucleus and other organelles such as mitochondria, sites of energy production. Through the study of yeast, scientists have identified the engines, gears, cogs, and brakes that govern the cell cycle. Many of these factors are disrupted by mutations that cause cancer, and progress in understanding human cancer syndromes is largely due to functional studies in yeast. With all six thousand yeast genes identified, scientists have set about assigning these genes into functional categories. The sequence of *S. cerevisiae* marked the first genome of a fully fledged eukaryote that arose about 1.7 billion years ago. It was the most complete and impressive genome sequence to date.

AS THE YEAST SEQUENCE WAS BEING COMPLETED, another major genome sequencing milestone was in sight: the culmination of a ten-year collaboration between the laboratories of John Sulston, director of the Sanger Centre, Britain's premier DNA laboratory near Cambridge, and Robert Waterston's group at Washington University in St. Louis. At first glance, a supple roundworm 1 millimeter in length that lives on a diet of *Escherichia coli* bacteria in petri dishes may seem an eclectic choice to join the security council. But the appearance of *Caenorhabditis elegans* on earth some 500 million years ago represents one of the turning points in the evolution of life on earth—a time when, as Nigel Calder wrote, "life had become subtle enough for the hierarchy of genes controlling genes to find ways of differentiating animal cells into muscle, skin, nerve, and so on, according to where they lived within a many-celled body."

In 1962, Sydney Brenner and Francis Crick, on the verge of unraveling the genetic code, had begun to discuss how to tackle the "new, mysterious and exciting" problems of biology: development and the nervous system. Brenner wrote to his boss, Max Perutz, outlining new areas that Cambridge's Laboratory of Molecular Biology (LMB) should move into. "As a more long term possibility," Brenner mused, "I would like to tame a small metazoan [multicellular] organism to study development directly."

Brenner already had a simple organism in mind: *Caenorhabditis briggsae,* a self-fertilizing hermaphrodite worm with fewer than a thousand cells, an elementary nervous system (just 302 neurons), a transparent body, short lifespan, and high fertility (producing two hundred to three hundred progeny). Brenner formally proposed to identify and trace the lineage of every cell of a close relative of *C. briggsae* called *C. elegans.* There were a number of technical reasons for the switch, but none as appealing as Brenner's quip that *C. elegans* was the more photogenic of the two.

Brenner entrusted this herculean task to John Sulston, who spent the better part of ten years hunched over a microscope, painstakingly charting the precise order of development of 959 cells in an organism the size of a comma during its two-week life cycle. Almost single-handedly, Sulston helped establish the nematode as one of the top five most valuable model organisms in biological research. It fills a critically important niche in evolution; as a multicellular organism, it can reveal clues to the process of development not possible with yeast. On the other hand, its rudimentary nervous system is far simpler than that of the fruit fly (which has more cells in its eye than the worm has in its entire body).

In 1983, Sulston and Alan Coulson embarked on another large-scale project, this time to produce a physical map of the six chromosomes in the *C. elegans* genome as a prelude to DNA sequencing. Many of Sulston's colleagues were surprised that he would embark on another daunting altruistic project, especially given the rudimentary state of sequencing technology at that time. Why not focus on a neat, tractable biological problem? He replied, "I have a weakness for grandiose, meaningless projects," and never looked back.

Together with Robert Waterston, another Brenner protégé, Sulston set out in 1989 (with the support of the Medical Research Council) to obtain the complete nematode sequence. In 1992, they published a report in *Nature* on the sequence of 100,000 bases of nematode DNA (by today's standards barely a morning's work for a bored technician, but at the time a major milestone), lending confidence to the possibility of eventually sequencing the human genome. Their success attracted the attention of American venture capitalist Frederick Bourke, who was flirting with the idea of setting up a company in Seattle with Leroy Hood to sequence the human genome. Sulston and Waterston even received a pledge from the financier that all of their data would be released publicly.

But Sir Aaron Klug, president of the Royal Society and the former director of the LMB, lobbied the Wellcome Trust to support the application of the nematode sequencing strategy to the human genome. "The work with the nematode illuminated the path forward while others were still discussing the route," says Klug. In May 1992, the Wellcome Trust, now the wealthiest medical charity in the world with assets of more than $20 billion, agreed to build a new DNA sequencing institute at a cost of $120 million, thereby ensuring that Sulston could continue his operation on home soil. The Sanger Centre, near Cambridge, England, was officially opened by Fred Sanger on October 4, 1993.

Throughout the 1990s, Sulston and Waterston turned their laboratories into two of the most efficient DNA sequencing laboratories in the world. Adopting an assembly-line approach that was alien to most biological research groups, shifts of technicians and researchers working around the clock tended to dozens of DNA sequencers to ensure that every last letter of DNA could be squeezed out of the machines. Two features draw the eye of first-time visitors to the Sanger Centre. One is the unusual staircase in the main foyer designed in tribute to the double helix, curving around in a clockwise direction, just as DNA is a right-handed helix. Above the receptionist's desk is an electronic display that counts the number of bases sequenced by the institute. By the end of 1998, it read more than 120 million, half of which were human, and the bulk of the remainder belonging to *C. elegans.* Eighteen months later, that number had soared over the 3 billion mark.

In December 1998, Sulston and Waterston finally completed their 15-year, $45 million endeavor to sequence the first animal genome. They had read 97 million letters of DNA, eight times more than the yeast genome completed just two years earlier. Many researchers considered the publication in *Science* slightly anticlimactic, because Sulston and Waterston had made a point of releasing their sequence data on a daily basis for the benefit of the international *C. elegans* community. In all, they counted 19,099 worm genes taking up about 25 percent of the worm genome. The architecture was beautiful to behold. Many of the essential genes clustered in the middle of each chromosome, suggesting that less critical, more rapidly evolving genes were dispersed toward the chromosome ends, which Waterston called "gene nurseries, or graveyards, or both." In the wake of the complete sequence, *C. elegans* has become an increasingly popular tool as researchers studying genes in other organisms discover related sequences in the worm. This is leading to important discoveries in fields ranging from the migration of nerves, the death of cells, the determination of sex, and the longevity of the organism.

One of the sobering lessons of the *C. elegans* genome project was that although 30 percent of the worm genes are related to human genes, most have never been seen before, which underlines how little we still understand about the secret of life. *C. elegans* has contributed to major advances in identifying genes important for human health, including Alzheimer's disease and the cell death pathways in cancer. Researchers have also identified genes that exert profound effects on the longevity and social interactions of worms, which also have highly related genes in humans. Sulston admiringly refers to the nematode as "a microcosm of humanity," an allusion reiterated by Bruce Alberts, president of the National Academy of Sciences, who says, "We have come to realize humans are more like worms than we ever imagined."

IN MANY RESPECTS, the success of the Sulston-Waterston partnership and the *C. elegans* genome project provided a perfect model for the Human Genome Project. Sulston showed that international teamwork and high-throughput sequencing could work. He also demonstrated the value of daily deposits of sequence data, allowing researchers instant access to valuable gene sequences. As the Sanger Centre took on more responsibility for the human genome, Sulston, a bearded, self-deprecating scientist, had to assume the uncomfortable mantle of linchpin of one of humanity's greatest scientific odysseys.

The son of a vicar, Sulston is an ardent socialist who is happiest pottering about in his garden or his laboratory. It was difficult to imagine anyone less suitable for holding the mantle of de facto spokesperson for the British genome project, but over the past few years, Sulston has gradually warmed to his new mandate, with the British press eager to anoint an authority to counter the perceived threats of patent-grabbing American corporations. Indeed, the British press has largely portrayed the "race" for the sequence as a transatlantic ideological battle of wills between Venter and Sulston, not Venter and Collins.

It is a role that Sanger and his colleagues have warmed to with considerable aplomb. In part, this stems from a sense of pride in upholding the unsurpassable reputation of Britain—some might say Cambridge—as the cradle of modern genetics. But more important, it reflects Sulston's passionate belief in the free trade of genetic information. Genome sequencing for commercial gain is "totally immoral and disgusting," Sulston says, whose disdain for Venter's company is palpable. "I find it a terrible shame that this important moment in human history is being sullied by this act," he said in 1999. Genome information should not be patented, he argues, because "it is intrinsically a part of every human being, a common heritage in which we should all share equally." But Sulston firmly believes that Venter wants "to establish a monopoly position on the human sequence." His final verdict is that "Craig has gone morally wrong."

Despite the controversy, Sulston waxes lyrical when asked to assess the significance of the genome project. "Galileo, a great hero of mine, took us away from the idea that we were the center of the universe. The theory of evolution took us away from the conviction that we were a unique life form. And this work will eventually tell us what makes our brains work and therefore our minds. It will tell us what we are." He is also an enthusiastic advocate of the potential of genetics to help society. "Genetic engineering is no more than a branch of surgery," he insists. "If we can fix a cleft palate genetically before birth, rather than surgically after birth, then sure, why not?" These views run against the grain of popular opinion, at least in Britain, where concerns about genetic technology are fueled by government ineptitude and tabloid sensationalism. Recently Sulston found himself criticizing Prince Charles for demonizing modern genetics research in the wake of the genetically modified food debacle. These are important issues, but in his late fifties, Sulston is keen to return to the relative serenity of laboratory life. In October 2000, he handed the reins of the Sanger Centre to Allan Bradley, another scientist with strong roots in Cambridge and an expert in mouse genetics and chromosome engineering.

Sulston and Waterston's magnificent collaboration gave hope for the ramping up of the sequencing of the human genome, but their success also underlined how far the Human Genome Project still had left to go. The worm genome was less than 100 million bases in length, a mere one-thirtieth of the amount of human DNA. Indeed, by the halfway mark of the fifteen-year genome project, a comparable amount of human DNA had been sequenced, but this was just 3 percent of the total human genome. For many, it was a poor return for an investment of around $1.8 billion. Exasperation with the perceived lack of progress by the public HGP was growing. The eminent American scientist Steve McKnight, chair of biochemistry at the University of Texas and cofounder of the biotechnology company Tularik, assailed the public genome effort as "a disaster—no cohesion, no focus, no game plan. Typical government work." Sydney Brenner criticized the central management of the HGP and its insistence on mapping before sequencing. "When it was the human genome initiative, it was really great," he says. "But once it got consolidated into this managed project, it became a bit like Stalinist Russia: if you're not with us, you must be against us."

Freeman Dyson has expressed the view that the decision to finish genome sequencing by 2005 was not driven by the needs of science and medicine. He suggests that it became "politically imperative" to persevere with existing sequencing technology, when a better strategy might have been to use existing technologies to focus on the important parts of the genome, and spend more resources to find faster and cheaper methods to sequence the rest. "In science," wrote Dyson, "to change the objectives of a program in the light of new discoveries is a sign of wisdom. In politics, it is a sign of weakness. Unfortunately, politics prevailed over science."

And it was no surprise that Venter should express an opinion. In early 1998, Venter derisively called the leaders of the sequencing centers the "Liars' Club" for their widely differing estimates of the final cost and timetable of the project. "They all have a different way of calculating their costs and the amount of sequencing they have actually accomplished." He complained that the public HGP still did not have any DNA fragments fully prepared for sequencing. Collins, in a line dripping with irony, fired back that Venter was probably concerned about keeping TIGR's sequencing machines busy.

He would soon learn just how busy Venter intended to keep them.

CHAPTER FIVE

The Circle of Life

Then felt I like some watcher of the skies
When a new planet swims into his ken;
Or like stout Cortez when with eagle eyes
He stared at the Pacific—and all his men
Looked at each other with a wild surmise—
Silent, upon a peak in Darien.

—"On First Looking into Chapman's Homer,"
JOHN KEATS

B Y 1994, TIGR WAS OPERATING AT FULL CAPACITY. Legions of white-coated technicians, most of them fresh out of college, tended to row upon row of Applied Biosystems sequencing machines, punching instructions into the computers adjacent to each machine, while others mixed fresh batches of DNA with hand-held pipettes. The scene bore a slight resemblance to the cotton mills of the early 1800s, as TIGR became the prototype for the genomics revolution, a facility dedicated to the manufacture of digital DNA code.

Day after day, TIGR churned out hundreds of thousands of letters of DNA sequence; the initial batch of thirty sequencing machines soon swelled to eighty. Above the factory floor, Venter entertained a steady stream of reporters and businessmen in his lavishly appointed office, relaxing on the burgundy leather couch, admiring the pristine model of the *Sorcerer* under glass and the numerous framed magazine covers, many featuring their host.

Venter's vision of a vast compendium of human ESTs, sampled from dozens of human tissues to increase the chances of detecting snippets of as many genes as possible, was rapidly coming to fruition. Other groups, notably in Japan and Colorado, were making respectable efforts, but they could only look on enviously at TIGR's resources and efficiency. By late 1994, TIGR's database had swollen to 150,000 ESTs corresponding to perhaps 35,000 different genes—between one-third and one-half of the total number in the genome.

Interest from the academic community, muted at first, quickly began to grow in anticipation of the large fraction of human genes accessible via computer. With Harold Varmus's decision not to seek patents on behalf of the NIH on Venter's original batch of ESTs, many researchers had assumed that TIGR's ESTs would be made freely available upon request. They were dismayed to learn that TIGR had imposed restrictions on access to the database because of Venter's commercial ties to its for-profit arm, Human Genome Sciences, which in turn had to honor a $125 million deal with pharmaceutical giant SmithKline-Beecham.

In fact, the terms were comparable to those imposed by other commercial outfits and academic centers. The chief bone of contention was that researchers had to allow Genome Sciences to review their scientific manuscripts at least one month before publication, allowing Haseltine to judge whether any newly discovered ESTs were worth patenting. Some organizations, such as the Howard Hughes Medical Institute, saw nothing wrong in this and readily signed the deal on behalf of about two hundred leading medical researchers in the United States. But other researchers felt that TIGR and Human Genome Sciences were taking advantage of their monopoly on an increasingly precious resource. Moreover, for-profit companies were denied access to the EST database completely.

THE POWER OF THE EST DATABASE was signaled dramatically in 1994, when one of the most respected cancer researchers in the United States made a spectacular discovery by perusing the TIGR database. By the early 1990s, Johns Hopkins University's Bert Vogelstein had pieced together a chain of genetic aberrations that inexorably turns healthy colon cells into proliferating, cancerous polyps. Many of these insights have stemmed from the study of inherited forms of colon cancer, which kills 45,000 people in the United States alone each year. In 1987, Vogelstein and Eric Fearon found that tumor samples from colon cancer patients were frequently missing a segment of chromosome 17. Further studies logged genetic aberrations on chromosomes 5, 11, and 18 at different stages of the disease.

In 1988, Vogelstein espoused his multistep model for colon cancer pathogenesis in the *New England Journal of Medicine*. Although none of the individual genes had been found, the model was profoundly important and elegantly confirmed by the genetic detectives over the next few years. The culprit on chromosome 17 was *p53*, which has been dubbed the "Guardian of the Genome" for its varied protective roles during cell division and is mutated in about 50 percent of all human cancers. Vogelstein's group also tracked down the gene mutated on chromosome 18. And in 1991 Vogelstein and Ray White, the University of Utah geneticist who had helped Botstein and Skolnick conceive the gene mapping revolution ten years earlier, collaborated to identify mutations among patients with an inherited colon cancer called familial adenomatous polyposis. The gene was dubbed *APC* (adenomatous polyposis coli). Vogelstein's findings rewrote the cell biology textbooks, and he can often be found sporting a T-shirt depicting the genetic pathway in colon cancer while pounding the keyboards for his part-time rock group, Wildtype.

But while Vogelstein was taking giant strides in elucidating the pathway of genetic aberrations that give rise to one form of inherited colon cancer, he was increasingly curious about another: hereditary nonpolyposis colon cancer. Finnish researchers brought to his attention a collection of about forty families with this disease, which differed markedly from familial adenomatous polyposis. These patients did not exhibit the proliferation of polyps before developing full-blown tumors. By the spring of 1993, Vogelstein and veteran Finnish geneticist Albert de la Chapelle had traced the location of the nonpolyposis cancer gene to chromosome 2 and were closing in on the errant gene.

Closer scrutiny of those *Science* reports localizing the gene to chromosome 2 revealed a telltale clue to its identity. The Hopkins group was relying on DNA markers called microsatellites: short stretches of di- and trinucleotides (such as CACACACA and CTGCTGCTG), the precise length of which varies in different people. Such a pattern was well known to Richard Kolodner, a geneticist at the Dana Farber Cancer Institute in Boston, who had seen it many times in yeast and bacterial cells lacking a gene called *MutS*. When he read Vogelstein's reports mapping the nonpolyposis gene to chromosome 2, he was in no doubt as to the gene defect. The MutS protein scans replicating DNA for errors in the newly formed base pairs and corrects them. Loss of MutS causes a massive increase in errors in DNA, particularly in stretches of microsatellite repeats.

Over the next six months, a frantic race evolved between Kolodner's small team and Vogelstein's army of workaholic postdocs and students.

Kolodner had fished out the human counterpart of the *MutS* gene but had no colon cancer family samples in which to test it; Vogelstein had the cancer families but had not made the connection with mismatch DNA repair until he was tipped off by Duke University researcher Paul Modrich. Soon, scientists were openly speculating that the nonpolyposis cancer gene was the human counterpart of a bacterial mismatch repair gene. Vogelstein's group wasted no time in tracking down Kolodner's sequence of the yeast mismatch repair genes, which allowed them to isolate the human equivalent called *MSH2,* sitting on chromosome 2, precisely where the faulty gene was predicted to reside.

But when Vogelstein called Kolodner in November to see where his new rival stood, he was stunned to find out that Kolodner had not only isolated the human gene but also identified putative mutations in colon cancer families. Worse, Kolodner had already submitted his paper describing the discovery to *Cell,* an eminent molecular biology journal based just across the Charles River in Cambridge. Vogelstein rushed his paper to the same journal, which eventually published it a fortnight after Kolodner's report, although news of both studies was released to the press simultaneously. The Hopkins scientist cast aside his disappointment and accepted an invitation from Francis Collins to appear jointly with Kolodner at a Washington press conference to announce the breakthrough.

Neither gene sleuth had much time to bask in the media spotlight. About 40 percent of the nonpolyposis families studied did not have a mutation in the *MSH2* gene, and there were at least two other DNA mismatch repair genes in yeast and bacteria that probably had counterparts in humans waiting to be discovered. Kolodner and his collaborators had been able to isolate one of them—the human *MutL* homologue, named *MLH1,* by first isolating the mouse gene and in turn using this to fish out the human gene. As Vogelstein's team was quickly learning, there was too little sequence similarity between the bacterial and human *MLH1* genes to use one to fish out the other directly. In desperation, Vogelstein looked for a shortcut to find the human *MLH1* gene.

Vogelstein's partner Ken Kinzler telephoned Venter and asked him if he had found any ESTs that looked like a bacterial DNA mismatch repair gene. Venter replied, "Yeah, we found three of them!" The only possible snag was that the rights to the genes belonged to Haseltine and Human Genome Sciences. In fact, Haseltine already appreciated the significance of the putative human mismatch repair genes and had even called Kolodner, his former Harvard colleague, in early January to examine them. But Kolodner already had the human gene in hand and so declined.

Undeterred, Vogelstein and Kinzler made the short trip from Baltimore to Gaithersburg to set up a brief collaboration with Venter and Haseltine. Vogelstein and Kinzler discovered mutations in the human DNA repair gene in some hereditary forms of colon cancer. In March 1994, just three months after the first nonpolyposis colon cancer breakthrough, the second gene was reported simultaneously by the groups of Kolodner and Vogelstein in *Nature* and *Science,* respectively. Among the authors of Vogelstein's report were Venter and Haseltine. It was a powerful testimonial for the utility of the TIGR sequence collection.

Vogelstein and Kinzler's savvy "clone by phone" strategy served as a wake-up call to genetics researchers, who realized that a few minutes perusing the sequences in the EST database could potentially save them months, possibly years, of tedious gene mapping and cloning experiments. Within weeks of the colon cancer gene discovery announcement, Venter was deluged with requests to search the TIGR EST database. When the gene for polycystic kidney disease, a common dominantly inherited disorder causing kidney failure in adulthood, was announced later that year after another tortuous positional cloning effort, the celebrations were bittersweet. TIGR had isolated several ESTs corresponding to the same gene, which, if they had been mapped, might have saved years of work. But a short time later, the EST strategy was again in the news as it netted another high-profile disease gene.

In 1991, British researcher John Hardy discovered the first example of an inherited mutation in a familial form of Alzheimer's disease. However, this gene, *APP*, accounted for only a small percentage of familial Alzheimer's cases. In June 1995, a Canadian team led by Peter St. George-Hyslop identified a new gene on chromosome 14, dubbed *S182,* which was mutated in a much larger proportion of Alzheimer's patients. But there were still many other affected families in which the disease did not trace to either *APP* or *S182.* The most intriguing of these were the Volga Germans—a group of families that had moved from the Hesse region of Germany to the Volga Valley of Russia around 1760 and later emigrated to the United States. A research group in Seattle, led by Gerry Schellenberg, had just mapped the errant gene in these families to chromosome 1.

The sequence of *S182* was a disappointment; it suggested that this new Alzheimer's protein resided in the cell membrane but divulged few other clues as to its likely function. Two of St. George-Hyslop's collaborators, Rudi Tanzi and Wilma Wasco from the Massachusetts General Hospital, decided to play a wild hunch: they wondered if the Volga German Alzheimer's gene might not be related to *S182*. Tanzi conducted a quick comparison of the *S182* sequence against the public EST database and found a striking match to an EST that had been deposited by Jim Sikela, one of the first champions of the EST method along with Venter. Moreover, Tanzi quickly found that it sat on chromosome 1, precisely where the mutation in the Volga German Alzheimer's families was thought to lie. Less than two months after the Canadian group's breakthrough, Tanzi and Schellenberg published the discovery of a single misspelling in this related gene in affected members of the Volga German pedigrees.

The sheer speed of progress in the field was stunning. Even more impressive, these findings have proven an important breakthrough in Alzheimer's research: *S182* and its close relative code for proteins, called presenilins, which are involved in processing of the APP protein and the production of protein plaques in the brains of Alzheimer's patients.

For Venter, the colon cancer and Alzheimer's disease discoveries provided an immensely gratifying vindication of his EST strategy. To make his point, he compares the routes to identifying two notorious neurodegenerative diseases. The search for the Huntington's disease gene took an international consortium ten years and cost an estimated $100 million before the gene was finally identified by positional cloning in early 1993. By contrast, a simple search of the EST database had netted a familial Alzheimer's disease gene in a matter of weeks (although finding the right gene to begin the search took longer). Moreover, as researchers mapped more and more ESTs to their respective chromosomes, the task of positional cloning—mapping disease genes and then methodically searching all the known genes in the region—would be greatly accelerated by the ready availability of a list of potential candidate genes. This is called, rather imaginatively, the "positional candidate" approach.

DESPITE THE GROWING POPULARITY OF ESTs, many researchers remained at best suspicious, at worst hostile, toward Venter and his sequencing operation. For three years, Venter had been portrayed as a conniving scientist, intent on locking up the patents on hoards of human genes for his own nefarious goals. Compounding the lingering resentment over the NIH's ill-conceived decision to apply for patents on Venter's first collection of ESTs, some researchers were frustrated by TIGR's refusal to grant unrestricted access to its EST database. But TIGR's deal with Human Genome Sciences gave Haseltine's company as much as one year to examine new DNA sequences to decide which ones might be worth patenting. In the eyes of many researchers, TIGR was deemed to be acting like Frankenstein. "They're coming after us with torches and pitchforks," said Leslie Platt, TIGR's chief operating officer in 1994. The stress finally got to Venter. That year, he flew home from Europe for an emergency operation after a sudden attack of diverticulitis. Doctors removed a small part of his intestines.

At an October 1994 conference in Washington, D.C., Michael Morgan, an executive with Britain's Wellcome Trust, hosted a closed meeting of genome leaders to discuss whether to use TIGR's EST collection as part of a large-scale project to map the location of human genes. Collins strongly opposed using Venter's ESTs for such a program and lent his backing to an initiative that had been put forward by the pharmaceutical giant Merck, which, like other big pharma companies, had been shut out of the TIGR database.

Alan Williamson, Merck's vice president, had conceived the idea of financing a separate program to identify ESTs in May 1994. Unlike TIGR's database, all of Merck's sequences would be made publicly available without delay and with no strings attached. Merck decided to give Washington University's Bob Waterston and Richard Wilson a grant of $10 million to produce hundreds of thousands of human ESTs over the ensuing two years or so. The cDNA clones would be provided by Bento Soares, a Columbia University professor and founding member of a small group of academic researchers who billed themselves the IMAGE (Integrated Molecular Analysis of Gene Expression) consortium. Merck's decision was not entirely altruistic, of course: Merck was intent on challenging SmithKline-Beecham's stranglehold on EST rights thanks to its exclusive $125 million deal with Haseltine's company. Among those who welcomed the Merck initiative, ironically, was Venter. "Imitation is the sincerest form of flattery," he said, knowing that his institute would enjoy the same access to Merck's publicly deposited cDNA sequences as any other researcher.

At the same time, Venter felt it was time to improve public access to the TIGR database. Under the new conditions, users of the TIGR database would have access to most sequences for free but would give Haseltine one to two months' notice prior to publishing data based on other ESTs and offer Human Genome Sciences commercial rights to any gene products (just as Vogelstein had done). Venter defended the conditions, which appeared to be in keeping with the terms offered by other biotech companies for access to proprietary information.

In 1994, Venter telephoned me to arrange a lunch meeting at the National Press Club in Washington, D.C. He wanted to discuss the prospect of submitting a paper to *Nature* unlike any it had ever published before. It was time to unveil details of the largest collection of human genes in history. It took more than a year for Venter's magnum opus to be reviewed, corrected, and ultimately accepted for publication.

Not everyone was excited. "If you publish this Venter stuff, I can promise you that nobody in the US genome community will ever send you anything ever again!" John Maddox shrugged off the telephoned threat by one of the leading geneticists in the United States and righteously defended his decision to publish Venter's report. "It would be a loss to the research community," he stated, "and a disservice to the authors of this material, if there were to be no record of it at this critical stage in the evolution of the human genome project." Nicholas Short, *Nature*'s biology editor, took up the baton. "For scientists," Short enthused in the accompanying press release, "the Genome Directory marks a turning-point in our understanding of the human genome and its contents. But for everyone, it is nothing less than the first atlas of ourselves."

In September 1995, *Nature* finally published "The Genome Directory." This momentous document, written by Venter, Fraser, Adams, Haseltine, and ninety fellow researchers from TIGR and Human Genome Sciences, described the burgeoning collection of ESTs and was published as a separate 379-page supplement, accompanied by contributions on mapping chromosomes from several public HGP groups. The sheer abundance of data in the Genome Directory was quite unlike anything that had gone before. Venter's team sampled genes expressed in thirty-seven human tissues ranging from the brain, liver, and heart to the greater omentum and the esophagus. Nearly 175,000 EST sequences were examined, yielding sequence matches ranging from the ABC1 family (proteins that control the passage of ions and small molecules in and out of cells) to the zinc finger family of proteins (key regulators of gene expression). Specially created software called the TIGR Assembler compared each sequence against the database to compile overlapping EST fragments into a single "tentative human consensus" sequence.

But Venter's analysis was not restricted simply to the TIGR/Human Genome Sciences collection of ESTs. Merck initiative had yielded nearly

75,000 ESTs by the spring of 1995, while other groups had identified a further 45,000 or so. With these sequences freely available in the NIH's EST database, Venter was perfectly at liberty to add these 118,000 ESTs to his own collection to provide an even more comprehensive analysis for his directory. Venter assembled the sum total of ESTs into 29,599 whole or partial genes, with another 58,000 ESTs unaccounted for. The EST sample provided a glimpse at the patterns of gene expression in every human tissue and organ and allowed Venter to classify in broad strokes the function of the sample genes. For example, 16 percent have a role in metabolism, 12 percent code for proteins that transmit and detect signals between cells, while 4 percent are required for cell division and DNA replication.

In conjunction with the landmark publication, Venter made all of his EST data publicly available in the TIGR Human cDNA Database, containing more than 345,000 ESTs. The only exception was for a small number of genes that Haseltine withheld from the publication, which briefly tarnished Venter's accomplishment. Nevertheless, in less than two years, TIGR had assembled the world's largest collection of sequenced genes, in a single stroke doubling the total amount of DNA sequenced by the rest of the world. Even by conservative estimates, it was reasonable to believe that TIGR had extracted roughly half of all human genes.

One vexing question remained, however. Just how big a contribution to the identification of the full complement of human genes was Venter's EST storehouse? The question hinged on knowing the precise number of genes, but scientists had spent decades bickering about that issue. Most textbooks settled on the unsubstantiated figure of 100,000, for no other reason than it was a convenient round number. Twelve months before the release of the Genome Directory, I invited Venter to ponder that exact question. After mulling over the startling discrepancy in various estimates put forward over the years and considering the latest data from his own EST and chromosomal sequencing work, Venter and his colleagues presented a consensus in a short article in *Nature Genetics*. The final prediction for the total number of human genes fell somewhere between 60,000 and 70,000.

Venter's article was highly persuasive, leading many writers to revise the textbook "100,000 genes" figure downward. Indeed, scientists have felt comfortable quoting a range from 60,000 to 80,000 for several years. Naturally, scientists with proprietary collections of gene sequences sharply disagree. Haseltine insists that the true figure is closer to 150,000.

Venter recalls that his benefactor, Wallace Steinberg, was so upset with his lowball estimate that he called up the TIGR chief to complain: "What the hell do you think you're doing, saying there are only 60,000 genes? I just sold 100,000 genes to SmithKline-Beecham!" Venter did not object too strenuously and appropriately dedicated the Genome Directory to the memory of Steinberg, who died a few months before it was published.

The lingering uncertainty over the precise number of genes stems in part from disagreements over the precise definition of a gene (many genes are expressed in slightly variable forms). Scientists recently revisited the question of how many genes there are in the human genome. A lottery has been organized at Cold Spring Harbor, with punters waging $1 bets on the likely final tally. Estimates range from about 40,000 to well over 120,000. The temptation to inflate the number of genes in *Homo sapiens* goes beyond appeasing anxious stockholders. "People like to have a lot of genes—it makes them feel more comfortable," says Sydney Brenner. "To be only eight times more complicated than [the bacteria] *E. coli* is an insult!"

EVEN AS VENTER WAS ANTICIPATING the publication of the Genome Directory, his interest in simply compiling more and more human cDNAs was beginning to wane. Human Genome Sciences was aggressively generating its own EST data and focusing more on selecting candidate targets for its drug discovery programs in cardiovascular disease, immunodeficiency, and several other disorders. Indeed, Haseltine was building a new manufacturing facility that would allow the company to purify these proteins, the next essential step in his ambitious plan to create a full-fledged pharmaceutical company.

Meanwhile, university researchers were systematically mapping the location of the ESTs to the human genome, providing a rough and ready guide to the position of genes along each chromosome. Although Venter recognized the value in systematically assigning the human ESTs to their rightful chromosomal locations in the genome to aid the identification of disease genes, that task would take years of work and require the collective efforts of many large genome centers.

Venter sensed the time was right to change tack. "Once we have the sequence data on the whole set of genes," Venter explained, "we have a chance to use the study of genomes to put together a complete genetic evolutionary comparison." But Venter was also shrewd enough to realize that sequencing fragments of genes barely scratched the surface of the information to be mined from the genome. The only long-term solution was to derive the DNA sequence of the human chromosomes themselves. Venter knew that TIGR could accomplish the desired accuracy of sequencing. Indeed, the institute had already proven it could sequence chromosomal DNA; just a few months before the publication of the Directory, Venter made another spellbinding announcement: the sequencing of the first free-living organism.

In late 1993, Venter was attending a genetics conference in Bilbao when he was approached by a lanky, white-haired scientist whom he vaguely recognized. It was Hamilton Smith, a biochemist from Johns Hopkins University who had shared the Nobel Prize in 1978 with his Hopkins colleague, the late Daniel Nathans, for the discovery that essentially launched the era of genetic engineering. Smith had characterized the first restriction enzyme, a natural protein in bacteria that specifically recognize and cleave small palindromic motifs in double-stranded DNA, for example GAATTC (turn the double helix 180 degrees and the opposite strand also reads GAATTC). In the late 1960s, Smith and his graduate student, Kent Wilcox, discovered that a drop of a cell extract from the bacteria *Haemophilus influenzae* could sever viral DNA into tiny pieces, leaving the bacterial DNA completely unaffected. Smith subsequently purified the broken DNA and identified the order of nucleotides at the frayed end of each DNA strand, providing the first molecular definition of the action of a restriction enzyme.

The discovery of this arsenal of DNA-cleaving reagents enabled researchers to cut-and-paste human DNA into bacteria, which were perfect for propagating genes. Because most restriction enzymes cut DNA in a staggered manner, much like a child's party scissors, the resulting "sticky ends" enabled researchers to join two foreign pieces of DNA that had been cut with the same restriction enzyme, using a ligase enzyme to glue the two sticky ends together. Smith's elegant biochemical detective work laid the foundations for the recombinant DNA era in the early 1970s, the DNA fingerprinting revolution of the 1980s, and the gene mapping of the 1990s. But when the call from Stockholm came in October 1978, nobody was more surprised than Smith and his family. His mother, on hearing the news on the radio, said to her husband, "I didn't know there was another Hamilton Smith at Hopkins." Smith wrote apologetically to the

Harvard researcher who had inspired his discovery, saying he should have shared the prize.

In the aftermath of the Nobel Prize, the almost reclusive Smith struggled to live up to his sudden celebrity, spend time with his five children, help his elder brother suffering from schizophrenia, and maintain a productive laboratory. He dabbled in computer programming and ill-conceived experiments, culminating in the ignominy of having his grant applications spurned by the NIH.

As they relaxed in a bar in Bilbao, Venter and Smith became acquainted. The two men were a study in contrasts, but the shy Smith found himself liking Venter's self-confidence, while Venter was impressed with Smith's quiet intelligence and ideas. As Venter puts it, "My guess is, we both wish we could be a little more like the other person." Within a couple of months, Venter had invited Smith to a position on the scientific advisory board of TIGR. In late 1993, Smith proposed a radical idea at a TIGR staff meeting. Sequencing technology was not yet at the stage that a single institution like TIGR could contemplate tackling the entire genome or even a single chromosome, but valuable lessons could be learned by sequencing a smaller organism. And so Smith daringly suggested that TIGR consider sequencing the complete genome of a bacterium.

The logical candidate in principle for such a program would have been *E. coli,* the bacteria that live by the billions in the human gut, occasionally causing serious outbreaks of food poisoning. Decades of research made it the most understood bacterium at the genetic level and one of the most understood organisms after humans. But *E. coli* was off the board: a low-key sequencing project had begun in the mid-1980s, although it was making slow headway and would not be finished for several more years.

Smith quickly suggested another possibility: the "pet bacterium" he had used 20 years earlier to study restriction enzymes: *H. influenzae.* Although in part a sentimental choice, *H. influenzae* is also of considerable medical importance. A large percentage of apparently healthy children carry the bacteria in the nasal passages, which can then spread to the ear, brain, and lungs, causing otitis media, bronchitis, and meningitis. There is growing evidence that *H. influenzae* is an important cause of childhood pneumonia in developing countries. But just as important perhaps, *H. influenzae* is a highly typical bacterium—average size and average DNA composition.

Ironically, Smith's idea met with an even cooler reception among his own small group at Johns Hopkins than it did at the TIGR meeting. Smith's team had already sequenced a small portion of the *H. influenzae* genome and was not enthusiastic about the time it would take to map out the circular bacterial chromosome, not to mention the tedium of sequencing roughly 1.8 million units of DNA. But Smith was not content to let the plan languish. He performed some computer simulations that suggested that an almost ridiculously naive shotgun strategy could work. Instead of spending months, possibly years, mapping the bacterial genome (that is, determining the order of overlapping fragments before sequencing them), Smith favored a much more brute-force approach. First, shear the bacterial DNA into thousands of random pieces; then sequence the DNA of each fragment; and, finally, use a computer program to align the overlapping fragments to produce a single, contiguous DNA sequence of the entire organism.

The TIGR team struggled to come to terms with Smith's eccentric sequencing strategy. But Venter liked it almost immediately, and Smith soon had TIGR's full support. Smith personally and painstakingly oversaw the shearing of the purified bacterial DNA into random chunks, selecting those smaller than 2,000 letters, which he then inserted into a different bacterial DNA to grow and purify before running the sequencing reactions. A team of eight TIGR personnel using fourteen ABI 373 DNA sequencers spent three months carrying out the 28,000 sequencing reactions, each about 500 letters long. Each letter of DNA was sequenced an average of six times.

The more daunting question was, Would the TIGR software be able to piece it all together? The TIGR Assembler program reduced the 28,000 individual sequences to 140 small sections, or contigs. Slowly but surely, the remaining gaps were closed.

As their sequencing work got off the ground in early 1994, Venter and Smith decided to apply to the NIH for a grant to fund the work. Given their recent history with the federal funding agency, neither man was terribly optimistic, but both wanted the satisfaction of proving the skeptics wrong. In August 1994, Venter finally received a pink sheet from the NIH rejecting his funding request; the reviewers had considered the task of identifying almost 2 million letters of DNA impossible. Venter and Smith laughed: their *H. influenzae* sequence was already 90 percent finished. (Funding was eventually obtained from the Department of Energy.)

It took several more months before the final pieces of the sequencing puzzle were put together. After thirteen months, history had been made at a cost of forty-eight cents a letter. It was not the largest contiguous sequence ever assembled—that honor belonged to the academic consortia sequencing chromosomes of yeast and nematode worms—but it was the most significant. Venter's momentous gamble had paid off. The first genome sequence of a free-living organism (as opposed to a virus) was in the books and on the web. He had deciphered the circle of life.

VENTER PRESENTED THE *H. influenzae* sequence for the first time in May 1995 at the American Society of Microbiology conference in Washington, D.C. The accolades flew thick and fast. Scientists lauded the result as a milestone and an incredible achievement. Even James Watson acknowledged Venter's achievement as "a great moment in science." In July, just two months before the publication of the Genome Directory, Venter's team published the first bacterial genome sequence in *Science* to worldwide acclaim.

The treasures of the *H. influenzae* genome lived up to everyone's expectations. The genome contains 1,830,137 letters of DNA. To its credit, TIGR did not try to conceal more than 100 instances where it had difficulty distinguishing the precise base. "Better to call a spade a spade," agreed one pair of commentators, "than to squint your eyes just right until it looks like a shovel (or a G)." The genome contains 1,743 genes—remarkably an average density of one gene every 1,000 bases. This meant that almost every DNA base coded for something important, with virtually no wastage or junk sequences. More than 1,000 genes were either identical to or resembled known genes from other organisms. These could quickly be sorted into families, providing immediate insights into biochemical pathways. Seventeen percent are involved in translating the instructions embedded in the DNA sequence into protein, 12 percent are required for transport, 10 percent are necessary for producing energy, and 8 percent to produce the outer envelope of the bacterial cell. However, about 40 percent of the genes were unrecognizable; they resembled no known genes, although about half of them were similar to *predicted* genes.

Bacterial geneticists like Oxford University's Richard Moxon, an early proponent of sequencing the *H. influenzae* genome, poured over the sequence, reveling in their new intellectual affluence now that all the genes were known. They traced the difference between virulent and non-virulent strains of the bacteria to a span of eight genes that were found only in the virulent type b strain. These genes encode proteins that help the bacteria adhere to their host cells. They noticed that the chromosome contained 1,465 copies of a short twenty-nine-letter DNA motif called the uptake signal sequence, with a conserved core that reads AAGT-GCGGT. The bacteria recognize and preferentially take up exogenous DNA with this sequence. Insights into the biochemical pathways used to generate energy and sustain life were immediately apparent from the presence or absence of key metabolic genes, and new clues regarding the bacteria's ability to adapt to changes in its environment came from the discovery that a handful of key virulence genes harbor short runs of a four-letter repeat sequence that deliberately introduce misspellings during DNA replication. This results in widespread variation in the protein sequences to help the bacteria cope with changes in the external milieu.

By the standards of pharmaceutical research, TIGR's expenditure on the sequence of the first free-living organism was trivial—a paltry $1 million or so. For that, TIGR had unearthed a treasure trove of riches for bacterial geneticists. The hundreds of mysterious genes for which there was no known function could be studied easily by systematically removing each gene and observing the effect on the bacteria. Even more ambitious, some researchers are purifying and crystallizing the corresponding proteins to glean something of their function from their three-dimensional shape. Finally, prospects for vaccine development to target the most virulent strains of H. influenzae have never before been more encouraging.

EVEN AS VENTER WAS BASKING in the acclaim of publishing the sequence of the first free-living organism, his team was putting the finishing touches to its second blockbuster and beginning work on a third. While the impact of the first microbial sequence was undeniable, that of Venter's next two sequences would be even greater, shaking the study of evolution to its very foundations.

In January 1995, working with Clyde Hutchison, a microbiologist from the University of North Carolina, Venter's team began sequencing the DNA of Mycoplasma genitalium. By contrast to H. influenzae, the rationale for choosing M. genitalium, a benign squatter in human genital tracts, was perfectly reasonable: just one-third the size of H. influenzae, it holds the smallest free-living genome known, a mere 580,000 bases of DNA. Compared to H. influenzae, the sequencing was straightforward, taking five TIGR personnel just two months using eight ABI machines to collect the raw sequence data and a few extra months to complete the assembly. For a total cost of just $200,000, Venter's team had identified 517 genes—480 coding for proteins and the remainder for RNA molecules.

Not surprisingly for bacteria that dwell in nutrient-rich mammalian cells, *M. genitalium* apportions its genetic resources quite differently from *H. influenzae*. For example, it gets by with just one gene involved in the biosynthesis of amino acids, whereas *H. influenzae* uses nearly seventy. And yet, surprisingly, close to one hundred genes in *M. genitalium* were unrelated to anything seen in *H. influenzae* or any other organism, befuddling researchers as to their function.

But Venter, Fraser, Smith, and their colleagues were not content to let matters rest there. If *M. genitalium* is able to sustain life with just a third of the genes of *H. influenzae* and many other bacteria, how many more genes might be dispensable? Could it be possible to define the minimum number of genes necessary to sustain life? With the smallest genome known fully catalogued, Venter now had the tools to address this controversial question. Using a top-down approach on the already threadbare genome of *M. genitalium,* Venter's group sought to disrupt as many genes as possible and examine the effect on viability. The trick was to introduce a mobile DNA element called a transposon, which insinuates itself randomly into the bacterial chromosome. The disruption of some genes would have little or no effect on the health of the cell, but when the transposon wormed its way into the middle of a critical *Mycoplasma* gene, then the bacteria would no longer remain viable. Similar experiments were also performed on *M. pneumoniae,* a nicely equipped version of *M. genitalium* with some two hundred additional genes.

It took over a year and more than 2,000 different insertion events, but finally a striking pattern emerged. Transposon insertions in ninety-three different *M. genitalium* genes seemingly had no effect on the health of the cells. Comparisons with *M. penumoniae* revealed a further group of 129 *M. genitalium*–related genes that were not required for viability. That left about 350 genes in *M. genitalium* that were never interrupted by the transposon in live cells, providing an upper limit for the "minimal genome." On the other hand, a lower limit for the number of essential genes could be derived simply by comparing TIGR's first two bacterial sequences, which showed that 256 genes were conserved, and thus likely to be essential. Thus, to a first approximation, Venter and Fraser estimated that somewhere between 250 and 350 genes constitute the skeletal sequence for life, and yet the function of about 100 of these genes remains shrouded in mystery.

Interesting though these results were, there was a much more direct method to ascertain the minimal gene requirement for life, which Venter drily explained: "One way to identify a minimal gene set for self-replicating life would be to create and test a cassette-based artificial chromosome, an experiment pending ethical review." Rather than remove genes to establish the minimal genome, Venter was contemplating a far more radical bottom-up approach that, if successful, would literally create life in the test tube. "The key point," says Fraser, "is that [the DNA] has to be self-replicating," but she admits there are many technological hurdles to overcome. The paleoanthroplogist Richard Fortey wrote, "That vital spark from inanimate matter to animate life happened once and only once, and all living existence depends on that moment." By building an artificial raft of genes, adding one gene after another to a synthetic chromosome placed inside a dormant bacterial shell, Venter might bring back that vital spark.

When Bill Haseltine was asked once whether he was playing God, he replied earnestly, "I wish we were." Initially Venter seemed to share that sense of giddiness. Presenting his plans at the American Association for the Advancement of Science annual meeting in early 1999, he casually invoked the creator of Frankenstein: "Shelley would have loved this!" he quipped. But the putative practical benefits of such a program, notably the creation of designer microbes to clean up toxic waste and oil spills or create renewable forms of energy, were lost amid the ethical controversy Venter had stirred up. Jeremy Rifkin, the prominent biotechnology gadfly, told the BBC: "This is a divide which takes us into a brave new world—a world in which scientists and companies can begin to create their own genesis. Do any of the scientists involved really have the wisdom to know how best to dictate the future evolution of life on this planet?"

Somewhat uncharacteristically, Venter opted for a more circumspect approach. Recalling the ethical firestorm in 1997 that followed the arrival of Dolly, the cloned sheep, and wary of the legitimate danger of custom microbes being adapted for nefarious purposes such as biological warfare, Venter called a halt to the experiment. Instead, he funded a panel of distinguished bioethicists to consider TIGR's proposal to build the minimal genome before the experiment got underway rather than after. The panel concluded that experiments minimizing the number of genes had no relevance to "the complex metaphysical issues about the status of human be-ings" and could find little fault in Venter's reductionist plans. "Too often," the panel wrote, "concern about 'playing God' has become a way of fore-stalling rather than fostering discussion about morally responsible manip-ulation of life." However, the panel added, "The temptation to demonize this fundamental research may be irresistible," especially in the absence of a standard definition of life.

Clearly some groups will condemn Venter's Promethean plans as an outrageous affront to the sanctity of life, a testament to the appalling hubris of modern science. The prevailing sentiment among religious leaders, however, is that while there is understandable discomfort with Venter's research plan, there is more to life than a jumble of genes acting in concert in a hollowed-out bacterium and that Venter's experiment should not be mistaken for the real creation of life. Whether the public re-action will be quite so restrained when a small DNA sequence provides that vital spark is another matter.

THE FIRST TWO BACTERIAL GENOME SEQUENCES convinced the world's scientific community that Venter's maligned sequencing strategy was the way to go—at least for bacteria. The Wellcome Trust and the NIH both poured millions of dollars into funding microbial sequencing work. As his technical wizardry rapidly became standard practice, Venter's next trick left the scientific world gasping in amazement as he conjured up a result that has literally rewritten the textbook of evolution and redrawn the tree of life.

This story has its origins in 1982, when a research submarine named *Alvin* combed the floor of the Pacific Ocean off the coast of Baja Califor-nia, in search of new forms of life. Two miles beneath the surface, *Alvin* rummaged around the base of a thermal vent known as a white smoker— a torrid emanation from the netherworld. The prize catch from the plu-tonic depths of the Pacific that day was a methane-producing microbe that thrived in temperatures around a balmy 85°C, 200 atmospheres of pressure, and no traces of oxygen. This curious organism, which had never been seen before, was named *Methanococcus jannaschii,* in honor of its chief waste product and the expedition leader, Holger Jannasch.

M. jannaschii is a peculiar beast, seemingly unsure of its own identity. At the cellular level, it looks like a typical bacterium, with its DNA unfet-tered by the walls of a nucleus—the central distinguishing feature of the two main domains of life, the eukaryotes (including all plants and animals) and the prokaryotes. But genetically, *M. jannaschii* is a representative of a distinct group called archaea (formerly called the archaeabacteria) that thrive under only the most extreme conditions on the planet. So hardy are these bacteria that they are favorites of the panspermia theorists—those who believe that life may have arrived on earth from space.

The genetic aloofness of archaea emerged from the trailblazing studies of Carl Woese, a physicist-turned-evolutionary-biologist at the University of Illinois. Starting in the late 1960s, Woese conducted a backbreaking survey of the genetic similarities and differences among dozens of different bacterial species. His rationale was simple. During evolution, mutations or changes inevitably accrue in the DNA of different organisms. By cataloguing these sequence changes and estimating the rate at which these changes take place, researchers can gauge the evolutionary relationship between different species. Organisms with many DNA changes in a given gene are more likely to have diverged earlier than organisms with only a small number of alterations.

Woese focused on a special RNA molecule that is a crucial component of ribosomes, the cell's protein-manufacturing machines. These ribosomal RNAs (rRNAs) are found in all species, befitting the critical importance of protein synthesis to sustain life. Initially (this was still years before Sanger and Gilbert perfected their DNA sequencing techniques), Woese had to rely on an indirect strategy to measure the sequence changes. So he used chemicals to carve up the rRNA molecules and monitored radioactive traces as he separated the resulting fragments through a gel.

In 1976, Woese began comparing the rRNA sequences from archaea and bacteria, and made a dramatic discovery. The rRNA of the archaea was so distinctive he simply could not classify them as bacteria. To Woese, the sequences were written in God's handwriting. They spelled an entirely new domain of life—a third domain, separate and distinct from prokaryotes and eukaryotes. Equally exciting, Woese felt the archaea might represent a vestige of the very first life forms on earth. It was the only organism capable of surviving in the hostile conditions that existed more than three billion years ago, reminiscent of the hot springs of Yellowstone Park—a "sulphorous surrogate of Hades" according to Fortey—eschewing oxygen in favor of carbon dioxide, nitrogen, and hydrogen.

Evolution is no stranger to controversy, but Woese's revolutionary notion that life has three domains, not two, and that two of these domains belonged to microbes was tantamount to heresy. His results, published in 1977, were widely reported in the popular press. Editorial writers proclaimed that Woese had revealed a third kingdom of life on earth, a strange alien organism that appeared to be neither animal nor vegetable. (In fact, the third domain Woese was invoking complemented bacteria and eukaryotes, which include all animals and plants.)

By contrast, the reaction of the established evolutionary experts was hardly fit to print. According to one observer, Woese's fanciful proposition "was greeted with wrath and ridicule, not to mention abuse." It was bad enough that Woese was an outsider practicing an abstruse technique that few others could master. Worse, he was essentially arguing that the alleged experts in the field of evolution had completely overlooked a huge limb of the tree of life. However, over the next few years, new studies of archaea biochemistry and physiology revealed marked differences in the properties of these organisms compared to bacteria. Incredibly, they seemed to have as much in common—perhaps more—with eukaryotes as with their single-celled compatriots.

Venter was fascinated by this controversy from both a scientific and a personal viewpoint. He could identify with Woese's plight—a researcher brave enough to risk ridicule by challenging evolution's central dogma while yearning for the respect and recognition of his peers. Venter knew that TIGR's sequencing prowess could settle this contentious debate once and for all by providing the tools to compare species at the nucleotide level, not gene to gene but genome to genome. And so, in collaboration with Woese, he selected the genome of *M. jannaschii* as the next contestant for the TIGR treatment.

Venter and Woese were not the first to embark on sequencing archaea. Back in 1993, a band of researchers formed a pressure group seeking support for archaeal sequencing projects named ARGO—the Archaeal Genome Organization—while hoping that their ignorance of Greek mythology would not come back to haunt them. They were later reminded that the *Argo* was the ship that carried Jason and the Argonauts in their quest for the golden fleece—a mission that was ultimately successful, though plagued by treachery and danger along the way. Several efforts soon got off the ground, but buoyed by a $1.5 million grant in early 1995, it was Venter and Woese who would claim the prize first.

Venter and Woese adopted the same proven sequencing strategy that had worked so well for the similarly sized *H. influenzae* genome. After dividing the genome into thousands of random DNA fragments about 2,000 to 2,500 letters apiece, they performed over 36,000 sequencing reads. The remaining gaps were filled in using larger clones to help orient the various fragments. It took close to a year, but Venter's group now had the complete DNA sequence of the first archaea. After a six-month grace period to allow Haseltine to preview the sequence data, the results appeared in *Science* in August 1996, twenty years after Woese put archaea on the evolutionary map. (Researchers were so desperate to see the archaea sequence that they raised little fuss about TIGR's obligation to Human Genome Sciences.)

The significance of Venter and Woese's sequence transcended the 1,664,976 letters of the circular chromosome, or even the two small extrachromosomal elements found in *M. jannaschii*. "It tells us things about life on this planet that would have seemed like science fiction even a few years ago," said Venter. The main chromosome was predicted to encode 1,682 genes, but remarkably, of these, only 11 percent matched genes from *H. influenzae* and 50 percent found no matches in the databases. The reason for such a low match was that the archaeal genes resembled those of animals and plants more than simple bacteria. Genes that make protein components of ribosomes (the protein factories), RNA polymerase, and other factors required for gene activation all resembled eukaryotic genes, not bacterial genes. The structure and replication of the archaea chromosome also bore the hallmarks of eukaryotic chromosomes. On the other hand, most genes involved in basic metabolism shared more in common with bacterial counterparts than eukaryotes.

The first archaea genome sequence vindicated Woese's long insistence that the archaea represent a third limb of the tree of life. According to this view, life traces back some 3.5 billion years to LUCA—the last universal common ancestor. LUCA divided into two nonnucleated cells, the bacteria and the archaea. Millions of years later, the archaea gave rise to cells with a nucleus—the eukaryotes. By a process of endosymbiosis ("endo" meaning internal, "symbiosis" meaning a mutually beneficial relationship), these cells took up small bacteria that serve as mitochondria (the energy-producing factories) and chloroplasts (sites of photosynthesis in plants), vital cogs in the evolution of animals and plants. Richard Fortey put it like this: "We are one tribe with bacteria that live in hot springs, parasitic barnacles, vampire bats and cauliflowers. We all share a common ancestor."

However, even Woese concedes that the tree of life is not quite this straightforward. The branches of the evolutionary tree are much more gnarled than we might expect. As we have seen, archaea possess bacterial genes, eukaryotes possess bacterial genes, and bacteria possess archaeal genes. This was borne home when Venter's group sequenced the genome of *Thermotoga maritima,* a rod-shaped bacterium first discovered in Vulcano, Italy, in an 80°C marine sediment. The sequence of this genome revealed a surprisingly archaea-like organization, with about a quarter of its genes related to archaea. This example, and many others, suggests that in addition to the vertical transfer of genes from generation to generation, a high degree of lateral gene transfer—the process by which bacteria can spread genes for antibiotic resistance—has occurred during evolution. It suggests that life may have evolved from a small population of primitive cells with shuffling genes. To some, this is an unwelcome complication in evolutionary theory. Observes evolutionary biologist Ford Doolittle, "It is as if we have failed at the task that Darwin set for us: delineating the unique structure of the tree of life." On the contrary, however, the sequencing insurrection that Venter sparked has yielded a far deeper appreciation of the origin of life and microbial evolution than anyone might have predicted just a few years ago.

Following *M. jannaschii,* several archaea genomes were sequenced in quick succession. Within two years, microbial DNA sequencing became so commonplace that publication of the sixth archaea sequence, reported by a Japanese group in 1999, was relegated to an obscure specialist journal called *DNA Research.* However, the sequence of *Aeropyrum pernix* was noteworthy for being the first representative of the aerobic archaea, or crenarchaeotes—virgin territory as far as genomic sequencing is concerned. (All previously sequenced archaea are members of the other major subdivision, euryarchaeotes.) The 1,669,695 bases of *A. pernix* seem to contain fewer genes that resemble eukaryotes and provide further evidence for the importance of the lateral transfer of genes in the early stages of evolution. As scientists continue to edge along the branches of the tree of life in search of new microbes to sequence, a clearing picture of the development of life on the planet will emerge, and with it a new appreciation of the true place of humans in the hierarchy of life on earth.

SINCE TIGR LAUNCHED THE MICROBIAL SEQUENCING REVOLUTION IN 1995, more than thirty microbial genomes have been published and a similar number are nearing completion. That number will double by the end of 2000, resulting in the known sequence of most of the important bacterial pathogens, and an increasing number of commercially useful microbes as well. The shotgun sequencing strategy that TIGR pioneered is now the textbook method of choice around the world. Venter's string of seminal microbial genome publications has made him one of the most highly cited researchers over the past five years (a distinction he shares with his TIGR colleague Mark Adams).

TIGR has sequenced the DNA of many medically important microorganisms, including *Helicobacter pylori,* which chronically infects more than half of the world's population and is responsible for most cases of stomach ulcers. Other organisms include those for tuberculosis, meningitis, cholera, the spirochetes that cause syphilis, and Lyme disease. Each of these achievements promises major advances in vaccine development. TIGR has also sequenced a handful of chromosomes of the malaria parasite and the mustard plant, a popular plant model organism.

Arguably the most interesting microbe sequenced by TIGR is a strain of radiation-resistant bacteria named *Deinococcus radiodurans,* whose exploits have earned it a spot in the *Guinness Book of World Records.* This remarkable microbe was first isolated in 1956 inside a can of meat that had supposedly been sterilized by gamma radiation. *D. radiodurans* (the name means "weird berry-shaped bug that resists radiation") can withstand exposure to 12 million rads of radiation, a dose thousands of times greater than required to kill a person. Such radiation bombardment slithers the bacterial DNA into dozens of pieces, and yet *D. radiodurans* is effortlessly able to piece together its genome, rather like coalescing drops of mercury. (In most organisms, a few DNA breaks are crippling.) Curiously, the *D. radiodurans* sequence—divided among four circular DNA molecules and encoding 3,187 genes—has not immediately provided the means to understand this repair mechanism. There are some clues, however. The bacteria are capable of pumping out small stretches of damaged DNA so there is no risk of their being reincorporated into new chromosomal material. It is also intriguing that each bacterial cell contains four to ten copies of the complete genome, prompting speculation that these extra copies function much like backup files in a computer, allowing intact copies of genes to be pasted into a damaged section of a chromosome.

But why did these extraordinary recuperative powers evolve in the first place? One intriguing possibility is that these unique abilities evolved to counter not massive doses of radiation (such levels would never be encountered on earth) but rather severe dehydration. Desiccating the bacteria induces DNA damage reminiscent of radiation. Indeed, *D. radiodurans* has been found in areas of Antarctica that have not seen rain in hundreds of years. Thus, resistance to radiation may simply be a serendipitous by-product of coping with prolonged periods of drought—encouraging to supporters of the panspermia theory that life was originally carried to earth through space. The *D. radiodurans* sequence should provide a boost for therapeutics to repair damaged DNA. New strains of the bacteria are already being used to clean up spills of toxic chemicals, and doubtless many more novel industrial purposes will be developed.

In August 2000, Fraser and her TIGR colleagues celebrated the completion of their twentieth microbial genome sequence—the El Tor strain of *Vibrio cholerae*, the bacterium that causes cholera. The sperm-shaped cholera bacterium causes a catastrophic loss of fluids from the body—perhaps half the total body content per day. The French called it *mort de chien* (a dog's death). The first of seven cholera pandemics occurred in 1817, when the disease spread from India throughout Asia, later spreading to Europe and North America.

Much to Fraser's surprise, cholera turns out to have two circular chromosomes (more than 4 million bases) and twice as many genes (3,885) as expected. The larger chromosome appears to contain most of the bacterium's standard housekeeping genes required for survival, whereas the smaller chromosome contains more host-related genes that were apparently captured during evolution. The decoding of this horrific organism's genetic code should reveal new clues to limiting further outbreaks and saving countless lives.

DURING THEIR FOUR-YEAR RELATIONSHIP, tales of dissent and disagreement between Venter and Haseltine were rife. Not only did the two men have strong powerful egos, but also their goals in collecting and distributing the EST data could not have been more different. Venter anxiously wanted to redeem himself in the eyes of the scientific community by providing easy access to academic researchers to the full collection of TIGR's database. Haseltine refused to allow immediate access, in the hopes of safeguarding his company's intellectual property.

Nevertheless, the news that Venter and Haseltine were suddenly dissolving their partnership was a surprise. When TIGR was founded, the original terms of the deal called for the institute to receive $70 million from Human Genome Sciences over a ten-year period. But in 1997, Venter waived the last $38 million that TIGR was owed. Despite the well-documented differences between Venter and Haseltine, it seemed like a steep price to pay to secure his freedom. Venter did not wait long to celebrate. His first action was to release reams of unpublished DNA sequence data into the public databases, to widespread acclaim from academics worldwide.

However, Venter was also mulling over his next—and most risky—venture. In the back of his mind, he suspected that shotgun sequencing might work for much larger genomes. As far back as July 1995, Venter was telling *Science*'s Rachel Nowak that the publication of the first complete bacterial sequence had "raised the ante worldwide for sequencing the human genome." Indeed, Venter had closed his epic *Science* article describing the first microbial genome sequence with one of those delicious understatements that scientists seem to reserve for truly classic papers. After reciting a litany of tips for would-be surveyors of bacterial genomes, Venter threw in a teasing line. "Finally," he mused, "this [shotgun] strategy has potential to facilitate the sequencing of the human genome."

Few people at that time took him seriously, but the astonishing success of the microbial sequencing projects suggested to Venter that his dream could work, if the right technology was in place. Meanwhile, Venter's confidante Hamilton Smith was in no doubt what should be done. "Sequence, sequence, sequence. That's what's needed to set the stage for the twenty-first century."

CHAPTER SIX

Treasures of the Lost Worlds

Purity of race does not exist. Europe is a continent of energetic mongrels.

—H. A. L. FISHER

ONE OF THE FIRST SIGHTS to greet visitors to Tristan da Cunha is a painted wooden sign that says simply: WELCOME TO THE LONELIEST ISLAND IN THE WORLD. On his first visit to this desolate 7,000-foot protrusion above the waters of the South Atlantic, Noé Zamel was too busy kissing the ground to notice. The seven-day, 1,700-mile voyage from Cape Town to Tristan da Cunha through the tumultuous seas known as the Roaring Forties tests the inner fortitude of even the most experienced sailors, let alone a sixty-year-old Brazilian pulmonary physician who was simply hitching a ride. Seas reminiscent of the "perfect storm"—the infamous coalescence of three weather systems off the Massachusetts coast in 1991—caused the *SA Agulhas,* a 100-meter oceanographic research vessel, to corkscrew through the water with a pitch of fifty degrees from vertical.

But despite the water that flooded his cabin and the constant fear that the boat would capsize, Zamel says his trips to the remote British territory midway between Cape Town and Buenos Aires were actually quite fun. The lure of Tristan da Cunha was not the island itself but its people—some 300 descendants of a handful of founders of European and African origin. The Tristanians enjoy a simple way of life—fishing, growing potatoes and cabbages, raising sheep—almost oblivious to the rest of the world. But now the world is coming to them, for hidden in the DNA of these ninety families is a clue that might unlock the mystery of asthma.

Zamel first became aware of the Tristanians in 1961. He was a resident at a London hospital at the time when the entire population of Tristan da Cunha was evacuated to Britain following an eruption of the island's volcano. The lava spewing out of a fissure near the base of the mountain submerged the local lobster factory but did little other damage; however, the 270 Tristanians were required to stay in Southampton for almost two years, during which time they were subjected to a battery of medical and psychological tests, from diet to dialect, anthropology to asthma. The tests confirmed earlier suspicions that half of the islanders had a history of asthma—the highest incidence of asthma in the world.

To understand how such a remarkably high frequency of a disease like asthma could arise on Tristan da Cunha, we should begin at the beginning. The island was discovered in 1506 by the Portuguese explorer Tristao da Cunha, but the first settlers did not arrive until 1810. Only one person survived by the time a British garrison was stationed there in 1816, to prevent its use by the French as a base to rescue Napoleon, who was in exile on St. Helena, an island one thousand miles to the north, following his defeat in the battle of Waterloo the previous year. When it withdrew the next year, three soldiers stayed, including the Scot William Glass, who eventually had sixteen children. Dutch, Danish, and English nationalities were among the other founding family members as the population slowly grew, arriving via the occasional shipwreck and a trickle of emigration from St. Helena. Among the latter were two sisters, who, it is now believed, introduced a genetic susceptibility to asthma to the island more than a century ago. This susceptibility subsequently spread through the inevitable cousin-cousin marriages as the population grew from about fifteen to over three hundred. Among the inhabitants today, there are only seven surnames.

The first serious case of asthma on the island was reported in 1910 by a missionary who documented the death of a twelve-year-old girl. Several other studies noted the incidence of the disease, culminating in the British study following the volcanic eruption.

In the early 1990s, Zamel, now at the University of Toronto, remembered the strange case of the Tristanians when he set up plans to study the genetics of asthma. His initial attempts to convince the islanders to participate were futile. First Zamel had to request the British Foreign

Office to transmit a telegram to Cape Town, which would convey the message to Tristan da Cunha, where it would finally be discussed at the quarterly council meeting. Twice Zamel was rejected; it was only after Zamel produced a video appeal for the islanders' cooperation that he was finally granted permission.

In September 1993, after Zamel had gratefully set foot on dry land, he surveyed the island. Nestling at the foot of the mountain decorated with grazing sheep was the settlement known as Edinburgh: a school, two churches, a post office, the fish factory, an administration building with a swimming pool nearby, a hospital, a pub housed in the village hall, and a supermarket. The islanders, Zamel noticed, speak an unusual English dialect smattered with Dutch idioms. They add an "h" in front of words beginning with a vowel, so that "asthma" sounds like "hashmere."

Zamel began a thorough clinical evaluation of the islanders, diagnosing asthma by measuring the airway hyperresponsiveness to methacholine and collecting blood samples for the genetic analysis. He found that nearly half of the islanders were allergic to house dust mites, and a quarter were sensitive to cat fur—though there had been no cats on Tristan da Cunha for twenty years because of toxoplasmosis. He also produced a complete genealogical record of the island's inhabitants tracing back to William Glass. Zamel fulfilled his commitment to collect ninety blood samples on his last day, despite having to self-administer morphine to cope with a kidney stone (which he finally passed two days en route to Cape Town).

Three years later, Zamel made a return trip to the island, this time with reinforcements. Accompanying Zamel was Carrie Le Duc, a scientist from a San Diego biotechnology company called Sequana Therapeutics, which was risking $10 million in an effort to find the Tristanian asthma susceptibility gene. Le Duc brought enough equipment to turn a room of the local hospital into a respectable cell biology laboratory, allowing her to freeze hundreds of blood samples to produce cell lines that would serve as an eternal source of Tristanian DNA. Using the time-honored positional cloning strategy, Sequana intended to map the region of the genome that harbored the tiny flaw associated with asthma and identify the gene. The hope was that this discovery would shed new light on the chemical responses that go awry in asthma and provide new clues for the development of drugs. Sequana licensed the rights to the gene discovery to the German pharmaceutical company Boehringer Ingelheim for $30.5 mil-

lion. It was a huge gamble: Could the gene that caused asthma on a desolate island near Antarctica possibly have relevance for the millions of
asthma sufferers around the world?

After the frozen cell samples were safely received back at Sequana's
headquarters, the staff, led by Jeff Hall, a former member of Mary-Claire
King's group that had mapped the breast cancer gene *BRCA1*, set about
analyzing the DNA with hundreds of polymorphic markers. Many markers showed hints of linkage to the putative asthma gene. However, the
strongest evidence implicated a segment of DNA on the short arm of
chromosome 11, in particular two adjacent genes.

Sequana's foray into the world of asthma genetics provoked howls of
dissent on two fronts. Academic researchers cried foul when Sequana refused to reveal its asthma discovery in the traditional manner of publication in a reputable science journal. In May 1997, Sequana issued a teasing
statement that it had finally discovered the gene mutation for asthma on
Tristan da Cunha. This example of "genetics by press release"—announcing a result without releasing the data for peer review—soon became
common practice among biotechnology and pharmaceutical companies
more interested in wooing investors than wowing investigators. (Many
such discoveries triggered milestone payments from partner pharmaceutical firms, and had to be publicly disclosed in order to avert potential
charges of insider trading.) The identity of the candidate genes, which
were initially dubbed *"Wheeze-1* and *-2"* but now go by the more formal
titles of *ASTH1I* and *ASTH1J*, became public only when the European
patent application was filed.

Meanwhile, environmental groups sharply criticized Sequana for
exploiting an isolated population for profit. A Canadian organization
called the Rural Advancement Foundation charged that Sequana was
"committing acts of genetic biopiracy and, in the process, violating the
fundamental human rights of the people from whom DNA samples are
taken." But Sequana countered that all of the Tristanians participating in
its study fully understood the goals of the research and had willingly
signed consent forms. Moreover, the islanders had been guaranteed free
medicine for life should the research prove successful.

Zamel and colleagues have conducted a separate study among 300
pairs of siblings with asthma in Canada. This study has reassuringly provided independent confirmation of an asthma susceptibility gene on
chromosome 11, although the evidence is not as convincing as that obtained using the Tristanian families. Zamel suspects the statistical differ-

ences may be due to the African influence from St. Helena in the early founders of Tristan da Cunha. But confirming the association in a more heterogeneous population of European ancestry suggests that the underlying trigger of asthma in the South Atlantic may be relevant to most cases of the disease worldwide.

BY CONTRAST TO THE NAVIGATIONAL ORDEAL facing visitors to Tristan da Cunha, the journey to Tangier Island takes a mere ninety minutes over the tranquil waters of the Chesapeake Bay. Despite its proximity to the U.S. mainland, the island is so isolated that most of the 750 inhabitants trace their ancestry to the first English settlers in the late 1600s. The guidebooks lure visitors to this small sand bar that bisects the eastern shores of Virginia and Maryland with stories of succulent soft shell crabs and by noting that the residents speak in a quaint Elizabethan brogue. A tour of the island lasts barely fifteen minutes in a golf cart, during which one cannot help but notice the repetition of family names on the gravestones: Crocket, Pruitt, and so on. As for those guidebooks, the crabs are indeed spectacular, but those hoping to hear snatches of ye olde English are inevitably disappointed. Perhaps the profusion of satellite dishes is to blame, but to the untrained ear at least, the inhabitants speak with nothing more distinctive than a strong southern drawl.

In the absence of a local doctor, residents take a ferry to the mainland for a doctor's appointment. In 1961, a five-year-old boy was referred to a well-known physician at the NIH named Donald Fredrickson (who would later become director of the NIH and president of the Howard Hughes Medical Institute). Fredrickson noticed that the boy's tonsils were grossly enlarged and a bizarre orange-yellow color, the result of a sheen of cholesterol. The boy's sister also had unusual tonsils and shared a number of other unusual traits, chiefly the virtual absence of high-density lipoprotein (HDL) cholesterol particles, the so-called good cholesterol. Fredrickson named the recessively inherited condition Tangier disease. Someone, perhaps when British troops were stationed on the island two hundred years earlier, had unknowingly introduced a genetic flaw into the island's gene pool. As the population grew and cousins married cousins, this flawed gene spread, until eventually a child was born who had inherited two copies of the gene—one from each parent. The defective gene clearly interfered with the body's normal mechanism for disposing of cholesterol.

Following the initial reports of Tangier disease, identical symptoms were reported in cases in Europe, Canada, and elsewhere, although worldwide there are no more than forty affected families. Nevertheless, several research groups were sufficiently motivated to hunt down the mystery gene. In 1999, three teams of researchers simultaneously identified the Tangier disease gene, an exciting discovery that should boost the development of cholesterol-managing drugs. Several different mutations were found in the Tangier gene, called *ABC1,* including the mutation in the original family from Tangier Island: two missing bases of DNA prematurely terminated production of the ABC1 protein less than halfway through its length, rendering it inactive. The original carrier of the mutation probably experienced only mild symptoms and would not have known she or he was ill.

Unlike the discovery of many other genes, which offer few clues to the function of the gene and the mechanism of disease, the sequence of the Tangier gene immediately revealed its link to cholesterol metabolism. A cursory glance at the sequence of the ABC1 protein revealed 12 stretches of conspicuously hydrophobic (water-repelling) amino acids, indicating that the protein sits lodged in the fatty membrane coat of cells, snaking in and out of the membrane. Furthermore, the sequence bared striking similarities to a well-known family of membrane proteins called the ABC transporters, many of which have been tied to genetic diseases including cystic fibrosis and adrenoleukodystrophy.

Thus, the normal function of ABC1 is to pump cholesterol out of cells, where it eventually forms HDL particles that are transported to the liver, where they are broken down. In patients with Tangier disease, the efflux of cholesterol from cells is blocked, causing increased cholesterol deposition inside cells and the breakdown of HDL, leading to premature atherosclerosis and coronary artery disease. Cholesterol deposits in other tissues cause other symptoms including nerve damage and blindness.

David Mangelsdorf's group at the University of Texas Southwestern Medical Center has shown that one of the major sites of action of the ABC1 transporter protein is in the cells lining the duodenum, where it serves to pump absorbed cholesterol back into the lumen of the intestine. Mangelsdorf has found that treating mice with a drug that boosts production of the Tangier protein dramatically inhibits cholesterol absorption, raising hopes that related drugs might be effective in preventing atherosclerosis and managing cholesterol levels.

There is a possibility, however, that by the time such a drug has been successfully developed, the tiny island that gave rise to Tangier disease may have been permanently swallowed by the rising waters of the Chesapeake.

FROM THE FROZEN TUNDRA of western Finland to the tropical islands of the Philippines, from the bedouin tribes of the Middle East to the Mexican-American natives of the Rio Grande, scientists are spanning the globe in pursuit of remote populations. The cultural isolation of these ethnic groups, whether primarily geographic or cultural, affords researchers a major advantage in trying to track the spelling errors in DNA that are associated with diseases such as asthma, obesity, multiple sclerosis, diabetes, cleft palate, psoriasis, and many others. In some cases, such as in Finland, the physical isolation of the population serves to insulate rare gene mutations. In other populations, such as the Dutch Amish of Pennsylvania, where consanguinity is still common, the chances of two identical recessive mutations being thrust together increases. In the offspring of two first cousins, for example, one-sixteenth of the DNA on average will be identical on both chromosomes, increasing the chances of the expression of a genetic disorder.

In recent years, the bedouin Arabs of the southern Negev region of Israel have become accustomed to the incongruous sight of a white van with the University of Iowa logo emblazoned on the side. Val Sheffield, a large man with dark curly hair, has been leading a major effort to study these Arab tribes, in which more than half of the marriages are consanguineous. As a result of this cultural isolation, the prevalence of some genetic diseases is unusually high. One disorder that particularly intrigues Sheffield is Bardet-Biedl syndrome, a curious constellation of features including mental retardation, retinitis pigmentosa, polydactyly, and obesity. Sheffield's group has traced three separate forms of Bardet-Biedl syndrome in three tribes to different chromosomes: 3, 15, and 16. Given the remarkable diversity of symptoms in this syndrome, the imminent identification of these genes will be greedily seized on by researchers in a range of fields. The most common genetic disease among the 90,000 bedouin Arabs is a recessively inherited form of deafness. Sheffield's team mapped the deafness gene to chromosome 13 before a British team subsequently revealed its identity as connexin 26, a protein constituent of gap junctions that permits the transfer of materials between adjacent cells. There are over sixty different forms of nonsyndromic hearing loss, but mutations in connexin 26 appear to be the most common.

Surrounded by the Baltic Sea to the south and west and the Arctic Ocean to the north, it is hardly surprising that the Finns feel they live "at the edge of the inhabitable world," as Leena Peltonen, one of the country's leading geneticists, puts it. Sandwiched between the culturally diverse, frequently feuding Swedes and Russians, Finns have remained isolated: few leave; even fewer arrive. The country was first populated about 10,000 years ago, after the last ice age, but numbers did not begin to grow appreciably until about 2,000 years ago. By the twelfth century, the population numbered 50,000 people, and by the late seventeenth century, it had grown to about 500,000. Many inhabitants left the coastal areas of "old Finland" to head inland, north and east, forming numerous isolated communities in the "new Finland." At least one-third of the population subsequently perished from the famine of the hunger years and a series of epidemics. The survivors became the founding members of Finland's present population of more than 5 million people, although many inhabitants live in small subisolates founded just twenty generations ago.

By the early 1970s, geneticists recognized that Finland was unusually fertile ground for the study of human genetics. All told, more than thirty genetic diseases are more common in Finland than in other populations, whereas other disorders that afflict large numbers of Europeans, such as phenylketonuria and cystic fibrosis, are scarce or absent. Geneticists attribute this unusual prevalence of so many disorders to a founder effect compounded by genetic drift. In other words, the introduction of a mutation by a single founder into a small, isolated subpopulation, followed by a period of rapid population growth (as well as some elements of inbreeding in isolated areas), provided ideal conditions for the spread and manifestation of the gene associated with disease.

With the cooperation of the population and excellent records kept by the state and the Lutheran church dating back almost 400 years, Peltonen, Albert de la Chapelle, and their colleagues have made great progress in identifying the genes for most of the relatively common recessive diseases in Finland. Some of these disorders, such as Imerslund-Gräsbeck syndrome and Mulibrey nanism, are obscure even by the standards of card-carrying geneticists. But all are important in their own right, including examples of blindness, deafness, and muscular dystrophy and neurological disease. The incidence of many of these diseases is clustered into discrete parts of the country, reflecting the stability of the population. For example, most cases of amyloidosis are grouped in the southeast, the vision disorder retinoschisis is most common in the southwest, and another serious vision loss disease, choroideremia, is most prevalent in northeastern Finnish Lapland. As predicted, in every instance where the defective gene has been found, a large majority of patients harbor the identical mutation, providing persuasive evidence for the passage and spread of an ancient mutation present in an early founder of the Finnish population.

While standard methods can be used to map the rough location of a disease gene by studying the segregation of polymorphic DNA markers in Finnish families, this still leaves millions of bases, and potentially hundreds of genes, to sort through. But in 1992, Eric Lander and de la Chapelle made the precocious suggestion that the exponential growth of a strain of bacteria in a flask could serve as a model for the rapid, almost exponential, growth of an isolated population such as in Finland carrying a mutant gene. In 1943, Luria and Delbrück had published a classic paper in which they derived equations to map mutations that arise in an exponentially growing bacterial population. Lander's mathematical dexterity enabled him to see the parallels with isolated populations such as those in Finland. "Although our populations are human, the situation is much the same," he observed dryly.

The test case was the recessive disease diastrophic dysplasia (DTD), which results in short stature and is unusually prevalent in Finland, where it affects about 200 people. Lander's collaborators, de la Chapelle's group in Helsinki, had mapped the *DTD* gene to chromosome 5, but that still left years of work ahead to sift through a region of perhaps 1.5 million bases. Applying Hawking's Rule, I will forgo the mathematical minutiae, but the result of Lander's calculations was that the *DTD* gene probably lay 60,000 bases from the most tightly associated DNA marker. After sequencing the DNA around this marker, Lander and colleagues soon identified the gene and the mutation responsible for the majority of cases of DTD in Finland. Much to Lander's chagrin, his fearless prediction was off—by just 10,000 bases!

In 1688, two Dutch immigrants—Gerrit Janss from Veldcamp and Ariaantje Adriaansse from Amsterdam—were married in Cape Town. This couple has the dubious distinction of having introduced a dominant mutation for the disease porphyria variegata (VP) to South Africa. Patients with VP, which is caused by a fault in the metabolism of the blood pigment haem, are sensitive to light and can suffer acute, potentially fatal neuropsychiatric attacks. They also pass urine that acquires a deep port color, for which the disease was named (after the Greek *porphuros,* meaning reddish purple). Today, more than ten thousand descendants of Janss and Adriaansse carry the same C-to-T spelling mistake in the gene for an enzyme called protoporphyrinogen oxidase—the VP gene—accounting for virtually all cases of the disease in South Africa, but none in other countries.

VP is also known as the royal malady in view of tantalizing evidence that King George III, America's last monarch, and other descendants of Mary, queen of Scots, suffered from the disorder. If the madness of George III was caused by a genetic disorder, the trait disappeared in subsequent generations, just as a more famous royal malady, hemophilia, was making its appearance in the House of Windsor. (Queen Victoria passed a defective gene for a blood-clotting factor to three of her children, inflicting the disease on the Russian and Spanish royal families, while King Edward VII and his descendants were spared.)

Suspicion as to the root cause of the madness of King George centers not only on his five serious bouts of delusion after the age of fifty, but also on the unusual color of his urine. Experts have clashed over whether his urine's complexion—"a pale blue ring upon the glass"—was indicative of VP or an unrelated disease. Proof that King George's insanity was attributable to an inborn error of metabolism could, in principle, be obtained by forensic DNA analysis. However, the chances of such an intimate probe of the royal family's DNA taking place seem rather slim.

THE SUCCESS OF FINDING DISEASE GENES among the geographically isolated populations of Finland or various island communities, or culturally isolated populations such as those in the Middle East, the Mormons of Utah, and the Amish of Pennsylvania, has prompted the search for other populations with similar advantages. The challenge is to see whether the methods that have proved so successful for tracking individual Mendelian genes in Finland and elsewhere can also be applied to more common disorders such as heart disease and diabetes, where the relative contributions of genetic and environmental factors are unclear. The key to studying these remote populations lies in the variations between DNA sequences of different individuals and races. Between any two people there are on the order of 1 million differences in the letters of the human genome—perhaps half a dozen on average in a single gene. Recall that these variable bases are known as SNPs. By tracking the inheritance of thousands of SNPs in families affected with a genetic disease, researchers can quickly pinpoint a SNP that is in close proximity to the errant gene. Biotech and pharmaceutical companies have seized on this paradigm and are stockpiling huge collections of SNPs to aid the search for common disease genes. Already, companies are reporting success in identifying important disease genes. The French company GenSet reports it has located a gene for schizophrenia, and Glaxo Wellcome has pinpointed three genes using SNPs for migraines, diabetes, and psoriasis (although the details are closely guarded for the time being). In less than a decade, gene hunting has progressed from an expensive, arduous trawl through uncharted chromosomes to a rapid, highly automated process where the limiting factor is merely gaining access to the relevant populations.

About 900 adults live on Norfolk Island, which lies about 1,000 miles off the east coast of Australia. About 75 percent are direct descendants of the British mutineers who took command of the *Bounty* and cast Captain Bligh adrift. Many of the men settled down with Tahitian women, but many of their descendants are strangely prone to cardiovascular disease. Researchers believe that the Polynesian predisposition to cardiovascular disease, coupled with a similar susceptibility among British descendants, will shed light not only on genetic factors for cardiovascular disease, but also on hypertension, migraines, and other conditions.

On the Micronesian island of Kosrae, Rockefeller University researcher Jeff Friedman is collaborating with local authorities to conduct a major study of genetic factors in obesity. Following Friedman's dramatic discovery of leptin in 1994, the big question was whether mutations in the human leptin gene might cause forms of obesity similar to that seen in mice. Despite an intensive search, only a small number of humans appear to carry mutations in the leptin gene, although it seems that most cases of morbid obesity in humans are due to abnormalities that affect leptin secretion or sensitivity. To uncover other genes involved in human obesity, Friedman has initiated a collaboration with the Department of Health on the Micronesian island of Kosrae, where the population, mostly of Micronesian and Caucasian descent, shows a high rate of obesity. Friedman and collaborators have completed a full medical evaluation of twenty-five hundred adults on the island, and hope soon to locate genes that contribute to obesity, diabetes, and heart disease.

Although not as isolated as other populations discussed in this chapter, 176 Mexican American families living near the Rio Grande in Starr

County, Texas, are heavily afflicted by type II (non–insulin-dependent) diabetes. After considerable success in identifying genes for relatively rare inherited forms of diabetes, notably MODY (maturity-onset diabetes of the young), Graeme Bell of the University of Chicago has led a large group trying to unmask a gene that predisposes to this disease, which afflicts 16 million Americans alone. Several years ago, Bell's team traced their candidate gene to chromosome 2 and, after an arduous search, recently identified what they suspect is the key gene that codes for an enzyme called calpain. How this discovery ties into diabetes pathophysiology remains to be fully explained.

WHILE SOME SMALL POPULATIONS are only too eager to cooperate with commercial companies, the example of French DNA discussed earlier shows just how political and controversial is this issue. But the French scandal has nothing on one of their fellow European nations. Nowhere else is the potential bounty and controversy from mining the genome in isolated populations more intense than in Iceland, the world's oldest democracy.

Iceland is about the same size as the state of Virginia, and yet the population—just 275,000 people—would fit comfortably into three football stadiums. The country was officially settled in 874 A.D., when Ingolf Arnarson fled the tyranny of King Harald Finehair and sailed westward from Norway in search of a new land in the North Atlantic that a countryman had discovered a few years earlier and named Iceland to discourage rampant immigration. Iceland became the first nation in Europe to convene a parliament (the Althing), in 930, in stark contrast to the feudal systems elsewhere. Among its other social distinctions, Iceland abolished slavery in 1117, ended illiteracy in the eighteenth century, gave women the vote in 1915, and elected the world's first female head of state.

The stereotypical picture of the Icelandic people differs little from the observations of Jules Verne almost 150 years ago, in *Journey to the Centre of the Earth:* "The men looked robust but clumsy, like fair-haired Germans with pensive eyes, conscious of being somewhat apart from the rest of mankind, poor exiles relegated to this land of ice and whom Nature should have made Eskimos, seeing that she condemned them to live on the edge of the Arctic Circle." W. H. Auden did not care for the capital, Reykjavik, which he called "Lutheran, drab and remote," but the country

itself was beautiful—"a holy land, with the most magical light of anywhere on earth."

Iceland's future is intimately tied to its remarkable history. In 980,
Eric the Red was banished from Iceland for murder and sailed west again,
settling on Greenland. His son, Leif, would establish the first settlement in
Newfoundland, 500 years before Columbus. Quickly brushing aside a few
clans of Irish monks who had previously settled in Iceland, the founding
group of a few thousand Vikings rose to about 70,000 by 1100. While the
Norse constituted the majority of the population, a healthy minority was
of Celtic stock, including women brought as slaves. But every time the
population threatened to explode during the past millennium, a natural
disaster appeared as if on cue to purge the population. In the fifteenth century, it was the black plague, which wiped out two-thirds of the island's
60,000 people. An outbreak of smallpox in the early 1700s plummeted the
population to about 30,000. And then in 1783, an enormous eruption of
the Lakagigar volcano produced one of the largest lava flows on record
and spewed so much suffocating sulfur dioxide and ash that a quarter of
Iceland's population died in the ensuing famine.

In the entire history of Iceland, only about 800,000 people have
been raised there. As there was virtually no immigration to the island (at
least not before World War II), most Icelanders trace their roots back to
the early Viking settlers, although historical and genetic evidence points
to a significant Irish contribution to the nation's gene pool. How much
exactly is uncertain, but some Icelanders say they are a cross between the
most boring people in the world and the most drunk. For these reasons,
Iceland is a geneticist's gold mine, its genetic heritage poised to do for one
island in the North Atlantic what North Sea oil did for another. In medical terms, Iceland seems an unremarkable place for such a bonanza. The
frequency of most genetic conditions is roughly the same as in the rest of
Europe and the United States. Degenerative arthritis is slightly more
common in Iceland, but the country has not seen the same increase in
prevalence of asthma and type I (juvenile) diabetes as western Europe.

It is the size and structure of the population, not the traits themselves,
that has geneticists so excited about Iceland. Geneticists studying heterogeneous populations in the United States or Europe can collect large
numbers of families with a genetic disease, but there is no guarantee that
the underlying mutation is the same in people from different ethnic backgrounds. In Iceland, the situation is very different. The homogeneity of

the population, descendant from the original Viking settlers, and the influence of natural selection on the population over the past millennium, makes it likely that individuals suffering from hereditary forms of Alzheimer's disease or breast cancer or depression probably have the same Viking mutation. Moreover, because of the wealth of genealogical information on the population, the task of tracking the location of the faulty gene is made much easier.

The awesome power of genetics in Iceland is graphically illustrated by the example of breast cancer. Following Mary-Claire King's success in mapping the location of the first breast cancer gene, researchers set out to look for other breast cancer susceptibility genes. They knew that such genes must exist, because many of the breast cancer families King had collected over the years did not harbor a mutation in the *BRCA1* gene. What had taken King nearly two decades for *BRCA1* took just two years for *BRCA2*. Improved technology played a part, but the crucial factor was the interrelatedness of the Icelandic population that allowed researchers to trace virtually every instance of breast cancer on the island to a man named Einar, who lived over four hundred years ago. "The single, ancient *BRCA2* mutation was a golden key," says King. "Iceland is just an amazing place to do genetics. The population there is like a gift from heaven."

One of the most famous characters in Icelandic lore is Egil Skallagrímsson, born in 910, whose résumé features turns as a scaldic poet, murderer, drunkard, miser, and farmer. According to Egil's saga, his hair-trigger temper was evident as early as age six, when chasing another boy in a game of Icelandic rules football, "Egil ran up to him and drove the axe into his head right through to the brain." In a later adventure, Egil and his men were challenged to a drinking contest by a man named Armod. When Egil realized he couldn't keep going any longer, he grabbed Armod, pressed him against a pillar, "then heaved up a vomit of massive proportions that gushed all over Armod's face, into his eyes, nostrils and mouth, and flooded down his chest so that he was almost suffocated." Later, he gouged out Armod's eye with his finger, leaving it hanging over his cheek. But Egil was also a famous poet, and in one famous adventure saved his life after being captured by King Erik Bloodax by composing a lengthy rhyming poem to the king.

Egil's place in Icelandic history is secure. Now, a thousand years and more than thirty generations later, a direct descendant of Egil has returned to his native land to claim his place in history. If successful, he will have composed the ultimate saga of Viking history.

EVEN IN A CROWDED HOTEL LOBBY, it is impossible to miss Kári Stefánsson. He stands 6 feet 5 inches, and his spiky white hair adds another couple to that. My first meeting with him began inauspiciously. He grunted a curt acknowledgment, his gaze fixed intently into the distance, and appears determined to live up to the press reports of his austere demeanor. But then, to my great relief, he removes the earpiece of his cell phone and I realize he was conducting an international conversation. All is not lost.

"I need some coffee" is his first intelligible comment, issued in a tone that won't take no for an answer. Stefánsson was jet-lagged, having flown to Washington from Iceland the previous day to deliver an 8:00 A.M. lecture at the meeting of the American Association for the Advancement of Science. His irritation was exacerbated by having to answer another round of critical questions and by an aching leg injury sustained in a basketball game with his teenage daughter. There is a more pervasive source of fatigue, too: much like Venter, Stefánsson has come under fierce criticism since he began pursuing his genetic dream in 1997. But once his caffeine levels rise and his focus is directed to his company and his homeland, he quickly warms to the task. Each answer becomes a passionate monologue, sprinkled with passages from Auden and Pope, and peppered with occasional expletives.

In 1997, after living in the United States for twenty years, Stefánsson quit his job as professor of neurology at the Harvard Medical School to return to Iceland. Part of his motivation was nostalgia. "I had forgotten how magical the summer is," said Stefánsson wistfully. "To have two and three months when the only night is a little purple dot on the horizon and the birds stop singing for a couple of hours."

However, there was a bigger reason—12 million reasons, to be exact. That was the amount of venture capital Stefánsson had secured in three months to realize his dream of building a biotechnology company to sift through the genes encoded in Viking DNA. Stefánsson had watched with interest the efforts of American researchers in the late 1980s and early 1990s to use isolated populations such as the Mormons in Utah and the Dutch Amish of Pennsylvania to attempt to map genes for bipolar affective disorder and other neurological disorders. Slowly he realized the immense potential of his homeland for similar studies. During a visit to Iceland in 1995, to prepare DNA from multiple sclerosis patients, his plan began to crystallize.

The mission of deCODE Genetics would be to uncover the genes for multiple sclerosis and dozens of common genetic diseases. In February 1998, Stefánsson signed a deal with Hoffman-LaRoche worth a staggering $200 million over five years, granting the pharmaceutical giant licensing rights to deCODE discoveries in twelve different genetic disorders. In addition to the financial windfall, the deal guarantees free drug treatments to the Icelandic people. "I thought it would be interesting to coerce 'big pharma' to recognize the contribution of the population," Stefánsson grins.

Stefánsson's most important asset is the Islendingabók—a digitized record of the unparalleled genealogical wealth of the country spanning 1,100 years. Iceland has no prehistory; its first settlers were literate and wrote extensively. As far back as the twelfth century, the genealogies of 430 early Icelandic settlers were recorded in the Landnámabók. After two centuries of settlement, Icelanders wrote down what is widely considered the pinnacle of medieval literature—the sagas. These famous stories, first written in calf's blood on vellum, contain extensive family histories of the central characters, not unlike passages in the Bible. Despite the well-chronicled brutality of Icelandic life in the Middle Ages, the Sagas surpass all other records of life that emerged in the following centuries from more civilized cities of continental Europe, including Chaucer and Dante. A more contemporary perspective is offered by the Icelandic pop star Björk, who proclaims, "We were the first rappers of Europe." Throughout the Middle Ages, Icelanders continued to record their family histories, a keen tradition among poor, isolated peoples where there was little to do but know the affairs of one's neighbors.

The Islendingabók database depicts the family trees and other genealogical information of well over 75 percent of the 800,000 Icelanders who ever lived and for whom records exist. (The total number of Icelanders who ever lived is probably about 1.2 million; however, if one includes the high infant mortality rate during the country's history, the total number of live births over the past thousand years may exceed 2 million.)

Thordur Kristjansson, one of the creators of the database, offers me a demonstration of the genealogy database. Typing in a single name, such as Stefánsson's, immediately reveals his family tree. Above and to the left of the blue square that represents Stefánsson is another blue square, designating his father, Stefán, a well-known radio personality and author. To his right, a red circle represents Sólveig, his mother. Above his parents is an increasingly dense pattern of blue squares and red circles, a complete representation of Stefánsson's family for six generations. It is only when one clicks on the symbol for Ingveldur Guonadottir, born in 1776, that part of the trail runs cold: the branches on Ingveldur's paternal side are bare. But on her maternal side, the line of descent continues unabated, back through the age of Shakespeare and the Norman conquests until finally the familiar name of Egil Skallagrímsson appears. I remark to Kristjansson how extraordinary this is, but he just laughs. "It's nothing, we can all trace our ancestry back to Egil!" He shrugs, and by *all*, of course, he means the entire Icelandic population. Kristjansson types in his own name, and in just a few keystrokes, we are effortlessly transported back one thousand years to another ancestor, known as "flatnose" to his friends.

If Stefánsson had been content to use the existing genealogical information and rely on patient referrals from the Icelandic medical community, there would have been little controversy. But his ideas went much, much further than that. Stefánsson proposed that deCODE Genetics build a powerful new database that would contain a wealth of medical data for the entire Icelandic population. To complement the genealogical data on every living Icelander, deCODE would use detailed medical histories, including patient diagnoses, visits to their general practitioner and the hospital, and cost of treatment—information that has been contained within Iceland's nationalized health system since 1915. Combining these clinical data with the unparalleled family histories of his countrymen could yield extraordinary opportunities for unearthing disease genes, especially when combined with DNA genotypes for most members of the population. In return for privately financing and constructing the database to the tune of $150 million, deCODE would be granted an exclusive license for the resulting data. In effect, deCODE Genetics would gain exclusive marketing rights to a country's complete genetic heritage.

On March 31, 1998, a bill to create the Icelandic Health Care Database was introduced into the Icelandic parliament. It provoked outrage among politicians and scientists, even those who had welcomed the formation of deCODE just two years previously. Opponents were appalled by the prospect of Stefánsson's owning a monopoly on Iceland's genetic research, collecting personal medical information without the individual's explicit consent, the threat it posed to competitive funding for academic researchers, and the massive fortune he was poised to acquire. (Indeed, some estimates put deCODE's potential value at more than twice Iceland's gross domestic product.) A vigorous opposition movement was born, charging that Iceland's very sovereignty was at stake.

One of the chief concerns about the deCODE proposals was privacy, although confidentiality is a less contentious issue in Iceland than it would be in most other democracies. Iceland's nationalized health system means that concerns over the loss of private health insurance are virtually irrelevant, and with every citizen required to carry a national identity card there are few worries about government intrusions on privacy. DeCODE says it guarantees patient confidentiality by adopting a rigorous encryption procedure. Once a month, the updated genealogy database is sent to the Data Protection Center, where the information is encrypted. Similarly, patient records submitted by physicians are also codified so that deCODE scientists have no knowledge of individual identities. Moreover, Stefánsson points out that the penalty for breaking the confidentiality is two years' imprisonment.

Fierce public criticism has erupted over the government's apparent kowtowing to deCODE in negotiations for the health-care database, including unsavory allegations of bribery, which Stefánsson adamantly denies. DeCODE is permitted to gather the information without consent, but Icelanders are allowed to exclude themselves from being in the database. For the time being, most of deCODE's samples come from physicians studying particular diseases, with whom deCODE is collaborating. "There are those that claim that this constitutes coercion," says Stefánsson, "but most people simply want to participate. I can see nothing wrong with that." In the future, however, Stefánsson wants to get as many people as possible to give blood to deCODE for DNA typing. He thinks that 75 percent is a realistic target. DeCODE has already genotyped more than twenty thousand people, and only 5 percent of the population have declined, which has little effect on the statistical analysis of family data.

Some of Stefánsson's severest critics have founded a rival biotechnology company called UVS. The name is derived from three witches—Urdur, Verdandi, and Skuld—who determine the fate of man, according to Icelandic legend. Stefánsson is unimpressed: "This is just an attempt to cash in on our sort of success," he says. "But I'm glad they did this, because it helps me in many ways." A recent public opinion poll voted deCODE the most popular company in Iceland by a margin of four-to-one. "The people still support me, and the doctors think I'm the devil," says Stefánsson, "so it's time to just let the future happen."

The Icelandic government is firmly behind Stefánsson. One government minister said "it would be unethical not to try to use the unique resources that we have in Iceland to try to improve the health of Icelanders, and hopefully of others." But aside from the vast medical benefits that might stem from deCODE's research, Stefánsson cites another reason that his venture must thrive. "We have the potential to have a significant impact on the Icelandic economy if we do well. We have repatriated some of the best biologists coming out of Iceland. It is terribly important for a small society like this not to lose its best educated people . . . the consequences of losing the brightest people, even for one generation, are significant and lasting." Indeed, deCODE has lured many expatriates back to Iceland, and currently employs over 350 people. In July 2000, deCODE Genetics went public on the U.S. stock exchange, resulting in a valuation well in excess of $1 billion. Some critics point to the company's registration in Delaware as evidence of foreign control over Iceland's genetic heritage, but Stefánsson will have none of it. "The Delaware registration was simply because of the marketing considerations. How much more Icelandic can we be? Every employee works here, and everything we do is rooted in Icelandic culture and Icelandic heritage."

Inspired by the potential of the deCODE database, several other European countries are planning similar ventures, with some modifications. Projects are underway in Estonia and Sweden, but of particular note is the ambitious $60 million plan, hatched jointly by the British Medical Research Council and the Wellcome Trust charity, to launch the U.K. Population Biomedical Collection. Hailed as the world's largest clinical trial, it will involve 500,000 Britons who will provide medical records, lifestyle details, and DNA samples. Unlike Iceland, the British project will allow volunteers to opt in. All will be middle-aged adults, forty-five to sixty-five years old, in the slightly perverse hope that they fall sick within ten years or so.

The British genomics company Gemini, based in Cambridge, is collaborating with researchers in Newfoundland to form a new organization called Newfound Genomics. Together they will study the genes of the Newfounders using much the same approach as deCODE has popularized in Iceland. Some 25,000 Britons settled in Newfoundland over the past few hundred years. The Newfies, as the residents of Canada's poorest province (and butt of Canadian jokes) are known, suffer from several disorders, including rheumatoid arthritis, diabetes, and psoriasis. And there will be many others, although few will match the scale of the deal signed by the French company GenSet, to study the rural population of China for health-related genes.

THE EXTRAORDINARY CIRCUMSTANCES and controversy over the launch of deCODE Genetics has somewhat overshadowed the company's research efforts, which are currently aimed at identifying forty major genes, including thirty-nine disease genes. Twelve of these disorders are the subject of the $200 million deal Stefánsson struck with Hoffman-LaRoche, including Alzheimer's disease, heart disease, strokes, emphysema, and schizophrenia. Already, the investment seems to be paying off.

One month before making a major announcement at a conference in Zurich in March 2000, Stefánsson told me of his company's success in locating a gene for stroke. Through collaborations with physicians across the country, deCODE collected and analyzed the DNA of more than 2,500 stroke victims, focusing on 1,200 patients from 180 large families. At this point, Stefánsson narrowed the diagnostic criteria, focusing on about 50 families in which strokes occurred earlier and more frequently than in the general population. By screening the Icelandic DNAs with an arsenal of 1,000 DNA markers spaced across the twenty-three pairs of chromosomes, deCODE scientists were able to pinpoint one small chromosomal segment in common among the stroke patients. "It's the first ever mapping of common stroke," says Stefánsson, "the first ever success in using this approach in a complex disorder."

Using similar approaches, deCODE has made strides in other areas, tracing mutated genes for preeclampsia (high blood pressure in pregnant women), the skin disorder psoriasis, osteoarthritis (the most common form of arthritis, afflicting some 40 million people worldwide), and two genes for multiple sclerosis. In August 2000, deCODE announced a potential breakthrough in mapping a gene for Alzheimer's disease. Unlike previous gene associations that are confined to early or late onset, deCODE considered a broad definition of dementia in order to spot potential hereditary links. Progress is also being made in hunting genes involved in hypertension and endometriosis. In all of these examples, the complete genome sequence will facilitate the final steps from linked DNA markers toward identifying the precise mutations, but it is not a serious impediment. Once a gene is mapped, a team of young bioinformatics experts produces a dense cluster of markers in the vicinity of the gene to narrow down the critical region. In this way, the amount of DNA sequencing of suspected genes in patients and healthy controls is kept to a minimum.

There is also news of a major discovery in osteoporosis, which Stefánsson says he wants to confirm in a second, unrelated population before publishing. While we are talking at his Reykjavik home, Stefánsson is interrupted by a call from his research director, Jeff Gulcher, on his cell phone. The deCODE president tells him to call back in an hour and starts laughing. "Jeff is incontinent with excitement over the work with osteoporosis," Stefánsson explains. The reason is that "we have found a SNP in the coding region of a receptor that changes very dramatically an amino acid. He just wants to talk about it, but I don't want to talk about it. This is so clear, there's nothing to talk about!"

The fortieth project at deCODE is of universal interest. Geneticists have long dreamed of searching for gene variants associated with old age, and there is excellent evidence from experiments with fruit flies and nematodes that mutations in certain genes can significantly extend an organism's life span. There have been hints that polymorphisms in humans may affect aging, but nobody has been able to assess the possibility in a homogeneous population like Iceland. In the genealogical database, however, Stefánsson had a tool that might reveal genetic clues to longevity. By searching for the 1,500 people who have lived to be more than ninety years old, Stefánsson finds a significant degree of clustering within families. Moreover, people who had a sibling over the age of ninety were more likely to reach the same age than those who did not. "Not only is longevity familial," concludes Stefánsson, "it is genetic, and the genetic mechanism is very simple"—perhaps as simple as a single gene.

The software tools, databases, and financing that deCODE has amassed—not to mention the invaluable resource of Viking DNA—suggest that it will play a major role in the identification of common disease genes in the next few years. Mapping those genes in an isolated population will be the easy part, but like so many others, Stefánsson is anxiously waiting for the definitive genome sequence to be available. "What Celera and the Human Genome Project are doing is just creating the cable to download information from the human genome," Stefánsson says, "nothing more, nothing less. I don't give a damn about this cable as long as someone else hooks them up."

Mention of Celera inevitably steers the conversation toward its president. Unprompted, Stefánsson volunteers the view that Venter "is a bit of an asshole," but in a manner that suggests he reaches that opinion about most people. "I don't dislike him at all," he adds quickly, with a flicker of mutual respect for his fellow genetic age gladiator. There is a key difference, however, Stefánsson points out with a final lunge: "We're in a position to turn the sequence into knowledge about health—he is not."

/

CHAPTER SEVEN

Prize Fight

The world continues to offer glittering prizes to those who have stout hearts and sharp swords.

—LORD BIRKENHEAD

LIKE MANY OTHER SCIENTISTS, James Weber, the director of the Marshfield Medical Research Foundation in Wisconsin, was frustrated over the slow progress of the public genome project. His forte—mapping genes that cause inherited disease—is a vital if tedious enterprise, but his successes frequently left him with the daunting task of trying to characterize dozens of genes in millions of bases of DNA to find the disease gene in question. How much easier it would be if the DNA sequence were sitting in a database, ready to download and search for the crucial mutation.

Around 1993, Weber began mulling over a bold new approach to sequencing the human genome that might dispense with the painstaking process of mapping and sequencing preordered fragments from each chromosome. Instead of sorting the encyclopedia of life into chapters before reading the text, why not, Weber reasoned, rip out all the pages, decipher the text of millions of random pages, and then use high-powered computational tools to assemble the full sequence? Weber contacted Gene Myers, a computer specialist at the University of Arizona in Tucson. Myers's chief claim to fame to that point was helping to write an indispensable piece of software called BLAST, which finds sequence alignments in DNA and proteins. Myers ran some computer simulations that showed that the shotgun approach could indeed work for genomes vastly bigger than bacteria. Furthermore, it could finish the sequence years sooner than the existing clone-by-clone method, and potentially save hundreds of millions of dollars in the process.

Weber sent off a grant proposal to the NIH, seeking $12 million to run a pilot shotgun sequencing project. If his strategy could realize just a 1 percent savings in annual medical health costs (some $35 billion in the United States), Weber argued, the project would more or less pay for itself. In February 1996, he was invited to present his plans before a group of about forty genome leaders who had gathered for a retreat in Bermuda. The reaction was cool. Weber had no track record in running a DNA sequencing center, and conceptually his proposal seemed naive and unworkable.

"They trounced him," says Myers, "they said [the sequence] would be full of holes, a 'Swiss-cheese genome.' " Weber takes a more sober view. "Part of the opposition," he recalls, "stemmed from the science. These large genome centers were well established to sequence cosmids and BACs, and to change to something radically different would have meant entirely overturning their labs." One of the participants at the meeting was Craig Venter, but he did not comment on Weber's proposal, which was duly rejected.

Disappointed, Weber and Myers wrote up an account of their proposal. Weber brought the article to the attention of Richard Gibbs, director of the Baylor Genome Center and an editor of *Genome Research,* an increasingly popular journal in the genomics community. The proposal was eventually published one year after Weber's presentation in Bermuda. For the first time, the rank-and-file researcher had a chance to assess Weber and Myers's provocative ideas.

The Human Genome Project needed a radically new approach, Weber and Myers argued. The true motivation for the project should be the sequencing of all genes and their regulatory regions, identifying polymorphisms, and uncovering the unexpected. It should not be "the accomplishment of some arbitrary, mythical goal of 99.99 percent accuracy of a single artifactual (in places) and nonrepresentative copy of the genome." Weber and Myers's bottom line was simple: "We should generate as much of the critical sequence information as rapidly as possible and leave cleanup of gaps and problematic regions for future years."

Immediately following the Weber-Myers manifesto was a rebuttal written by Phillip Green, a highly respected bioinformatics expert at the University of Washington. In the 1990s, Green had written some of the most influential computer algorithms for analyzing the genome, in particular two programs called PHRED and PHRAP, which became the standard tools for interpreting DNA sequence data. A reserved figure on

the genomics circuit, it was not Green's style to publish opinionated articles in science journals, so his courteous but scathing criticism of the
Weber-Myers shotgun sequencing strategy carried considerable weight.
Green exposed a series of potential scientific and economic flaws in
the Weber-Myers proposal. The most serious problem was that the crucial finishing stage—filling in the thousands of gaps in the DNA sequence
that would be left after the initial shotgun sequencing—would be expensive and was destined to fail. Moreover, that failure would be discovered
only late in the project. Myers's computer simulations were "greatly over-
simplified," and it was naive to predict the success of shotgun sequencing
in humans by extrapolation from the method's successes in bacteria. The
human genome is a thousand-fold larger and studded with repeated tracts
of DNA that would confound the computer alignment. Finally, Green
questioned the organizational aspects of such a monolithic undertaking:
"It is not clear how one would deal with hiring, training, and laying off
the relevant people on the massive scale required. It also is quite unclear
how the project could be distributed among several laboratories. . . .
There is no reason to switch."

Green's deconstruction of the shotgun proposal left the two young
geneticists with few sympathizers. But among the few scientists who
would give some credence to their plan was Craig Venter. Shortly after
Weber's Bermuda presentation, Venter, Smith, and Leroy Hood had published their own recommendations of how the human genome should be
sequenced. Like Weber and Myers, Venter and colleagues proposed that
the tedious chore of mapping DNA fragments along each chromosome
before sequencing was dispensable. Venter had always placed a premium
on computational analysis. Myers's simulations had suggested that a variation of Venter and Smith's shotgun approach had a realistic chance of sequencing the human genome, although it was 2,000 times larger than a
typical bacterial genome. Venter was up for the challenge and, thanks to
the dissolution of his ties with Haseltine, free to give it a go. In the waning
months of 1997, only the lack of a breakthrough in DNA sequencing efficiency was holding Venter back from an assault on the human genome.
That was about to change.

IN FOSTER CITY, CALIFORNIA, Michael Hunkapiller gazed admiringly
at the result of ten years and some $100 million of investment. The future
of DNA sequencing had arrived, and it bore the logo of Hunkapiller's

company, Applied Biosystems. Venter is hardly exaggerating when he says "there would be no genome field at all if it wasn't for Mike Hunkapiller."

In the early 1980s, Hunkapiller was a key figure in the heaving laboratory of Leroy Hood at the California Institute of Technology. Hood's mission was to build an automated DNA sequencing machine that, while still employing Fred Sanger's chain termination method, would take the drudgery out of DNA sequencing. Sanger's original system had relied on a radioactive trace to distinguish the various DNA fragments, which could subsequently be detected by exposing to an X-ray film. Hood, Hunkapiller, and colleagues developed an alternative system using fluorescent dyes to label each of the four letters in DNA a different color. As the DNA fragments run past a laser at the bottom of the gel, the wavelength of light emitted by the dye indicates the identity of each successive letter in the sequence of that fragment. A computer records the traces of all four dyes and "calls" the sequence automatically. Hood's prototype machine had many advantages. One was that the use of fluorescent dyes dispensed with the need for radioactivity. Moreover, since the four DNA letters were uniquely labeled, the mixture of DNA strands from a sequencing reaction could be run on one lane of the gel, thereby increasing capacity.

Hood pitched his plans to produce a commercial sequencing machine to nineteen different companies—and was summarily rejected by all of them. So in 1983 he decided to start his own company, named it Applied Biosystems, and entrusted Hunkapiller to run it. For the next three years, ABI worked in tandem with Hood's lab to refine and improve the first automated DNA sequencing machine. (Both men are listed in the patent as inventors of the machine.) By 1986, the 377 automated sequencer was ready for testing. One of the first people to test it was Craig Venter, who arranged to test a prototype in his NIH laboratory (the first agreement of its kind between the NIH and a private company). In 1988, Applied Biosystems licensed the rights to the machine from Caltech, and the ABI 377 was on the market.

In 1993, Applied Biosystems was bought by the Perkin-Elmer Corporation, a sixty-year-old instrument manufacturer producing a diverse range of products ranging from bombsights to chip-making machines. Hunkapiller became the president of the company. In 1995, Tony White, the new president of Perkin-Elmer, set about rejuvenating the company by selling the struggling analytical instruments business and even the company's name. The sale was finalized in 1999, and White's slimmed-down company became known simply as PE Corporation (now known as Applera Corporation). By dispensing with the analytical instruments division, White could now focus entirely on biological instrumentation and resources, including software, reagents, and, of course, DNA sequencing technology. The transformation was an unqualified success, with revenues of more than $1 billion in 1999.

Throughout the 1990s, Hunkapiller had introduced a series of useful improvements into the DNA sequencer, for example increasing its sample capacity and accuracy. The machines were immensely popular, with more than 6,000 sold from 1987 to 1997. But from the beginning of the human genome endeavor, the project leaders knew that wholesale improvements in sequencing technology were essential to finish the sequence on time. Even before producing their first DNA sequencer, Hunkapiller and Hood had toyed with a radical idea to improve the speed and efficiency of DNA sequencing. Instead of using a thin slab of a gel-like material to separate the different lengths of DNA, Hunkapiller tried to use ultrathin capillary tubes filled with a polymer solution, each tube no wider than a human hair. The system showed promise, but it was difficult to standardize four reactions running in four tubes simultaneously. This problem inspired the development of dyes to mark each base of DNA, enabling Hunkapiller to shelve the capillary idea temporarily.

A decade later, Hunkapiller was ready to combine the capillary tube method for separating DNA fragments with the fluorescent dye technology for detecting them. There were many advantages to the capillary system: the tubes could tolerate higher temperatures than conventional slab gels, so DNA samples could be separated in half the time. Moreover, the new machine could have samples loaded automatically, used small volumes of reagents, and dispensed with the cumbersome task of pouring the slabs of gel for DNA electrophoresis. Finally, the new machine could process several DNA samples per day.

The finished product was dubbed the ABI PRISM 3700. Considering its $300,000 price tag, the 3700 is rather anonymous in appearance—a beige box on wheels with a tinted lid. One somehow expects a few more bells and whistles. Even the mechanics look deceptively simple. On the left, a robotic arm is suspended over a plastic tray containing ninety-six small wells storing DNA. On the right are the copper-colored capillary tubes, through which the DNA is separated. Once the robot has loaded each DNA sample into the appropriate tube, the samples migrate through the capillaries until they pass through the laser detector that tracks the fluorescent tags distinguishing each letter of DNA.

With ninety-six lanes and the capability of reading 550 bases in a single lane, the PRISM 3700 dramatically increased the sequence capacity over its predecessor, cutting sequencing time by 60 percent and labor costs by 90 percent, and producing sequence about eight times faster. Requiring only about fifteen minutes of human intervention every twenty-four hours (compared with eight hours for the original machine), the PRISM 3700 can produce as much as 1 million bases of DNA sequence per day. Finally, here was a machine capable of providing the increase in sequence capacity required to complete the human genome on time, or perhaps even sooner.

But if his new toy finally provided the means to sequence all 3 billion bases of the human genome, Hunkapiller wondered if perhaps his company could go one step further, to exercise control of the sequencing data instead of merely selling the machines to the public consortium. If a few hundred machines could sequence millions of letters of DNA every day, then it would require only a couple of years or so to sequence the 3 billion letters of the human genome. There was one snag, however: PE lacked experience in sequencing strategies. "We didn't have skills in-house, and we knew Craig Venter was interested," Hunkapiller recalls. In January 1998, Hunkapiller invited Venter and Adams to California to inspect his new sequencer. Venter was suitably impressed by the prowess of the PRISM 3700, but totally unprepared for Hunkapiller's indecent proposal: the chance to team up to sequence the entire human genome. Venter's immediate response was along the lines of "You guys are certifiable!"

Despite the unequivocal success of shotgun sequencing for bacterial chromosomes, not even the pioneer of that technique could be certain it would work for the human genome, which is 1,000 times larger. But the more he rolled the figures over in his mind, the more he liked the idea. By the end of the day, Venter was convinced that it was going to be a major breakthrough. Hunkapiller's recollection is that "we spent a few days working through the math and came away thinking maybe it's doable." Although Hunkapiller had conceived the idea and White was willing to bankroll it, the project was now thrust squarely onto Venter's shoulders. He knew that this was the ideal moment to leave the security of his non-profit institute and embark on the biggest gamble of his career.

The launch of the new company was planned under a thick veil of secrecy. Venter invited his senior aides at TIGR, including Mark Adams, Anthony Kerlavage, and Granger Sutton, to join him in the new venture, which would become one of two distinct subsidiaries of PE Corporation (the other being Hunkapiller's Applied Biosystems). As for who would succeed Venter as the president of TIGR, the logical choice was Clare Fraser, Venter's wife, and, more important, the director of many of the institute's microbial genome sequencing projects. In early May, James Watson telephoned Michael Morgan at the Wellcome Trust, warning him that there were rumors that Venter was going to announce a major breakthrough at a press conference in the next few days.

ON MAY 8, 1998, just a few hours after a meeting with NIH director Harold Varmus, Venter and Hunkapiller sat down with Francis Collins at the United Airlines Red Carpet Club at Washington's Dulles Airport, and proceeded to break the news of their new genome sequencing venture to the genome chief. Venter proposed that his new company and the public Human Genome Project share their data—and presumably the credit once the sequence was complete. Neither Collins nor Varmus was exactly thrilled about the prospect of a private outfit's muscling into the long-standing public effort to sequence the human genome, no matter how quickly they could do it, and share the kudos that they had worked so hard to earn. A stunned Collins told Venter he needed more time to consult with his genome chiefs.

Venter had no intention of waiting to see how the NIH reacted, nor did the PE board, which formally approved the business plan that afternoon. Two days later, on May 10, the bombshell Venter had dropped exploded over the front page of the Sunday edition of the *New York Times*. Venter had forewarned veteran *Times* science reporter Nicholas Wade, who had regularly covered Venter's research during his tenure at TIGR.

In his story, Wade outlined how the new company, as yet unnamed, would complete the human genome sequence by the year 2001, four years ahead of the public Human Genome Project's timetable. It would combine Venter's sequencing savvy with about three hundred of Hunkapiller's new PRISM 3700 sequencers to become the largest producer of DNA sequence in the world—over 100 million bases per day. This torrent of data would be handled by an $80 million Compaq terabyte computer. The total cost of the operation was estimated to be $200 million to $500 million. Venter later said his assault on the entire human genome could be described as "the full Monty."

Contrary to the terms of the Bermuda agreement, under which the genome centers pledged to release DNA sequence data every twenty-four hours, Venter said he would release the sequence data free of charge every three months and insisted that they would not be held hostage. However, pharmaceutical companies and other parties interested in gaining immediate access to a custom-built genome database that would facilitate the search for specific genes and drug candidates could purchase a license. As for gene patenting, Venter said he would only seek patents on 100 to 300 human genes, less than 1 percent of the total genome. He had no intention of building a pharmaceutical company, but rather saw his venture as an information company, analogous to the Bloomberg News Service or the Lexis–Nexis news retrieval service. In addition to the raw sequence, Venter's database would provide a rich source of information on gene identification, DNA variants, medical relevance, and comparisons with other species.

But running throughout Wade's piece was a sense that the new firm posed daunting obstacles for the public HGP to continue on its current course. If Venter and Hunkapiller had been sincere in seeking to forge an alliance with the leaders of the public HGP, there were few signs in the *Times* article. "If successful," Wade suggested in just the second sentence, "the venture would outstrip and to some extent make redundant" the $3 billion public HGP. Wade also indicated that Collins and Varmus were willing to seek ways to integrate their programs with Venter's company, with one possibility being to refocus the efforts of the public genome centers on sequencing the mouse genome. Finally, he suggested there might be tough questions from Congress regarding the financing of the public HGP if it was going to be scooped by Venter's company. Wade's article drew a swift rebuttal from Varmus. In a letter sent to the *New York Times* the same day that Wade's article appeared, the NIH director protested that the success of Venter's new entity was not a "fait accompli" and that the feasibility of his shotgun approach "will not be known for at least 18 months."

Three months later, Venter revealed that his company would be known as Celera Genomics (from the Latin *celeris,* meaning quick or swift). PE Corporation would own 80 percent of the company, Venter 10 percent, with the remainder divided among other staff and advisers. The company's slogan was forcefully direct: "Speed matters. Discovery can't wait." Celera's simple logo was composed of two intertwining strands, a subtle play on the iconic double helix, forming the figure of an outstretched person.

SOME OF THE INFERENCES IN WADE'S *New York Times* story about the potential redundancy of the public genome effort may have dismayed its leaders, but at least he refrained from using the four-letter word that would hound the scientists on the public side for the next two years. However, many other reputable news organizations were only too happy to play the "race" card. "On your marks. . . ," exclaimed the British weekly magazine *New Scientist*. "The race to read the book of life is about to start in earnest, and tensions are rising." *Science News* said, "Consider it modern biology's equivalent of the fabled race between the tortoise and the hare." Even the *Wall Street Journal* used the term, although it saw the race more as one between Venter and other biotechnology companies. These stories set the tone for much of the subsequent press coverage of Celera and the public Human Genome Project, despite increasingly fervent objections from the protagonists.

On Monday May 11, Venter, Hunkapiller, Varmus, Collins, and Ari Patrinos, director of the Department of Energy's Office of Biological and Environmental Research, appeared together at an extraordinary press conference in Washington to discuss the new venture and the impact it would have on the public Human Genome Project. The NIH leaders were politely supportive, and Varmus acknowledged that the new firm would "move things along more quickly." Collins seemed satisfied by Venter's pledge to release the "basic sequence" data four times a year, but said it would be "vastly premature" to change direction of the genome program, and told reporters to read Phil Green's critical assessment of the original Weber-Myers shotgun proposal. Venter saw it somewhat differently and reiterated his position that the genome centers should consider focusing more on sequencing the genomes of mice and other experimental organisms. Meanwhile, PE chairman Tony White said he envisioned Celera's becoming "the definitive source of genomic and related medical information . . . [courtesy of] a genomics sequencing facility with an expected capacity greater than that of the current combined world output."

The following morning, Collins held an emergency breakfast meeting with senior genome center scientists and staffers at the Cold Spring

Harbor National Laboratory on Long Island, to decide how best to protect the $1.9 billion invested over the previous ten years in the public
HGP. That morning's *New York Times* featured another bracing piece by
Wade, in which he referred to "the takeover of the human-genome project" by Venter's new company and again questioned the funding and direction of the public HGP: "It may not be immediately clear to members
of Congress that having forfeited the grand prize of human-genome sequence, they should now be equally happy with the glory of paying for
similar research on mice."

As the conference participants gathered on the pristine campus of
the Cold Spring Harbor Laboratory, Venter's shocking proposal domi-
nated the conversation. Scientists eagerly seized on the perceived flaws in
Venter's proposal. The whole genome shotgun sequencing strategy had
never been tried on DNA remotely close in size to the human genome.
The heavily touted PRISM 3700 sequencers were untested and might
prove unreliable. The final sequence would be full of thousands of gaps
and an unacceptable number of base-calling errors, rendering it virtually
illegible. The specter of rampant commercial genome exploitation and
gene patenting went against the core beliefs of most academic researchers,
who steadfastly believed that the human genetic code should be free for
everyone. And if Venter managed to overcome these obstacles, what was
his business model? If he was sincere about releasing the sequence for free
every three months, how was he going to make money?

These were all entirely valid questions, but most scientists knew they
could not afford to underestimate Venter's vision and resourcefulness.
Anyone else who had made the claim that he or she would almost single-
handedly sequence the human genome would have been dismissed as a
laughing stock. But as Venter had conclusively demonstrated twice before—with the development of the EST strategy and the shotgun sequencing of microorganisms—he had a knack of proving the skeptics
wrong. Nevertheless, the hundreds of assembled scientists were in no
mood to forfeit the genome project to anyone, especially Venter. They
angrily labeled Venter the "Bill Gates of Biotech," threatening to gain almost total domination over the human genome sequence just as Mi-
crosoft had (at least until government threats to break up the company)
over the software industry. James Watson, the president of the Cold Spring
Harbor Lab, went much further: he compared Venter's assault on the
genome project to Hitler's annexation of Poland and asked whether
Collins intended to be Churchill or Chamberlain.

But the most important reaction that week came from Michael Morgan, the genetics program director for the Wellcome Trust, the world's richest medical philanthropy. Addressing the packed auditorium, Morgan brought news that the trust had decided to double support for Britain's flagship DNA sequencing institute, the Sanger Centre, bringing the total up to about $350 million. Despite appearances to the contrary, Morgan stresses that the move had been in the works for six months. Discussions about a second five-year budget for the Sanger Centre had begun in 1996, with the initial plan that the Wellcome Trust and the MRC should each spend about $170 million each, only for the MRC to pull out. At the end of 1997, the Wellcome Trust decided to make up the shortfall. "It was coincidental," Morgan explains, "that two days after Craig made his announcement, the trust was sitting down to consider the Sanger application that the budget should be increased. My concern at the time, having heard the announcement, was that it would persuade the governors that they needn't put the money in. Instead, they took an entirely radical view, and the rest is history."

This funding increase would allow the Sanger Centre to handle at least one-third (or even more) of the public HGP's sequencing effort, which had long been the goal of John Sulston. Morgan also announced that the trust would legally challenge any patent applications on DNA sequences it considered to be unmeritorious. The timing of Morgan's resolute response could not have been bettered. "To leave this to a private company, which has to make money, seems to me completely and utterly stupid," declared Morgan. His talk was greeted with a standing ovation, although he jokes, "They weren't cheering me!"

By the end of the Cold Spring Harbor meeting, there was a sharp escalation in the level of vitriol emanating from the public program leaders. Alluding to the gaps that would tarnish Venter's final sequence, Sulston remarked, "I really don't see this as being any great advance whatsoever. We are going to provide the complete archival product and not an intermediate, transitory version of it." Sulston's close colleague, Washington University's Robert Waterston, said Venter's version of the sequence would be equivalent to "an encyclopedia ripped to shreds and scattered on the floor."

Venter took the stiffening of the public leaders' resolve in stride. "If I were on the other side of this, I would feel upset and threatened, too," he admitted. But he was frustrated, angered, and conflicted by the criticism too. The hand wringing over the thousands of gaps was frivolous, said

Venter, because the final sequence would be 99.99 percent complete, and the holes would be left in regions of the chromosomes that did not contain genes.

But there were also harsh words from his peers in the biotechnology sector. Incyte Genomics' Randy Scott dismissed Celera's chances of commercial success. "It's really Craig Venter going after the Nobel prize," he surmised. Bill Haseltine did not doubt his former partner would succeed, but felt the plan to sequence every letter of human DNA was a little like cosmology—fascinating from a scientific standpoint but of minimal commercial value. Nevertheless, it made more sense to shift the sequencing of the human genome to the private sector than to drain limited federal resources. In an op-ed piece in the *New York Times,* Haseltine urged the NIH to refocus its priorities: "The era of government-sponsored big science, in which a few laboratories receive as much as $10 million a year to analyze mostly junk DNA, while scientists doing disease-related research beg for financing, should end. Let private companies and charitable foundations finish the job of sequencing the human genome. National pride should come from conquest of disease, not winning a race that is not worth winning."

But some scientists applauded Venter's plan to sequence the genome privately, because it would speed up the generation of the human sequence, which at that point was only 4 percent complete. "Once the genome initiative got consolidated into this managed project, it became a bit like Stalinist Russia," Sydney Brenner told me. "If you're not with us, you must be against us. Leaving aside the egomania, I think Venter was right to create Celera. Rebelling against the kind of monolithic structures that have now become the genome program, I think, is fairly healthy. [But] the consequences were predictable."

One month after their stunning announcement, Venter, Hunkapiller, and four of Venter's senior aides published a manifesto of their plans in *Science.* The prose had the dry, emotionless tone so characteristic of scientific publications, but the numbers leaped out of the page: 230 ABI PRISM 3700 DNA sequencers . . . 1,000 samples . . . approximately 100 million bases of raw sequence . . . and about 200,000 sequencing reactions per day . . . 99.9 percent coverage of the genome . . . 10X sequence coverage . . . less than one error every 10,000 bases.

Venter and his colleagues endeavored to allay any misconceptions about his plans for disseminating the sequence data. "An essential feature of the business plan is that it relies on complete public availability of the sequence data," they wrote, which would be released at least every three months. They did not intend to seek patents on primary human genome sequences, but would focus on 100 to 300 "novel gene systems." Venter's team tried to end on a conciliatory note:

> It is our hope that this program is complementary to the broader scientific efforts to define and understand the information contained in our genome. It owes much to the efforts of the pioneers both in academia and government who conceived and initiated the HGP [Human Genome Project]. . . .
> We look forward to a mutually rewarding partnership between public and private institutions, which each have an important role in using the marvels of molecular biology for the benefit of all.

Science magazine has a circulation of about 180,000 copies a week and a readership of well over half a million people, and yet it is not clear that anyone read Venter's article. If they did, few were willing to take him at his word.

On June 17, the House Subcommittee on Energy and the Environment convened a meeting to discuss the meaning of Venter's entree into the Human Genome Project. The most colorful testimony came from Maynard Olson, the director of the University of Washington Genome Center in Seattle. Peering over his thick-framed glasses, Olson chided Venter's May announcement as merely "science by press release" and challenged him to "show me the data!" Olson predicted that the Celera's assembled sequence would contain "over 100,000 serious gaps in the assembled sequence," a far cry from the 5,000 that Venter predicted. He challenged the House panel to safeguard "the development of a sequence that will meet the test of time."

With considerable self-restraint, Venter refused to take the bait and ignored Olson's jibes. Venter urged Congress "not to cut the funding for the Human Genome Project but actually to consider increasing it." And he issued a line that has become a mantra for all human genome researchers: "The acquisition of the sequence is only the beginning."

In his testimony, Collins defended the record of the genome program, and, using language the members of Congress could understand, checked off the challenge ahead of Venter: "In order to meet the standards adopted by the international genome community, the sequence produced must have four characteristics—the four "A's"—of the Human Genome Project. The sequence must be *accurate* [99.99% or better] . . . the sequence must be *assembled* . . . [the] sequence must also be *affordable* . . . finally, high-quality, finished human DNA sequence must be *accessible.*"

Collins played down any sense of rivalry between the two sides by stressing that he looked forward to cooperating with Venter's private initiative. There was even an attempt at humor. "Notice Dr. Venter and I seem to have worn the same clothes today without intending to," he told the panel. "We are intending to be partners in every possible way . . . this is not a race."

Collins's assertion that the public alliance could work in harmony with Celera seemed a little at odds with the real world. To judge from the slurs in the press, the emergency meetings, and the increases in funding, it was going to be a bitterly contested race with huge stakes. Just one week before the congressional hearing, Collins had offered a far different characterization of Venter's plan, telling *USA Today* that the end result of Venter's approach would be like an issue of *Mad* magazine. (Collins has since said he regrets the quote.)

THE MOST EFFECTIVE WAY FOR VENTER TO STEM the constant criticisms would be to produce a model sequence that would demonstrate the effectiveness of shotgun sequencing on a sizable genome, as a dry run for the human genome. With some initial concerns about the performance of the new PRISM 3700 DNA sequencers abating, the only question was whether the shotgun strategy, so effective on small bacterial genomes, could be scaled up by a factor of one thousand. As the genomes of the yeast and nematode model organisms were already complete, Venter selected the most famous animal model of all—the fruit fly, *Drosophila melanogaster*—and pledged to do so ten times faster than Sulston and Waterston had done for the nematode worm. With a genome less than 5 percent the size of human, the fruit fly was an ideal bridge between the millions of bases in bacterial genomes and the billions in the human genome. The strategy would be the same: sequencing millions of DNA fragments using hundreds of the 3700 DNA sequencers, followed by computerized assembly into the final sequence. But the fruit fly carried significance far beyond validating shotgun sequencing. *Drosophila* geneticists like to brag that "a fruit fly is a human with wings"—one of the most important model organisms in biology.

The fascination with fruit flies was lyrically captured almost fifty years ago by Curt Stern, one of many brilliant students of the Nobel laureate Thomas Hunt Morgan:

> For more than 25 years I have looked at the little fly *Drosophila* and each time I am delighted anew. . . . I marvel at the clear-cut form of the head with giant red eyes, the antennae, and elaborate mouth parts; at the arch of the sturdy thorax bearing a pair of beautifully iridescent, transparent wings and three pairs of legs; at the design of the simple abdomen composed of a series of ringlike segments.

In 1866, the same year that the Moravian monk Gregor Mendel published his classic breeding experiments on garden peas, Thomas Hunt Morgan was born in Lexington, Kentucky. Four decades later, Morgan was a respected embryologist at Columbia University, hoping to identify the physical basis of hereditary factors. He began experiments on dozens of different species, but the species for which he would become famous was lured to the top floor of Columbia's Schermerhorn Hall by some bananas placed on the windowsill outside his laboratory.

The fruit fly has many advantages, chiefly a life cycle that is both short (one new generation every twelve days) and fecund (the average female lays some 1,000 eggs). The fruit fly is also inexpensive to grow, requiring little more than a steady supply of half-pint milk bottles, which Morgan's students pilfered from local stoops and the student cafeteria. One day early in 1910, Morgan observed through his simple hand lens a male fly with startling white eyes rather than their normal brilliant red color. The trait, not surprisingly, was dubbed *white,* and *Drosophila* had taken its first flight to fame. When Morgan visited his wife in hospital recovering from the birth of their third child, she asked after the white fly. Morgan launched into a detailed account before remembering to ask, "And how is the baby?"

Back in the cramped confines of the "fly room," Morgan bred his white-eyed male with a normal red-eyed female, and found that the progeny all had red eyes. But when he crossed the "children" (as he liked to call the progeny), the white-eyed trait reappeared in males. This clear "sex-limited" pattern of inheritance reminded Morgan of color blindness, which affects only males. Morgan knew that *Drosophila* contain four pairs of chromosomes, including a pair of sex chromosomes (female flies have two X chromosomes, whereas males have an X and a Y). Morgan deduced that sex-limited factors such as *white* must reside on the X chromosome, manifesting themselves only in males with one X chromosome. That summer, Morgan published his white-eye results and wrote about additional sex-linked mutants he had found in a letter to one of his collaborators in which he deadpanned, "They may throw some further light on the process of heredity." Morgan also suggested that the sex-linked traits did not always segregate together because the chromosomes shuffle material during meiosis. This shuffling process, known as crossing over, is the key catalyst of evolution.

Morgan's fly room contained eight working desks crammed into a space no bigger than a New York studio apartment, woefully equipped, and desperately filthy (legend speaks of cockroach infestation and marauding mice). The human inhabitants, by contrast, were always immaculately attired in waistcoats and ties, and the intellectual atmosphere was intoxicating. Students worked long into the evenings, loudly discussing experiments and exchanging ideas. "I sometimes wondered how any work got done at all, with the amount of talk that went on," said Alfred Sturtevant, one of his students. Leading the banter was Morgan. Although a shy man (his students would resort to hidden cameras to photograph him at work), Morgan had a keen sense of humor. He once advised two good friends after the birth of their first child, "I have no fatherly advice to offer, except not to call the little girl Drosophila. We have resisted the temptation ourselves three times." Another time Morgan was asked to write a letter of recommendation on behalf of one of his students for a teaching post at a midwestern university that emphasized the student's religious persuasion. Morgan promptly had a photograph taken of himself and two students praying in the laboratory to enclose with his letter.

Fruit fly aficionados like to jest that Morgan's three greatest discoveries were his students: Alfred Sturtevant, Calvin Bridges, and Hermann Muller. In the five years from 1910, Morgan's Raiders, as they were called, carried on Mendel's legacy, combining the rigorous statistical methods of Mendel with Morgan's meticulous microscopic analysis. In 1911, Morgan explained his concept of crossing over to his favorite student, Sturtevant, arguing that the closer two genes were on the X chromosome, the less frequently they would separate during meiosis. But Sturtevant saw something else. With half a dozen factors known to sit on the X chromosome, an examination of the frequency of crossing-over between them would yield their locations with respect to each other—in other words, the first simple genetic map. That night, Sturtevant cast aside his homework and pored over the breeding records of well over 25,000 flies—the children of specific matings between flies with pairs of sex-linked traits. By comparing the frequencies with which these traits showed up in the progeny, Sturtevant could estimate the relative proximity of one X-linked gene to another, charting the distances in terms of the frequency of recombination. By daybreak, Sturtevant had sketched a line with the locations of five fruit fly traits on it, much like this:

Yellow	white	vermilion	miniature	rudimentary
body	eye	eye	wing	wing

In what his boss called "one of the most amazing developments in the whole history of biology," Sturtevant, still only a teenager, had produced the first map of genes on a single chromosome. In the ensuing publication, Sturtevant wrote, "We have no way of knowing if chromosomes are of uniform strength, and if there are strong or weak places, then that will prevent our diagram from representing actual relative distances—but, I think will not detract from its value as a diagram." Indeed it did not. Sturtevant's genetic map helped lay the foundations of the genome project seventy-five years later.

Sturtevant's inspired construction of the first genetic map by itself would have ensured Morgan's place in history, but it was merely the beginning of a wave of discoveries that engulfed the fly room for the next few years. Sturtevant and Bridges built maps of the other three chromosomes, containing dozens of genes, and proposed the notion that some traits arose from the combination of several different genes, with some modifying the action of other—in other words, polygenic inheritance. Muller showed that crossovers could occur in pairs—a double crossover—shuffling the genes along a chromosome prior to the formation of sperm and egg cells. And Bridges observed that occasionally the chromosomes failed to separate properly during meiosis, with the result that some egg cells received two X chromosomes instead of one. (The equivalent unbalanced segregation of human chromosome 21 causes Down syndrome.) By the fiftieth anniversary of Mendel's famous experiments, the chromosome theory of heredity was all but established, and a scientific revolution was underway.

In October 1933, Morgan received a telephone call and listened in shock. It was Stockholm calling: "I beg to inform you that the Nobel

Prize for physiology and medicine for 1933 has been awarded to you for your discoveries concerning the hereditary function of the chromosomes." The presentation speech that accompanied Morgan's Nobel Prize read in part: "It is often difficult to distinguish what is Morgan's work and what is that of his associates. But nobody has doubted that Morgan is the ingenious leader. . . . The results of the Morgan school are daring, even fantastic, they are of a greatness that puts most other biological discoveries into the shade."

Morgan's students continued to make spectacular discoveries. Bridges published beautiful cytological maps of the giant chromosomes from the fly's salivary glands, aligning the genetic and physical map of a chromosome for the first time. Meanwhile, Muller successfully induced mutations in Drosophila by bombarding flies with X rays, for which he won the Nobel Prize in 1946. Muller's presence at Indiana University in 1951 attracted a young student named James Watson. In honor of Morgan, the unit of genetic distance between markers along a chromosome is called the centimorgan.

Interest in Drosophila waned before and after the discovery of the double helix. Bacteria and their viruses became the popular laboratory organisms for establishing the nature of the hereditary material, elucidating the genetic code, and understanding the rudimentary mechanics of gene regulation. But with each passing landmark, the prospect of understanding at the molecular level how more complex organisms develop became more feasible. In 1979, David Hogness's group at Stanford University isolated a gene complex called bithorax, a homoeotic mutation (one in which one body part is substituted for another) with two sets of wings. At the same time, two young German scientists, Christiane Nusslein-Volhard and Eric Wieschaus, set out in search of mutants that affected the development of the Drosophila embryo. In 1980, they published a classic paper in Nature describing fifteen different mutants altering the segmental pattern of the fruit fly larva. Their work was rewarded with the 1995 Nobel Prize—the third for pioneering genetic studies in fruit flies in the past century—and helped restore Drosophila to its rightful position as the preeminent organism to root out genes involved in growth and behavior, development and sexual orientation, and aging and addiction. Drosophila was a natural candidate for genome sequencing.

The sequencing of the Drosophila genome began in 1994 under the auspices of two groups: the Berkeley Drosophila Genome Project, headed by Gerry Rubin, and the European Drosophila Genome Project. The fruit fly contains 180 million bases of DNA, but about one-third is made up of highly repetitive tracts of DNA called heterochromatin, which has no known purpose and is exceedingly difficult to sequence. Thus, the researchers would concentrate on the 120 million bases known as the euchromatin, which contains the protein-coding genes. Initial support was lukewarm, says Rubin, because the fruit fly community had been spoiled by decades of classical studies on *Drosophila* variants and the availability of tools that made gene identification and manipulation easier in fruit flies than virtually any other organism. By the spring of 1998, the two *Drosophila* sequencing efforts had finished a respectable 30 million bases, or 20 percent of the genome, but researchers were now pining for the sequence, having seen the profound benefits emerging from the yeast genome.

Venter was fully aware of this, and thus approached Rubin at Cold Spring Harbor in 1998 to pitch an extraordinary offer. Celera would test its shotgun sequencing strategy on the *Drosophila* genome, offering Rubin and the entire *Drosophila* community the prospect of the complete genetic script of the fruit fly two years ahead of schedule. The cost would be borne by Celera. It was, says Rubin, "an offer I could not refuse." Many of his colleagues were appalled that Rubin would agree to collaborate with Venter, but Rubin was simply being practical. With the Berkeley project unlikely to finish before the end of 2001, Celera's help in completing the sequence would be a bounty to fruit fly researchers around the world and save taxpayers an estimated $10 million to $15 million in the process.

OF ALL THE CONCERNS RAISED by Venter's announcement—the threat to the established genome project centers, the quality of the raw sequence—the most contentious issue concerned the thorny issue of gene patenting. Venter stated that he was interested in patenting at most 300 human genes, but the cut-off was entirely arbitrary. If Venter succeeded in sequencing the genome as quickly as he projected, potentially years ahead of the public program, he would have a golden opportunity to apply for patents on perhaps thousands of valuable genes. These in turn could be licensed to pharmaceutical companies for vast sums of money. But Venter argued that the rapid release of the basic sequence would halt the aggressive speculation by certain corporations on human genes. He told the June congressional hearing:

By making the sequence of the entire human genome available, it makes it virtually impossible for any single organization to own its entire intellectual property. It eliminates the entire speculative nature that is currently associated with patenting DNA sequence information and requires that researchers understand the biology of a sequence before they file a patent application. Our actions will make the human genome unpatentable.

Gene patenting is a particularly sensitive issue in Europe, where the prevailing opinion is firmly that the human heritage cannot be monopolized. Sulston believes passionately that the public must take full responsibility for the genome, rather than entrusting it to the private sector. An editorial in the *Guardian* stated, "Entrepreneurs have a vital role in any economy in starting companies, but not in making money out of something that most people in the world justifiably regard as public property. All research on human genes should be released to the world without patent protection." Such sweeping statements, however, tend to overlook the important role that patent protection plays in medical research. Without such protection, the pharmaceutical industry would have no incentive to invest the hundreds of millions of dollars and years of research it customarily takes to bring a drug to market. Nor are patents the sole province of private industry. Since the early 1980s, researchers receiving funding from the NIH were required to apply for patents on potentially important gene discoveries.

A striking example is afforded by the *BRCA1* breast cancer gene, identified by Myriad Genetics in 1994. The ownership of the rights to such an important, commercially lucrative gene belonging to an upstart biotechnology firm provoked consternation among many researchers and academics. But there have been more than 700 research papers on *BRCA1* since its discovery, mostly from researchers at universities. Indeed, the most prescient advances in our understanding of the normal role of the BRCA1 protein in repairing damaged DNA have come from academic researchers undeterred by Myriad's ownership of the gene patent.

Nevertheless, there is growing concern that patents on genes used in genetic testing are barring people from potentially life-saving information. "We would have to refuse some women this valuable new [breast cancer] test," says John Burn, director of the Northern Genetics Service in the United Kingdom, which issues about two hundred breast cancer gene tests a year. "This is exactly the type of situation which will arise if genetic information continues to be patented." There has been a similar outcry in the United States over the insistence of Miami Children's Hospital to charge royalties for a gene test for Canavan's disease, a fatal neurological condition prevalent in the Ashkenazi Jewish population. In suburbs of New York City, the hope of replicating the successful screening for Tay-Sachs disease by identifying carriers of the disease before marriage is in jeopardy because of the gene fees.

In many discussions of the genome project, Celera has been portrayed as the most egregious offender in the dispute over DNA patenting, with slightly ludicrous charges that Venter wanted to patent humanity or the complete genome. Although Celera certainly has the potential to apply for huge numbers of human gene patents, it is not, and most likely never will be, the leading collector of human gene patents. The most notorious landlords of the genome are companies such as Incyte Genomics and Human Genome Sciences, which have been awarded hundreds of patents each in the past few years for human genes (ESTs) that have been characterized by virtue of their expression in different tissues. Incyte currently leads the pack with more than five hundred patents—twice as many as its nearest rival, SmithKline-Beecham—in many cases for genes that have been predicted entirely by sequence analysis, and with thousands of applications pending. Indeed, some biotech entrepreneurs consider Celera's late entree into gene patenting to be almost irrelevant. Haseltine smugly suggests that "when the Human Genome Project sequence is completed, it will be a rather meaningless footnote to a revolution that is well under way."

One disgruntled citizen voiced her protest about the gene-patenting stampede in rather bizarre circumstances. Donna MacLean, a British waitress and poet, filed patent application GB0000180.0, in which she sought to patent . . . "Myself." MacLean wrote, "It has taken 30 years of hard labor for me to discover and invent myself, and now I wish to protect my invention from unauthorized exploitation, genetic or otherwise. I am new: I have led a private existence and I have not made the invention of myself public. I am not obvious." MacLean said she was outraged by the "grasping, greedy atmosphere" surrounding human genome research and decided to see if she could be granted protection for her own genes. The case is pending.

• • •

THE WELLCOME TRUST'S TIMELY DECISION to boost funding for the Sanger Centre was a huge morale booster for the beleaguered public consortium. But Collins had to decide what changes, if any, should be made to the American genome centers, which still had the responsibility of producing two-thirds of the genome sequence. There were still more than a dozen U.S. laboratories handling the sequencing, with differing degrees of productivity and cost. Furthermore, only about 5 percent of the human genome had been sequenced. At a meeting in Virginia shortly after the May 1998 Cold Spring Harbor meeting, many of the principals in the genome program considered the suggestion of producing a "rough draft" of the sequence. This was a radical departure from the long-term goal of the Human Genome Project, which was still to produce a near-letter-perfect assembly of all 3 billion bases in the human genome. Collins quickly reassured his colleagues in the community that this was "not a substitute for finishing the whole thing." But the availability of at least 90 percent of the sequence by the end of 2001, even in less than perfect shape, would yield immediate dividends, most notably for researchers hunting for disease genes.

The plan was controversial. Some scientists thought rushing to cobble together a rough draft would distract from the Human Genome Project's ultimate goal of producing a gold-standard sequence. But others welcomed the idea as a boon for gene identification. At the genome center's advisory council meeting that September, the working draft strategy was formally approved. Collins hailed the plans as "big and ambitious, even audacious . . . this is not a time to be conservative, cautious, or coast along." Tens of millions of additional dollars would be set aside to support the sequencing effort.

Six months later, the timetable had been shifted forward again. With the availability of additional funds and increased productivity of all the public genome centers thanks to the introduction of the new PRISM sequencers, the NIH and the Wellcome Trust halted the pilot phase of the sequencing program and signaled the start of the all-out campaign to complete the sequence. The new deadline for the project's completion was spring 2000, eighteen months earlier than the previous forecast. "The best service to the scientific community," explained Eric Lander, "is to deliver the draft sequence rapidly, and then to circle back and perform in the course of another year-and-a-half at most the finishing on that sequence." Collins vigorously denied that the rough draft was a response to Celera's timetable—he termed the new plan "action, not reaction"—and insisted that discussions had preceded the launch of Celera.

For the working draft, each genomic DNA fragment would be sequenced only four times on average, not ten as planned for the final sequence, which would result in more errors in the sequence. However, work would continue to produce the final gold-standard sequence by 2003, two years ahead of schedule.

But it was clear to Collins and his advisers that pumping extra money into the current loose alliance of laboratories was not going to guarantee success. Having presided over a pedestrian, loosely organized confederation of laboratories across the United States, Collins had to make the difficult decision to redirect the bulk of available funds toward the three most productive genome centers. Collins announced that he was funneling over $80 million over five years to just three centers: those headed by Robert Waterston (St. Louis), Eric Lander (Cambridge, Massachusetts), and Richard Gibbs (Houston). Meanwhile, the Wellcome Trust announced it was putting a further $77 million into the Sanger Centre. Together with the Department of Energy's Joint Genome Institute, the lion's share of the final sequence would be produced by five genome centers— henceforth known as the G-5.

Now it was the turn of the public Human Genome Project leaders to hold out an olive branch to Venter, offering him to join their alliance, but there was no chance of that happening. The extra resources being marshaled on behalf of the public project were evidence of how seriously Venter's new venture was being taken. Not surprisingly, however, Venter rebuffed the offer, dismissing Collins's reworked schedule as "nothing to do with reality," accusing him of "putting humanity in a Waring blender and coming up with a patchwork quilt." That did not sit well with Collins, who shot back that Celera's program was the "Cliffs Notes version of the genome."

Billing himself the "operating manager and field marshal" of "Team Sequence," as he called the reshaped transatlantic alliance, Collins's hopes of delivering the sequence more or less contemporaneously with Celera rested in the hands of five major genome centers. About half of the sequence would be produced by the close partnership of the Sanger Centre and Washington University in St. Louis. The John Sulston–Bob Waterston alliance, which had famously completed the sequence of the *C. elegans* genome, could rightfully claim to be the most successful sequencing operation in the public genome alliance.

The team at Baylor College of Medicine in Houston, led by the Australian Richard Gibbs, would concentrate on three chromosomes, including the X. Just over 10 percent of the genome would be handled by the Joint Genome Institute in California. German, French, and Japanese groups would handle a further five chromosomes, including the two smallest, 21 and 22. A few months later, Collins rewarded three additional centers at Stanford University, the University of Washington, and a Massachusetts biotech company, Genome Therapeutics, which would contribute to chromosome sequencing and technology development.

Chromosome Assignments
for Public DNA Sequencing Centers

Laboratory	Director	Chromosomes
Sanger Centre	John Sulston	1, 6, 9, 10, 13, 20, 22, X
Washington University	Bob Waterston	2, 3, 7, 11, 15, 18, Y
Baylor College of Medicine	Richard Gibbs	3, 12, X
Joint Genome Institute (DOE)	Elbert Branscomb	5, 6, 19
Whitehead Institute / MIT	Eric Lander	17 [+others]
University of Washington	Maynard Olson	7
Genome Therapeutics	David Smith	10
France	Jean Weissenbach	14
Germany	Andre Rosenthal, Helmut Bloecker, Hans Lehrach	8, 21
Japan	Yoshiyuki Sakaki, Nobuyoshi Shimizu	8, 18, 21, 22

Last but not least, the Whitehead Institute's Eric Lander bagged his personal favorite chromosome—17, which had been his lucky number since he became fascinated by its unique mathematical properties—and helpfully "whatever needs to be done." With an extraordinary résumé, Lander had earned the right to select his own projects. His early interest was in mathematics; he won the Westinghouse Science Talent competition at age seventeen (for a paper on quasiperfect numbers), a Rhodes scholarship to Oxford, followed by a MacArthur "genius" award. In the early 1980s, while teaching mathematics at Harvard Business School, Lander began moonlighting at MIT, learning molecular biology from, among others, his younger brother Arthur, a neuroscientist. In 1989, Lander joined the faculty of the Whitehead Institute, affiliated with MIT, and was later appointed director of the Whitehead Institute Genome Center. Lander sensed that his mathematical expertise could be a powerful tool in the booming field of human genetics, which was searching for ways to map disease genes. He has applied those skills to help determine appropriate criteria for forensic DNA fingerprinting and mapping of genes for complex traits, while constructing a comprehensive genetic map of the mouse.

In the past few years, the Whitehead Institute Genome Center has become the flagship of the U.S. genome effort, with Lander supervising an army of more than two hundred researchers and technicians and producing about a third of the human sequence. Despite his entrepreneurial flair (in 1993, he cofounded Millennium Pharmaceuticals), Lander's loyalty in the genome race lay firmly behind the public genome project—or as he jokingly refers to it, "the Forces of Good."

VENTER SET UP HIS NEW CORPORATE HEADQUARTERS in a 200,000-square-foot facility on the northern outskirts of Rockville, Maryland, just eighteen miles from the White House and a few subway stops away from Collins's office at NIH. The office building formerly belonged to a government defense contractor and was sandwiched between the electric company and a housing development. It was hardly an exotic location, but Venter did not have the luxury of time. In August 1998, his team was moving in.

As each $300,000 sequencing machine was installed and checked, Celera scientists began naming each machine after a famous science-fiction character, affixing the new monikers with masking tape. There were Leia and Skywalker from *Star Wars,* Ripley and Hicks from *Alien,* and many others. The machines were stacked neatly in rows, patiently awaiting the first samples. On another floor, generic laboratories were constructed, where Smith and colleagues would caringly prepare the human DNA for sequencing. The monthly electricity bill to run the sequencers, computers, and air conditioning was $100,000.

By early 1999, the data center was operational. Compaq had built one of the two largest civilian computers in the world, rivaled only by an Intel-built machine used by the U.S. government for classified modeling of nuclear explosions. The computer contained some 800 Alpha EV6 and EV67 processors with 64-bit architecture and over 80 terabytes of memory (equivalent to five times the Library of Congress) housed in a specially designed quarters about the size of a basketball court. The computer was capable of 1.3 trillion floating-point operations per second, reducing functions that used to take years to a mere fifteen seconds. Over 400 miles of fiber-optic and copper cable were laid to handle the data. Even so, Venter estimated that it would take three months for this gargantuan computer to assemble the billions of bases of raw data into the finished human sequence. In one corner of the data center were servers dedicated to Celera's pharmaceutical clients. In another wing of the building, Venter built a command center reminiscent of the bridge of the Starship Enterprise. The curved wall contains two giant screens and a bank of more than thirty other television monitors, providing instant checks of the flow of sequence information and genome assembly.

In addition to his cadre of TIGR loyalists, Venter recruited other leading scientists, including Sam Broder, former director of the National Cancer Institute, who said he was in charge of "medical stuff." The scientific advisory board included luminaries such as Arnold Levine, the Rockefeller University president who discovered the notorious cancer gene *p53;* Victor McKusick, the Johns Hopkins geneticist who founded the catalogue of human genetic disorders, and Mel Simon, the Caltech researcher who developed BACs, a key resource for handling DNA in large-scale sequencing projects.

The key recruit was the felicitously named Gene Myers, the Arizona bioinformatics specialist who had publicly raised the notion of sequencing the human genome by shotgun sequencing, only to be dismissed by the majority of geneticists. Myers begged Venter for the opportunity to design the computer algorithms to assemble the millions of random DNA sequences into the book of man. Indeed, so anxious was Myers to join Celera that Venter jokingly recalls, "It didn't put him in a very good negotiating position!" Shortly after Myers naively agreed to join Celera for his academic salary, he called Venter and asked sheepishly, "What are stock options?"

Myers is a flamboyant character, usually attired in a stud earring and a scarf, but he did not care for the simmering tension between the public and private camps. He was consumed with the enormity of the task in front of him—a problem tantamount to reassembling the complete text of the Bible from ten copies that have been torn to shreds.

The general consensus was that the task of assembling the human genome would be a million times more difficult than bacteria. Not only is the human genome 1,000 times larger than the typical bacterial genome, but the greater proportion of repetitive DNA would confound attempts to assign each sequence to its proper place in the genome. Myers was confident that he had developed a strategy to discriminate the thousands of repetitive tracts of DNA scattered throughout the genome that threatened to sabotage the shotgun sequencing approach. Whether his confidence was well placed would become clear soon enough.

COLLINS'S DECISION TO PUSH for an expedited rough draft sequence initially angered some of his European collaborators, who felt excluded from the strategy deliberations. But after private meetings at the May 1999 Cold Spring Harbor meeting, all parties in the public HGP signed on. "The unanimous opinion of the international sequencing community is that keeping genetic information in a form that is not accessible [as Celera is doing] will harm research," said Collins.

Ironically, the key ingredient of the public Human Genome Project's ramp up in production would be hundreds of PRISM 3700 sequencing machines—the same machines adopted by Celera and manufactured by its sister company, Applied Biosystems. "The day we announced Celera," said PE Corporation chairman Tony White, "we set off an arms race, and we were in the arms business. Everyone, including the government, had to retool, and that meant buying our equipment." With Celera planning to use up to 300 PRISM 3700 sequencers, the public genome centers had no choice but to follow suit. By March 1999, Applied Biosystems had received more than 500 orders, including 36 from the Sanger Centre. Orders quickly followed from all of the G5 centers except the JGI, which opted to acquire similar machines from Amersham Pharmacia. By the end of 1999, Applied Biosystems had shipped 1,000 PRISM 3700 sequencers, placing them in 250 different laboratories.

Although there were a few technical issues to be ironed out, the genome center chiefs were almost as enthusiastic about the new sequencer as the PE executives. Baylor's Richard Gibbs called it "a superb example of engineering." Ironically, this posed a potential problem for PE, which felt the need to fulfill Celera's initial order first but also wanted to impress academic groups. While anxiously waiting for their shipments of the machines, Venter and Lander, who had ordered 300 and 120, respectively, hounded Hunkapiller to complain that he was favoring the other side.

Once installed at Celera, Adams, Smith, and Venter set the PRISM

3700 machines to work. In the summer of 1999, they successfully read the total DNA sequence of *Drosophila melanogaster*—120 million letters. By September, the sequencing phase was declared complete. It had taken just four months, or one-tenth of the time required to sequence the previous largest genome—just as Venter had predicted. Whole-genome shotgun sequencing, the strategy the experts had predicted was doomed to failure, looked as if it had been an unequivocal success.

There was further good news for Venter in the closing months of 1999. Celera's stock, which had debuted inauspiciously in May at less than $20 a share (placing the company's worth at some $400 million), began to show signs of life, pushing above $50 per share, even though the company had posted a loss of $40 million for the 1999 fiscal year.

CHAPTER EIGHT

The Story of Us

> *What a piece of work is man! How noble in reason! How infinite in*
> *faculty! In form, in moving, how express and admirable! In action*
> *how like an angel! In apprehension how like a god! The beauty of*
> *the world! The paragon of animals! And yet, to me, what is this*
> *quintessence of dust?*
>
> —WILLIAM SHAKESPEARE, *Hamlet*

IN 1975, GENETICIST MARY-CLAIRE KING published an astounding paper in *Science* magazine. Working under the guidance of the late Allan Wilson (the father of molecular anthropology) at the University of California, Berkeley, and using the crude comparative techniques available at the time, King discovered that humans and chimps are much more closely related at the genetic level than previously thought. Despite having diverged about 5 million years ago, King only found a 1 percent difference between the proteins and genes of the two species. "I kept thinking the project was a disaster because I couldn't find any differences," King recalls, "but Allan kept saying, 'This is great—it shows how similar we really are to chimps!' He turned straw to gold, and I wrote a perfectly reasonable dissertation that landed on the cover of *Science.*"

While King went on to make her name by mapping the *BRCA1* gene and using DNA fingerprinting to help reunite kidnapped children with their families in Argentina, other groups confirmed and extended her findings by showing that humans and chimps are more than 98.5 percent identical at the DNA level. (A detailed recent study of 10,000 bases of junk DNA on the X chromosome, for example, showed a difference in the corresponding sequence between humans and chimpanzees of just 0.94 percent.) By contrast, the DNA of chimpanzees and gorillas is just 97 percent identical, indicating that humans have more in common with chimpanzees than chimpanzees do with gorillas. Humans and chimpanzees last shared a common ancestor 5 million years ago; gorillas split from this branch of life more like 8 million years ago.

You do not need a Ph.D in anthropology to conclude that there are some fairly significant physical and behavioral differences between *Homo sapiens* and *Pan troglodytes* (or even *Gorilla gorilla*). Humans prefer to walk upright on two feet, have large opposable thumbs, and have dispensed with a thick matte of body hair. Female chimpanzees develop breasts only when nursing; the males have bones in their penises. We have acquired the gifts of language and consciousness, the mental ingenuity to sequence our own genetic makeup (and, as one early skeptic noted, the gullibility to pay for it!), and the impudence to believe that the hallmarks of humanity might be explained by a few typos within it.

Of course, it will not be that easy. But there is every reason to expect that the key to understanding how these distinguishing characteristics evolved will emerge from a scrupulous comparison of the full sequence of the human and chimp genomes. That astonishing prospect is just a few years away, as the chimpanzee genome will almost certainly be one of the first animal genomes selected for sequencing after the human genome project is finished. "Knowing the genetic basis of these changes," says *Nature*'s emeritus editor Sir John Maddox, "will give us a more authentic history of our species and a deeper understanding of our place in nature."

The most obvious difference between the genomes of humans and chimpanzees is in the number of chromosomes. Whereas humans have twenty-three pairs of chromosomes, chimpanzees have twenty-four pairs. Eighteen pairs appear virtually identical, but millions of years ago, a pair of chimpanzee chromosomes fused end to end to become what we now refer to as human chromosome 2. It is not yet known what beneficial consequences, if any, this cytogenetic union had on primate evolution (although author Matt Ridley impudently suggests that, following Pope John-Paul II's decree that an "ontological discontinuity" during primate evolution gave rise to humankind, this chromosomal junction marks the location of the genes for the soul).

There are several other significant changes between the two species that can be observed by "painting" the chromosomes with fluorescent DNA probes specific for particular chromosomes. Thus human chromosome 9 is larger than the chimpanzee chromosome 9, but human chromosome 12 is marginally shorter, and there have also been rearrangements on chromosome 4. David Nelson and Elizabeth Nickerson, of Baylor College of Medicine, have found that a leukemia-related gene called *AF4* has moved to a different location in chimpanzees compared to all other apes, which might partially explain why chimpanzees are less susceptible to leukemia.

These chromosomal fusions and rearrangements are obvious places to start the search for physiologically relevant DNA differences between humans and chimps, but a more fruitful search may be to identify critical DNA sequence changes that alter the expression of influential genes, such as those that code for transcription factors that regulate gene expression. A total difference of 1.5 percent between the genomes of humans and chimpanzees corresponds to about 50 million sequence differences, but most of these will exist in junk DNA. The number of significant differences in the two codes that affect the structure and/or expression of a specific gene will be just a fraction of that number. In Denver, a biotechnology company called GenoPlex has begun to search systematically for such sequence changes in the hope that they may reveal clues to the behavioral differences between humans and chimpanzees. The company was founded by two University of Colorado researchers, Tom Johnson, an expert on the genetics of aging, and James Sikela, one of the early advocates (with Venter) of EST sampling. One strategy being used by GenoPlex is to compare the sequences of thousands of potentially interesting human and chimp genes, searching for genes that contain significant numbers of amino acid changes in the corresponding proteins. In this way, the company says it has identified human genes involved in memory and HIV susceptibility. It stretches credulity, but GenoPlex scientists believe that the cognitive differences between the two species may come down to as few as fifty genes.

So far, only one significant biochemical difference between humans and chimpanzees has been described at the DNA level, but it provides a fascinating glimpse into the consequences of a tiny glitch in the genetic code. Nearly all cells in the body are decorated on their surface with glycoproteins, special proteins that have been modified with long chains of sugar molecules. One of the most common carbohydrate additions is a sugar molecule called sialic acid. Chimpanzees possess an enzyme that adds an oxygen atom to sialic acid molecules, producing a hydroxylated form called Neu5Gc. Humans, by contrast, do not express this modified form of sialic acid. The reason became apparent in 1998 when Ajit Varki, a researcher at the University of California, San Diego, and colleagues compared the sequences of the human and chimp forms of the sialic acid-modifying enzyme. It turns out that the human gene (which resides on chromosome 6) has fallen fallow, rendered inactive by the loss of ninety-two bases, in all human populations. By contrast, chimpanzees, and indeed all other African apes, possess a fully functional version of the gene.

The structural alteration of such an abundant molecule is likely to have a host of physiological consequences. For example, it provides a rational explanation for the failure of kidney transplants in the early 1960s using chimpanzee organs, which would instantly be recognized as foreign by the human immune system. Many pathogenic viruses and bacteria latch onto the glycoproteins decorating the cell surface, and the gene deletion may partly explain the increased resistance of chimpanzees to infection or in controlling the way cells interact and communicate with one another. Curiously, the mutation removes the last traces of the hydroxylated sialic acid from the human brain, although the significance of this is not yet known. A Japanese group is engineering mice that lack this enzyme to see if there is any effect on the animals' growth or behavior. "Maybe their mice will speak," says Varki dryly.

Could the profound differences between humans and our closest primate relatives really come down to a handful of genes? I have already described a Mexican family with an atavistic mutation that leads to a dense mat of hair covering their bodies. Variations in single genes that code for transcription factors can cause polydactyly, the development of an extra finger, so changes in the timing or location of expression of such genes could have a major effect on limb development. The relative resistance of chimps to diseases such as AIDS could also be attributed to alterations in a single protein receptor. And Nelson speculates that a single genetic change could expand the number of neuronal precursors that would, at least in part, explain the larger brain of human beings. The ultimate proof of the function of such genes would come from observing their effect in monkeys, which is a technically routine task, although nobody foresees wading into that ethical quagmire for some time.

The recent rapid increase in the speed of genome sequencing means that a systematic comparison of the genomes of humans and chimpanzees is within reach. "Until we have a detailed understanding of the genetic differences between ourselves and our closest evolutionary relatives, we cannot really know what we are," say Morris Goodman and Edwin McConkey, advocates of launching a human genome evolution project once the human sequence is complete. A primate genome project would help determine uniquely human behavioral and anatomical traits; shed light on human susceptibilities to diseases such as AIDS, cancer, and malaria; and raise public awareness of a potentially endangered species, a concern underlined by the extinction of Miss Waldron's red colobus in West Africa. Among many prominent scientists backing the primate genome project plan are Mary-Claire King and Francis Crick. Venter and Collins are both enthusiastic in principle. Venter indicates that the chimp genome will be an early priority for Celera, and Collins has already embarked on studies of chimpanzee DNA polymorphisms using gene chips. But Bob Waterston sees problems in comparing the human and chimp sequences. "I'm discouraged," he admits. "You can't do experiments on either organism — and we shouldn't."

DNA HAS BEEN CALLED the greatest archaeological excavation of all time, and for good reason. The DNA sequence is the ultimate arbiter of human identity, finding myriad applications in modern criminal investigations, rewriting presidential history, reuniting families torn apart by kidnapping, tracing the roots of ancient civilizations, and providing unequivocal evidence of human origins. Even before the complete sequence of the human genome was catalogued, scientists used variations in the sequence to remarkable effect, enhancing our understanding of the origins and global migration patterns of human populations over the past hundred thousand years.

Most molecular anthropology studies have focused on two unique chromosomes that are inherited in highly complementary ways. The DNA for which we have most information is a tiny circular chromosome (just 16,600 bases) found in the mitochondria, the cellular compartments that produce an essential chemical called ATP, the cell's internal fuel. Mitochondria are derived from bacteria that developed a symbiotic relationship with ancestral cells billions of years ago, which is why they possess a tiny chromosome. Because sperm do not contribute any mitochondria during fertilization, in any organism the mitochondria are derived exclusively from the mother. Hence, mitochondrial DNA provides a pristine record of the maternal line of descent. Your mitochondrial DNA is derived exclusively from your mother, hers from your grandmother, and so on, reaching back hundreds of generations.

Unlike the twenty-three chromosomes that we inherit from each parent, mitochondrial DNA is not subject to the shuffling and exchange of genetic material between pairs of chromosomes that precedes the production of sperm and egg cells (meiosis). The only source of variation is mutations, which gradually accumulate in the mitochondrial DNA sequence over time. By comparing the number of sequence changes in different DNA samples and estimating the rate with which such mutations accrue, molecular anthropologists can estimate the relatedness between any two samples and the approximate date at which any two samples diverged.

The other staple of human evolutionary studies is the Y chromosome, which bears the male sex-determining gene, and is passed only from father to son. Thus, studies of variations in the Y chromosome shed light on the passage of genes through the paternal line, complementing the study of maternal mitochondrial DNA. Elegant studies by the Whitehead Institute's David Page and Bruce Lahn of the University of Chicago have retraced the incredible evolutionary journey taken by the Y chromosome over the past 300 million years, transforming a formerly well-endowed chromosome to a shriveled excuse that possesses a mere thirty or so genes. However, some of these are fairly important, for they include the gene that triggers the male sex-determination pathway *(SRY)* and a family of genes required for sperm development.

Page and Lahn propose that about 300 million years ago, what is now the mammalian Y chromosome was hijacked by a mutation that led to the existence of the male sex-determining gene. (In our reptilian predecessors, sex was determined not by genes but by external factors such as temperature.) Over millions of years, the Y has been subjected to waves of gene rearrangements and deletions, as unneeded genes were discarded. As a result, the similarity between the X and Y chromosomes (the twenty-third pair of human chromosomes) has waned to such an extent that the two chromosomes are unable to exchange material during meiosis, except for two small regions at each tip. Thus the only significant source of variation in the Y chromosome is mutations that arise in the DNA sequence. By identifying specific segments of DNA that are prone to variation, researchers can compare the pattern of these sequences, or haplotype, building an evolutionary picture of paternal descent that nicely complements that of mitochondrial DNA.

IN 1856, A MAJOR ARCHAEOLOGICAL DISCOVERY was made in the Feldhofer cave in the Neander Valley, near Dusseldorf, Germany. The skullcap and partial skeleton unearthed by lime miners that summer's day were the founding fossils of the hominids we call Neanderthals—*Homo sapiens neanderthalensis*. The Neanderthals, stooped yet powerful with a large brain, receding forehead, and strong brow ridge, lived in western Asia and Spain for hundreds of thousands of years. About 40,000 years ago, modern humans arrived in Eastern Europe, cohabiting with the Neanderthals for 10,000 years or more until the Neanderthals disappeared. Ian Tattersall of the American Museum of Natural History argues that the key to the success of our immediate forerunners was the use of their brains: "Art, symbols, music, notation, language, feelings of mystery, mastery of diverse materials and sheer cleverness: all these attributes, and more, were foreign to the Neanderthals and are native to us." Perhaps the most important of these qualities was language. Imagine how utterly confused the Neanderthals must have felt to hear the first stuttering syllables of language uttered by modern humans. The English novelist William Golding came close: "The sounds made a picture in his head of interlacing shapes, thin, and complex, voluble and silly, not like the long curve of a hawk's cry, but tangled like line weed on the beach after a storm, muddled as water."

Although archaeological evidence shows when the Neanderthals disappeared from Europe, it remains unclear whether they were evicted by modern humans or evolved into *Homo sapiens*. In 1997, a team of investigators led by Svante Pääbo, a leading molecular anthropologist now at the Max-Planck Institute in Leipzig, reported that they had isolated traces of intact Neanderthal DNA sufficient to determine its sequence. According to calculations by Thomas Lindahl, DNA exposed to the elements would decompose within about 10,000 years. But if the DNA is attached to a material such as hydroxyapatite, found in bones and teeth, it might survive intact for two or three times that time.

Having rehearsed the extraction of ancient DNA on a Bronze Age specimen a few years earlier, Pääbo painstakingly sandblasted a few grams of the arm bone of the original Neanderthal skeleton, taking every precaution to avoid contamination with foreign DNA. The trace quantities of the fossil DNA they recovered were amplified using the polymerase chain reaction to the point that there was enough intact material to sequence 379 bases of the Neanderthal mitochondrial DNA. (One reason that this technical tour de force was possible might be that the fossil came from the cold north of Germany; efforts to obtain DNA from bones of the same era from the Middle East have not proven so successful.) This stretch of human mitochondrial DNA sequence typically varies at eight locations among different individuals, but Pääbo found that there were twenty-seven differences between the Neanderthal sequence and a standard human mitochondrial DNA sequence. This strongly suggests that Neanderthals did not contribute any DNA to the current human gene pool and that Neanderthals and humans diverged some 500,000 years ago. The cover headline accompanying Pääbo's article in *Cell* emphatically declared, "Neanderthals Were Not Our Ancestors."

Even Pääbo's supporters might have had difficulty reaching such a blanket conclusion based on a sample of one, but three years later, a team led by William Goodwin, of the University of Glasgow, provided indispensable verification of Pääbo's findings. Goodwin extracted and sequenced DNA from a 29,000-year-old Neanderthal fossil recovered from the Mezmaiskaya cave in the northern Caucasus in southern Russia, nearly 2,000 miles to the east of the Feldhofer cave. The resulting sequence, obtained by amplifying DNA extracted from a rib bone, differed in twelve positions (3.5 percent) from the original Neanderthal specimen, but in twenty-two positions with a reference human sample. This confirmatory result effectively lays to rest the possibility of either "Neanderthal" sequence being due to contamination or an artifact and proves almost beyond doubt that humans are not direct descendants of the Neanderthals. Rather, Pääbo believes that humans and Neanderthals last shared a common ancestor roughly 500,000 years ago (although the true figure could be anywhere from 300,000 to 700,000 years ago).

But others feel that the cohabitation of modern humans and Neanderthals for hundreds of thousands of years must have led to some interbreeding. Their case was bolstered by the 1999 discovery of the skeleton of an unusual four-year-old child in Portugal displaying a complex mosaic of modern and Neanderthal features, notably a human-like jaw with teeth and lower leg bones more characteristic of Neanderthals. The authors claimed that their specimen "resulted from inbreeding between indigenous Iberian Neandertals and early modern humans dispersing throughout Iberia sometime after 30,000 years ago." It has even been suggested that interbreeding was common. "To their contemporaries, the Neandertals were just another group of Pleistocene hunter-gatherers, fully as human as themselves." This controversy is unlikely to be resolved unless DNA analysis can reveal the presence of "human" sequences in a specimen that meets the anatomical criteria of a Neanderthal, or vice versa. Tattersall praised the Portuguese study as "brave and imaginative [and] courageous," but his opinion was that "this is simply a chunky

Gravettian child, a descendant of the modern invaders who had evicted the Neanderthals from Iberia several millennia earlier." More sequence analysis on additional bone samples should eventually resolve this fascinating chapter of human evolution.

THE ANALYSIS OF NEANDERTHAL DNA is arguably the most dramatic breakthrough in evolutionary genetic research, a field that effectively began on New Year's Day 1987, when Allan Wilson, Rebecca Cann, and Mark Stoneking published a landmark paper in *Nature* that forever changed our view of human evolution. Wilson conducted a thorough comparison of mitochondrial DNA sequences from 147 people representing five geographic populations: African, Asian, Australian, Caucasian, and New Guinean. He focused on mitochondrial DNA because it accumulates mutations much faster than nuclear genes and is passed down exclusively through the maternal line. Wilson's group was able to survey sequence variation at about 9 percent of the mitochondrial chromosome by measuring the cleavage of the DNA with a battery of restriction enzymes.

By comparing the number of sequence differences among each of the five major population groups, Wilson could group the sequences based on the degree of variation among them. Two sequences that differ by just a few bases must be more closely related than samples containing many more discrepancies. Put another way, the fewer the number of sequence changes, the more recently the two lines likely diverged. The crux of Wilson's paper was a horseshoe diagram in which the mitochondrial DNA samples were grouped according to their relatedness with each other. The sequences fell into two large groups: the African sequences in one and all five populations in the other, suggesting that the most ancestral sequence arose in Africa. By contrast, all of the non-African populations had multiple origins. Finally, by assuming that the mutations accrue at a constant rate of 2 to 4 percent per million years, Wilson's group came to a dramatic conclusion: "All these mitochondrial DNAs stem from one woman who is postulated to have lived about 200,000 years ago, probably in Africa."

Wilson's findings were heralded as evidence for human evolution from a single woman—African Eve. This is not to say that there was only one woman alive at that time; more likely, there was a small population of a few thousand people, but the progeny of only one woman successfully thrived. Wilson's classic paper has been subjected to a plethora of criticism over the past decade, and yet its central conclusion has survived remarkably intact. "The earliest divergences in the current mitochondrial gene pool occurred in Africa," says Svante Pääbo. Subsequently, he says, "some but not all, mitochondrial lineages took part in the colonization of the rest of the world." Interestingly, analogous studies performed using markers on the Y chromosome produce remarkably similar results, agreeing to a reasonable approximation on both the date (up to 200,000 years ago) and the location (Africa) of the earliest ancestor. But the comparisons of data from mitochondrial DNA and the Y chromosome point to some interesting differences. If anything, "Y-chromosome Adam" lived somewhat later than "mitochondrial Eve."

Over the past few years, studies of markers from the Y chromosome have also shed light on the migration of human populations, particularly since the identification of scores of new variable DNA markers on the Y chromosome. For example, at one position on the chromosome, designated M42, most of the world's population, including Africa, has a T. But a rare form, with an A at this location, is found in regions of Africa, most notably 15 percent of the Bushmen, or Khoisan. Another marker, studied by Michael Hammer's team at the University of Arizona, also has an A in its most ancestral state (including chimpanzees). Again, it is most prevalent in the Khoisan, whereas the rest of the world's population shows a G. (Hammer's study also points to the likelihood that some Asians returned to Africa carrying a new mutation on the Y chromosome about 100,000 years ago.)

These findings are strongly supported by results from Douglas Wallace, a prominent mitochondrial geneticist at Emory University in Atlanta. Over the past few years, Wallace's team has systematically catalogued the diversity of mitochondrial DNA sequences in the world's populations. These results suggest that mitochondrial Eve had eighteen "daughters," each with a distinct mitochondrial DNA sequence that spread to different regions of the globe. Each mitochondrial DNA pattern, or lineage, is tagged by a letter. American Indians have lineages A–D. There are seven European lineages that fall into two groups, H–K and T–X. The ancestral Asian lineage is M, which split into E–G as well as A–D. The main African lineage is termed L, with three branches L1–L3. Wallace recently compared the mitochondrial DNA sequences of seventy-four Khoisan-speaking individuals from the northwestern Kalahari Desert in southern

Africa and found that this population represents one of the most ancient African populations, reinforcing the age of African mitochondrial DNAs of about 145,000 years.

Comparisons of mitochondrial DNA and Y-chromosome variations lead to some surprising conclusions about the migratory habits of men and women in human history. A recent study by Luca Cavalli-Sforza's group at Stanford finds that there is a much greater degree of variation of Y-chromosome markers between populations than for mitochondrial DNA. One attractive explanation for these surprising differences is that of patrilocality: women tend to travel farther to join their husbands, with a consequent blurring of mitochondrial DNA variation between populations compared to men. This hardly fits with the popular image of marauding male savages (think Ghengis Khan) terrorizing the people of Europe and Asia in recent history. But the sequence analysis suggests that if men did travel extensively in search of a wife (or two) over recent millennia, they were more likely to bring them home.

Despite the tremendous usefulness of studying mitochondrial DNA and Y-chromosome variation, such studies inevitably provide a biased view of human history as, by definition, they can reveal information only about female or male origins and migrations, respectively. With improvements in sequencing technology, scientists are now generating valuable information regarding the variation of nuclear genes. For example, Pääbo's group has sequenced more than 10,000 bases of junk DNA from the X chromosome from sixty-nine racially diverse individuals and compared the pattern of variations. This segment of DNA, which seldom recombines and lies over 1 million bases from the nearest gene, is unlikely to be influenced by any selective pressures caused by changes in neighboring genes.

There is more sequence variation found among different African representatives than non-Africans—twenty-four different base changes compared to just seventeen in the rest of the world. Together, this is further evidence in favor of an African origin for the polymorphism seen in this small stretch of the X chromosome, but probably applicable to the entire human genome. Pääbo's results put the date of the most recent common ancestor at 535,000 years ago (albeit with a 20 percent margin of error), which agrees well with the shorter estimates calculated for mitochondrial DNA and the Y chromosome. (The calculated times back to a common ancestor are expected to be three to four times longer in studies based on

genes on the X chromosome or autosomes, because of the larger number
of chromosomes being studied compared to mitochondrial DNA or the
Y chromosome.)

Studies by Cavalli-Sforza and others show that there is a much
higher degree of genetic variation among individuals of a given race—on
the order of 80 percent—than between different populations. Moreover,
Pääbo's group has found that the three subspecies of common chimpanzee exhibit about four times as much sequence diversity as a group of
seventy unrelated humans from different populations. This suggests that
most of the variation in the human gene pool predates the dispersion of
humans out of Africa and the evolution of racial groups, and that humans
evolved relatively recently from a small founding population of a few
thousand people living in Africa around 200,000 years ago.

THE SAME PRINCIPLE WILSON USED to establish the concept of a mitochondrial Eve is also being used to shed light on the more recent migration of populations around the world. One of the leading exponents of
this research is Bryan Sykes, a geneticist at Oxford University's Institute
for Molecular Medicine, who gained notice a few years ago by performing DNA tests on a tooth from a 9,000-year-old corpse found in the
Cheddar Caves in Somerset, England. Remarkably, the mitochondrial sequence of Cheddar Man matched that of Adrian Targett, a local school-
teacher. Sykes has applied mitochondrial DNA sequencing to study
human variation and migration, showing that Polynesians are related to
people from Southeast Asia. Sykes is now expanding his study to the
length and breadth of Europe. By sequencing 500 bases of mitochondrial
DNA collected from 6,000 people who volunteered samples from across
the continent, Sykes has been able to categorize the sequences into seven
well-defined clusters—the same European haplotypes mentioned earlier,
designated H–K and T–X. But in a cute touch of anthropomorphizing,
he has created thumbnail sketches of the seven lucky mothers, or as he
fondly calls them, the Seven Daughters of Eve. "Every time a European
takes a breath, he or she is using the same genes to metabolize the oxygen
as one of those seven women," Sykes says excitedly.

Ursula, the oldest, lived in northern Greece about 45,000 years ago.
Among her descendants, who moved west to France and Britain, and
south, was Cheddar Man. Xenia lived 25,000 years ago near the Black
Sea. Tara lived in Tuscany 17,000 years ago, her descendants (including

Sykes) pushing north through France to Britain and Ireland. Helena lived near the Pyrenees; her descendants spread throughout Europe. Katrine lived near Venice 10,000 years ago; her descendants live around the Alps. Valda was a Spanish woman alive 17,000 years ago; her descendants trekked north to Scandinavia, giving rise to the Lapps. Jasmine was born in Syria before the last ice age, and her descendants moved west to England and the Baltic Sea. Sykes has set up a company called Oxford Ancestors that allows individuals to submit DNA samples (collected by swabbing the inside of one's mouth with a cotton swab) to find which of the seven daughters of Eve they are most closely related to.

Sykes's company also offers people in Britain the chance to trace an association between their Y-chromosome haplotype and their surname. When Sykes compared Y-chromosome markers from more than sixty men sharing his surname, he found surprisingly that half carried a unique haplotype. His conclusion is that all of the British Sykeses are descended from one man who first took that surname about 700 years ago. The Sykeses that do not bear the distinctive Y markers could be explained by occasional incidents of nonpaternity, at a rate of about 1 percent every generation. "It's really quite low," says Sykes about the infidelity rate. "It essentially means that 99 percent of Mrs. Sykeses have been very well-behaved."

In addition, studies of polymorphic markers have shed light on more recent patterns of global migration. Several years ago, Cavalli-Sforza and colleagues conducted an exhaustive study of nearly one hundred markers in more than two dozen European populations. The frequency of these markers varies in a gradient across the continent, the respective patterns offering precious clues to the origin, age, and subsequent dispersal of different populations. An example can be seen in the prevalence of serious mutations, such as the most common mutation in cystic fibrosis patients, which is most frequent in northern Europe and Denmark and decreases toward the south. Cavalli-Sforza's classic studies reveal patterns of DNA polymorphisms that correspond to the Celts and the Basques, recognizable due to their relative isolation, both genetically and geographically. Gradients of polymorphisms correspond to the routes of neolithic farmers west and north, from the Middle East to Spain, Italy, and northern Europe. A second group represents the Lapps of northern Finland and Russia. A third migration may signify the nomads who domesticated the horse.

Not surprisingly, studies of the Y chromosome have also been fruitful. One example focuses on the origins of the Japanese population. Hammer has shown that Japan's current population is a mixture of two cultures: the ancient Jomon people, hunter-gatherers who trace their origins back 10,000 years, and the Yayoi, agriculturalists who arrived from the Korean peninsula 2,300 years ago.

But the most remarkable application of Y-chromosome markers is to Jewish populations in the Middle East and beyond. The Book of Exodus describes the sanctification of Moses's brother Aaron and his sons, "so that their anointing will make an eternal hereditary priesthood for all generations." Aaron thus became the first Jewish priest, or *cohen,* a tradition that has since been handed down from father to son. Hammer, Karl Skorecki, David Goldstein, and colleagues studied Y markers from three hundred Jews, including more than one hundred *cohanim,* and found that half of the Jewish priests shared the same genetic signature, compared to less than 5 percent in the lay Jewish population. Moreover, the origin of this chromosome dates back some 3,000 years, in agreement with biblical history.

This impressive confirmation of oral traditions pales beside the astonishing tale of the "Black Jews of Southern Africa," worked out by Tudor Parfitt, a fifty-five-year-old Welsh Protestant who has, to his embarrassment, been called the Welsh Indiana Jones. Parfitt, a Jewish studies scholar at the University of London, became fascinated by the stories of the Lemba, a Bantu-speaking tribe of about 50,000 people living in southern Africa and Zimbabwe, which claimed to be one of the Lost Tribes of Israel. The Lemba regaled Parfitt with tales of a long-lost homeland called Sena and a mysterious ocean voyage to their new country, as well as their devoted adherence to many Jewish customs, including the practice of circumcision, the observation of the Sabbath, and the ritual sacrifice of animals. In *Journey to the Vanished City,* Parfitt traces the Lembas' long journey to southern Africa in reverse, traveling to the ruins of Great Zimbabwe and to the east coast of Africa, where, he believes, the Lembas' ancestors arrived from south Yemen about 1,000 years ago. He eventually discovered a small village in South Yemen called Sena that is most likely the Lembas' former home.

Parfitt has followed up this remarkable journey in collaboration with Goldstein and Neil Bradman, a British publishing tycoon and founder of the Center for Genetic Anthropology in London. On a return visit to South Africa, he collected blood samples from the Lemba, politely telling the tribesmen that he had to be careful not to contaminate their "Lemba Y chromosomes" with his "Welsh Y chromosomes." The results of the

DNA studies were stunning: a significant portion of the Lemba Y chromosomes exhibit the characteristic genetic signature found in the *cohanim,* including more than 50 percent of the Buba, one of the 12 Lemba clans. These markers have also turned up in the Bene Israel, the oldest Jewish community in India. More recent studies have demonstrated surprising similarities in the Y chromosomes of Jews, Palestinians, and other Arab populations in the Middle East. Thus, DNA is proving to be an invaluable tool of Jewish history, and that of many other populations besides.

Parfitt's work demonstrates how DNA analysis can reveal extraordinary ties between Africa and the Middle East. Now some researchers are hoping to offer African Americans the chance to learn something of their ancestral roots by studying their DNA. Rick Kittles, a geneticist at Howard University in Washington, D.C., has amassed a database of 3,200 DNA sequences from forty populations in West Africa. Most Africans brought to the United States came from this region, particularly Ghana, Nigeria, Sierra Leone, and Angola. For about $300, Kittles offered to compare an individual's DNA with the database for clues to the homeland of his or her ancestors. In one example, a Washington television reporter learned that his father's ancestors hailed from Nigeria, his mother's from Somalia and surrounding regions. By contrast, Kittles's own paternal line bears the mark of German ancestry—the indisputable mark of a white slave owner subjugating or raping one of his slaves. (About one-third of African males have some white ancestry.) Kittles's fledgling program has received intense criticism accusing him of profiteering from others' misfortune. "We're talking about American slavery, forced migration, prisoners of war," says University of Maryland anthropologist Fatimah Jackson. "I don't think you ask the descendant community to pay to find out something that's their God-given right." But Kittles counters that many African Americans are desperate for information on their origins, ties that were usually severed during enslavement.

Many scientists are hoping that the genome sequence will provide the foundation for a more systematic and powerful analysis of human genetic variation: the Human Genome Diversity Project. This proposal aims to compare the genetic material of scores of remote, isolated native populations around the world, in the interests of understanding human history and migration—a genetic snapshot of man's astonishing diversity before some endangered populations disappear for good. However, the plans have stirred anger and resentment among representatives of many native populations, who hurl accusations of "immoral" and "racist" and suspect a conspiracy to exploit their genetic heritage.

DNA CLEARLY HAS IMMENSE POWER to reveal the migration of populations across the globe over the past 100,000 years, to much more recent sorties, records of which, for one reason or another, have succumbed to the ravages of time. But increasingly, DNA is also being used for the study of individuals rather than populations, finding a variety of powerful applications ranging from criminal investigations—helping to solve some of the most heinous crimes of the past century—to presidential probes, rewriting modern Russian and American history in the process.

Genetic fingerprinting was invented by Sir Alec Jeffreys at the University of Leicester in the mid-1980s. His discovery of highly variable repeat regions in human DNA was in many ways the equivalent of a genetic bar code. Sample four or five such regions and the chances were extremely high that no two unrelated people would show the same pattern of variants. The method was as good as fingerprinting, with the advantage that even the most minute amounts of material—from a single hair, a drop of blood, a soupçon of semen—could provide enough DNA for the analysis.

It did not take long for Jeffreys to demonstrate the awesome power of DNA fingerprinting. In 1987, Dawn Ashworth was raped and strangled to death in the village of Narborough, just outside Leicester, England. The murder was similar to another schoolgirl death three years earlier, and the police soon arrested a seventeen-year-old kitchen porter from a nearby hospital. But unable to prove his guilt, the police took blood samples from 5,000 local men in an effort to match one to evidence. The murderer, Colin Pitchfork, persuaded a friend to give a blood sample in his place, but police eventually learned of the switch, and Pitchfork confessed shortly afterward.

Jeffreys's team then went on to demonstrate the feasibility of typing DNA from skeletal remains. In December 1989, workmen in Cardiff, Wales, discovered the remains of a young girl, who had been murdered several years before, wrapped in a carpet. A medical artist used a plaster cast of the skull to make a model of the girl's face, which was quickly recognized by a Cardiff social worker as that of Karen Price, a young runaway. Jeffreys compared DNA from the victim's femur with samples from her presumptive parents, comparing the sizes of six different polymorphic microsatellite markers. Jeffreys declared a complete match, marking the first time that bone DNA analysis was recognized by the British courts. Jeffreys's group went on to make other notable forensic identifications, including an analysis of the remains of Wolfgang Gerhard, who had drowned in Brazil in 1979. Jeffreys's team extracted DNA from the deceased man's femur and concluded that he was in fact the Nazi "Angel of Death," Josef Mengele.

Mitochondrial DNA has proven invaluable in reuniting victims of Argentina's military junta in the 1970s with their families. In 1976, a military coup seized power in Argentina, with soldiers killing an estimated 9,000 young "subversives." About 200 orphaned or abducted young children were given to childless couples in the army, who passed them off as their own. After the military regime was thrown out in 1983, an organization called *Las Abuelas* (the Grandmothers) tried to reunite these children with their biological families. Mary-Claire King compared mitochondrial DNA samples of the children with their relatives and consequently reunited dozens of families. The first case was that of an eight-year-old girl, Paula Eva Logares, who was living with a former police chief and his girlfriend. The Grandmothers said she was kidnapped from her parents when she was less than two years old. King's group proved beyond doubt that Paula was indeed related to three living grandparents, who claimed her. King says, "When she went back to her grandparents' house, which she hadn't seen since she was two, she walked straight to the room where she'd slept as a baby and asked for her doll."

ON THE NIGHT OF JULY 16, 1918, three hundred years of Romanov rule in Russia were brutally terminated in a hail of bullets from a Bolshevik firing squad. A local newspaper reported that Czar Nicholas II and his family, who had been imprisoned in the Ipatiev House in Ekaterinburg, had been "shot without bourgeois formalities but in accordance with our new democratic principles." The bodies were doused with sulfuric acid to hinder identification and hastily buried in a shallow grave, the location of which remained a mystery for more than seventy years. After the location of the site was revealed by a Russian film maker in 1989, President Boris Yeltsin ordered the exhumation of the remains from the makeshift grave. One thousand recovered bone fragments were assembled into nine skeletons—five female, four male—corresponding to Czar Nicholas, Czarina Alexandra, three of the five children, as well as three servants and the family physician. (The bodies of two children were burned.) Experts disputed the authenticity and identity of the remains, with some arguing that Czarevitch Alexis and Marie were missing, others maintaining that Anastasia could not be accounted for.

In September 1992, a Russian DNA fingerprinting expert named Pavel Ivanov flew to London carrying small bone samples of the remains. He was met at Heathrow Airport by a BBC television director who had rented a funeral hearse, because he felt it would be "inappropriate to carry the Russian Imperial family in the [trunk] of my Volvo." DNA tests were performed in the laboratory of Peter Gill at the British Forensic Science Service. Gill's first test showed that five of the bodies were female, and three were siblings. Mitochondrial DNA testing showed that DNA from the putative remains of the Czarina Alexandra matched a sample provided by the duke of Edinburgh, her great nephew (both are maternal relatives of Alice of Hess, Queen Victoria's daughter). But finding a cooperative maternal relative of the czar was much more difficult. A swatch from a bloody handkerchief of the czar's was too contaminated, and the czar's nephew refused to cooperate because of the refusal of the British to offer Nicholas's family refuge after the revolution.

When samples were finally obtained from two more distant maternal relatives, the results were frustratingly ambiguous. The mitochondrial DNA sequences were identical except for position 16169. The sequence in the two relatives showed a T, whereas, curiously, the putative remains of the czar had a mixture of about 70 percent C and 30 percent T. Gill and colleagues argued that this was an example of heteroplasmy, a naturally occurring mixture of DNAs from the same person, and calculated the odds of the remains' belonging to the czar of almost 99 percent.

But with contamination still a possibility, one last test was called for. At the urging of the Russian Orthodox church, the body of the czar's brother, Grand Duke Georgij Romanov, who died of tuberculosis in 1899, was exhumed from its coffin in St. Petersburg. Ivanov took the duke's bone samples to the Armed Forces Institute of Pathology DNA Identification Laboratory in Maryland. Incredibly, Georgij's mitochondrial DNA also revealed heteroplasmy at position 16169. The chances of observing such heteroplasmy in the general population, coupled with the proven match between the czarina and a living relative, produced odds of 100 million to 1 that the remains were authentic. With an emotional President Yeltsin in attendance, the Romanov royal family was given official burial in St. Petersburg in 1998. There are still people who claim that one of the czar's children survived the 1918 atrocity, but the claims of the last pretender, a woman named Anna Anderson, have been cast out by DNA analysis.

"The longest-running mini-series in American history," as Eric Lander calls it, dates back to 1802, when the *Richmond Recorder* broke the story of an illicit relationship between President Thomas Jefferson and his mulatto slave Sally Hemings, the illegitimate half-sister of his wife, Martha Wayles. Suspicion that Jefferson fathered some of Hemings's seven children centered on several lines of circumstantial evidence, including the physical resemblance of Hemings's children to the president and Jefferson's presence at his Monticello home at the time the children were conceived.

The Jefferson saga aroused the interest of Eugene Foster, a semiretired professor living in Charlottesville, Virginia. Foster began soliciting volunteers from the Jefferson and Hemings families, looking for paternal relatives who could trace a direct male-to-male line to Jefferson and Hemings's son. It is estimated that there are more than 2,000 living descendants of Jefferson and his wife, but none carries his Y chromosome because his only legitimate son died in infancy. However, Foster was able to obtain blood from male line descendants of Field Jefferson, the president's paternal uncle. He also received the cooperation of John Weeks Jefferson, the great-great-grandson of Eston Hemings, Sally's youngest son.

In all, Foster collected nineteen DNA samples, which, securely packed in his carry-on luggage, accompanied him to England, where they were delivered to Oxford University geneticist Chris Tyler-Smith. When each DNA sample was typed for nineteen polymorphic markers on the Y chromosome, there was a complete match in the pattern of polymorphisms between the male descendants of Field Jefferson and a male line descendant of Eston Hemings. The same study also excluded Jefferson as the father of Hemings's oldest child, Thomas Woodson.

The discovery caused uproar among presidential historians, who have long debated the validity of the Hemings family's claims. The timing of the report was also remarkable. As Lander and historian Joseph Ellis wrote, the illicit relationship between the author of the Declaration of Independence and his young slave "reappears to remind us of a truth that should be self evident. Our heroes—and especially presidents—are not gods or saints, but flesh-and-blood humans, with all of the frailties and imperfections that this entails." Lander and Ellis were alluding, of course, to a more contemporary White House sex scandal.

The course of the soap opera known as the Lewinsky affair was to all intents and purposes sealed on August 3, 1998. That day, the Federal Bureau of Investigation (FBI) informed independent counsel Kenneth Starr that it had identified two semen stains on specimen Q3243—a size 12, dark blue GAP dress owned by White House intern Monica Lewinsky. That evening, in the Map Room of the White House, President Clinton's physician drew a small sample of blood, which the FBI labeled specimen K39. Exactly two weeks later, Clinton was informed that two tests on the DNA in samples K39 and Q3243 revealed a complete match, suggesting that K39 and Q3243 were identical "to a reasonable degree of scientific certainty." The odds that the donor of the DNA deposited on Lewinsky's dress could have been any Caucasian other than Clinton were put at a modest 1 in 7,820,000,000,000 (7.8 trillion). President Clinton was subsequently impeached by the U.S. Senate. Seeking to draw comparisons between the two presidential scandals, a letter to the editor of the *New York Times* sarcastically recommended that the independent counsel "be empowered to bring posthumous impeachment charges against our third President. No cost should be spared to remove the stain of immorality from our national record."

THERE SEEM TO BE FEW LIMITS to the intrusive reach of DNA fingerprinting, and few other places more eager to see how far it can go than in France. On June 8, 1795, a ten-year-old boy named Louis-Charles died in a Paris dungeon from tuberculosis, two years after his parents, King Louis XVI and Queen Marie-Antoinette, were guillotined during the French Revolution. While performing the autopsy, the royal physician smuggled out the boy's heart in his handkerchief and kept it as a memento of the monarchy (the young king's body was unceremoniously dumped in a mass grave). There followed an extraordinary chain of custody before it was presented to the basilica of the Memorial of France at St. Denis in 1975, where it has remained in a crystal globe, along with other royal relics. French historians have argued passionately about the identity of the heart. Many believed that the boy who died in the Parisian prison was an imposter, the real king having been whisked away by loyal monarchists. Scores of pretenders have laid claim to the throne over the past two hundred years.

There is little in the way of precedent when contemplating the dissection of a royal organ for DNA analysis. But in 1999, the urn was covered with a purple cloth and taken to a Parisian laboratory, where small sections of the young king's heart were sliced off with a razor blade, divided, and sent to the laboratories of Jean-Jacques Cassiman in Belgium and Bernd Brinkmann in Germany. Despite the atrocious state of the heart, the two labs were able to compare mitochondrial DNA sequence from the heart with DNA from hair samples of Marie-Antoinette and two living maternal relatives, including Queen Anna of Romania. Although the analysis does not definitively prove the heart belonged to Louis XVII rather than another close relative, the historical circumstances leave little doubt. The vestige of the uncrowned king now awaits a more dignified burial.

Other historical curiosities could also be resolved with the judicious application of DNA testing, although whether the cases merit such intrusion is debatable. The death of Napoleon Bonaparte in 1821 on the island of St. Helena was reportedly from stomach cancer, but there have been allegations that he was poisoned by the British. These received a scintilla of credibility when the FBI found traces of arsenic in a lock of his hair removed after his death, leading some imaginative historians to suggest that Napoleon's body was switched before his coffin was shipped back to France in 1840. When French authorities opened the casket, they found a bearded corpse (Napoleon was supposedly shaved after his death) and no sign of his silver spurs.

In the United States, the body of the man some think was really the outlaw Jesse James was recently exhumed for DNA testing. Most historic accounts say that James was shot to death in 1882, but others believe that James faked his death and lived under the alias J. Frank Dalton until he died at age 104. Under Dalton's name on his tombstone is inscribed in small letters, "Supposedly killed in 1882."

In other cases, DNA testing is being proposed to add a footnote to history by revealing the diseases that afflicted famous people. The best known example is that of President Abraham Lincoln, whose lanky gait is suspiciously reminiscent of Marfan's syndrome. In the words of the London *Times*'s Washington correspondent, Lincoln had "stooping shoulders, long pendulous arms terminating in hands of extraordinary dimensions, which, however, were far exceeded in proportion by his feet." The gene for Marfan's syndrome was identified in 1991, and it would be feasible to test Lincoln's DNA—extracted from blood, bone, or hair museum specimens—for the presence of a mutation. At the other end of the spectrum is the case of French artist Toulouse-Lautrec, a dwarf whom some scholars believe suffered from pyknodysostosis. Robert Desnick and Bruce Gelb, researchers at New York's Mt. Sinai Hospital, have identified the gene mutation in families with this disorder but so far, Lautrec's descendants have refused permission to have the body of the famous artist tested.

THERE IS NO DOUBT THAT FORENSIC DNA analysis has enormous potential to nab criminals and pardon the wrongly convicted. Famed attorney Barry Scheck, cofounder of the Innocence Project, calls DNA "the gold standard of innocence," which his team has used to win the freedom of more than seventy wrongfully convicted people in the United States, many including eight facing the death penalty. "For every seven people executed in the U.S., one person is exonerated on the grounds of innocence," says Scheck. Those odds, he adds needlessly, are appalling. Of course, the statistical power of DNA fingerprinting is effective only if the jury can understand the odds being presented, and there are no other mitigating factors. Scheck rose to stardom as one of the phalanx of attorneys defending O.J. Simpson, who stood accused of murdering his ex-wife and another man. Amid a flurry of accusations of racism and incompetence directed chiefly toward the Los Angeles Police Department, the potentially damning evidence contained in forty-five bloodstain samples presented to the jury was obscured. In her closing arguments for the prosecution, Marcia Clark stated that the chance that a person other than Simpson would have the genetic profile found in a blood sample at the crime scene was 1 in 57 billion (ten times the number of people on the planet).

In the United Kingdom, the birthplace of DNA fingerprinting, a national DNA database contains some 700,000 profiles and has been used to link evidence to 75,000 criminals, at a rate of 500 a week. But Scheck notes that only a handful of convicts have been exonerated using DNA evidence. In the United States, the FBI's Combined DNA Index System contains profiles of over 250,000 criminals and thousands of unidentified crime scene samples.

There are few limits, it appears, to the potential applications of DNA fingerprinting. In France, the body of the popular singer Yves Montand had to be exhumed for DNA analysis to resolve a paternity dispute. He was eventually exonerated. An even stranger case arose in 1996, when the normally sedate English village of Bruntingthorpe was rocked by a contemptible crime wave. When one of the residents refused to clean up after his or her dog fouled the sidewalk, the local parish proposed to create a DNA database of the thirty dogs in the village which could be used to trace the perpetrator. Alas, after a barrage of negative publicity, the idea was shelved.

But DNA fingerprinting of house pets has already left its mark in the annals of criminal law. In 1996, police investigating the murder of a young Canadian woman recovered a bloodstained leather jacket near her home that also contained several cat hairs in the lining. Following a subpoena to collect a blood sample from Snowball, the pet cat of the victim's estranged husband, DNA tests were performed by Stephen O'Brien at the National Cancer Institute. A complete match with the cat hairs in the jacket helped lead to the conviction of the suspect.

CHAPTER NINE

The Croesus Code

Know ye not that they which run in a race run all, but one receiveth the prize.

—I CORINTHIANS 9:24

Sequencing DNA now is one of the easiest jobs you could have besides sloppin' burgers.

—KARY MULLIS, Nobel laureate

BY THE AUTUMN OF 1999, the radical overhaul of the public genome sequencing strategy that Collins had engineered began paying dividends. The infusion of tens of millions of additional dollars enabled the centers to reach their potential, and the provision of hundreds of new automated DNA sequencers provided the needed boost in technology. The G-5 centers optimized their organization, recognizing that DNA sequencing is repetitive, tedious work, hardly the intellectual environment to inspire young, ambitious Ph.D. students. Finally, the decision to produce a rough draft meant that time was not wasted struggling to perfect stubborn stretches of DNA sequence.

In November, the public Human Genome Project finally had something tangible to celebrate. Thanksgiving Day came a week early, November 17, when a simple "G" earned the distinction of becoming the one-billionth letter in the public Human Genome Project sequence repository, GenBank. Transatlantic festivities took place a few days later, with the British science minister David Sainsbury presenting an award to John Sulston at the Sanger Centre, while Francis Collins hosted a celebration for hundreds of researchers at the headquarters of the National Academy of Sciences in Washington, D.C. The major genome center groups in the United States and United Kingdom participated via a live webcast. "The twenty-first century came about 6 weeks early!" proclaimed a beaming Donna Shalala, secretary of the Department of Health and Human Services, before handing out awards to representatives of the American G-5 groups. Shalala likened the accomplishment to a Toni Morrison novel: "What it lacks in structure, it more than makes up for in the richness of its content." It was left to Collins to close the ceremony by exhorting the researchers to push even harder. His final words: "One billion and counting—let's keep climbing!"

Meanwhile, Venter was hosting an unusual gathering of private and public scientists at Celera's headquarters. He had invited forty-five leading *Drosophila* scientists from around the world to an "annotation jamboree," where they would pore over the raw DNA data generated by Celera's phalanx of DNA sequencers and assembled into "scaffolds" by Gene Myers's software. For eleven days (and nights), experts on almost every facet of *Drosophila* genetics and biology scrutinized the sequences of thousands of new genes for clues to their function. The experts checked the predictions made by computer algorithms with names like Genie and Genscan, while the bioinformaticians worked through the night to refine their programs to improve the accuracy of annotation.

Identifying new genes in the *Drosophila* genome proved more difficult than for the yeast or nematode sequences: the genes are spaced farther apart and are interrupted by larger introns, and the characteristic sequence motifs that signal the start of a gene are less recognizable. The biggest surprise came in the final gene tally: just 13,601, compared to 18,000 in *C. elegans.* Why a primitive worm should have nearly 5,000 more genes than a fly is anyone's guess. Part of the difference appears to be in the size of different gene families in the two species. For example, *C. elegans* contains 1,000 olfactory receptors compared to about 60 in *Drosophila.* It is also clear that about one-third of *Drosophila* genes are alternatively spliced so as to give rise to more than one protein sequence, so *Drosophila* probably contains more than 20,000 different proteins.

Veteran *Drosophila* geneticists marveled at the productivity of the Celera collaboration, which had made the complete sequence of their beloved fruit fly freely available over the Internet. Work began on writing the papers that would pay tribute to the prowess of Venter's shotgun sequencing technology.

• • •

BENEATH THE HANDS OF GOD and Adam from Michelangelo's *Creation of Adam,* the cover of the December 2, 1999, issue of *Nature* proudly proclaimed "The first human chromosome sequence." One of the largest international collaborations on record—more than two hundred researchers from the Sanger Centre, working with groups at the University of Oklahoma, Washington University in St. Louis, and Keio University in Japan—described virtually the entire sequence of chromosome 22. Although representing just 1.1 percent of the total human genome, scientists hailed the feat as a major landmark in the history of biomedical research and a timely validation of the mapping strategy advocated by the public genome alliance.

In keeping with the international collaborative effort underlying the work, press conferences were held in England, Japan, and Washington, D.C. In Cambridge, the normally reserved Brits got quite carried away. Sanger Centre chief John Sulston suggested that the work was "as important an accomplishment as discovering that the Earth goes round the Sun, or that we are descended from apes." Michael Dexter of the Wellcome Trust compared the achievement to the invention of the wheel. In Washington, D.C., reporters crowded into a small room at the Willard Hotel, a stone's throw from the White House. This proud moment in honor of human technological prowess was marked by the inability to secure a reliable audio link to Cambridge, where project leader Ian Dunham was waiting to broadcast a statement. (At one point, the Washington press corps listened in puzzlement as Dunham and Sulston, unaware their voices suddenly could be heard 3,000 miles away, casually discussed the weekend's upcoming soccer matches.)

The American collaborators were as lyrical as their British counterparts. Bruce Roe, of the University of Oklahoma, was in philosophical mood: "A new era has dawned: we have fulfilled the dreams of Mendel, Morgan, Watson and Crick, and Sanger, as we now have the essentially complete structure of the first human chromosome." Collins conceded, "I don't often pick up a scientific paper and find myself getting chills, as I did when I saw this whole chromosomal landscape." He likened it to "seeing an ocean liner emerge out of the fog, when all you've ever seen before were rowboats." Mark Patterson, a senior editor with *Nature,* compared it to "seeing the surface or the landscape of a new planet for the first time. It's allowing us to say something about its geography [and] its history."

Unlike the somewhat arbitrary celebration to mark the one billionth base, there was nothing artificial about celebrating the first "operationally complete" sequence of a human chromosome, as defined by the terms of the Bermuda Accord. Dunham and colleagues reported the assembly of more than 33,400,000 letters of DNA, including one contiguous stretch of 23 million bases, the longest on record to that point. But Roe's "essentially complete" definition underscored one important fact. The sequence of the long arm of the chromosome contained a dozen gaps in the DNA sequence (amounting to just over 1 million bases) that could not be closed for technical reasons. The researchers had also deliberately ignored the stumpy short arm of the chromosome, which is made up of heterochromatin and contains few, if any, coding genes. Despite the inevitable gaps, however, the quality of the sequence was superb, with just one discrepancy in every 50,000 decoded letters. The published sequence is not of any one individual, but a composite assembly of sequence derived from several DNA sources. The final estimated cost was put at $15 million, less than 50 cents a letter. But just how much information could be gleaned from this unprecedented expanse of raw DNA sequence?

As the Sanger Centre sequencers read the script of the twenty-second chromosome, Dunham's team examined the sequence for the presence of genes. Previously identified genes were easily recognized by comparing the chromosome 22 DNA sequence with all available DNA and protein databases. To identify new genes, Dunham's team analyzed the sequence with a battery of computer software programs designed to estimate the likelihood that a given stretch of DNA encodes a gene. Even the best of these software programs is not perfect; as many as 30 percent of the gene fragments predicted by the program Genscan might be spurious, whereas 20 percent of bona-fide exons could be missed using computational methods. "As a result," Dunham and colleagues reluctantly concluded, "we do not consider that *ab initio* gene prediction software can currently be used directly to reliably annotate genes in human sequence."

Nevertheless, Dunham's team was able to apply other tricks to find genes lurking on chromosome 22. One strategy is to scan for stretches of DNA enriched in the letters G and C. Adrian Bird, at the University of Edinburgh, has demonstrated that these so-called CG islands are commonly found in the regions just in front of genes. The presence of a CG island—there are about 550 on chromosome 22—is an excellent indicator of a nearby gene. Another resource for cross-referencing was the EST database, which contains hundreds of thousands of gene fragments.

Amid the barren, repetitive stretches of DNA of no known function, Dunham's team confidently identified 545 genes: 247 were matches to known human genes, 150 were related in sequence to genes from other species (such as those of the mouse, fruit fly, and yeast), and 148 showed significant similarities to ESTs. There are also 134 pseudogenes: sequences that were once functional but are now unable to code for a functional protein because the sequence is disrupted. However, the sequence also revealed the staggering extent of so-called junk DNA. Some 42 percent of chromosome 22 is composed of more than 55,000 repetitive DNA elements.

The search for additional genes gave interesting results. The Genscan program, for example, predicted the existence of an additional 300 genes. Even allowing for the likelihood that many of these are false positives, there are probably at least 100 genuine new genes among this group. How many more is still unclear. The CEO of Incyte Genomics, Roy Whitfield, insisted that his company had assigned more than 1,000 genes to chromosome 22. Another Californian bioinformatics company, DoubleTwist, has used a combination of gene prediction algorithms to predict 1,485 genes and 2,700 alternative spliced products. The final number of genes may be close to 1,900. However, the Sanger Centre's Richard Durbin dismisses these inflated estimates as "bogus." In nine months since the chromosome 22 sequence was published, his colleagues have found fewer than 20 new genes.

Even a cursory glance at the public alliance's gene haul provides a breathtaking hint of the variation among them. The average size of genes on chromosome 22 is 19,000 bases: the shortest is just 1,000 bases long, whereas the largest (appropriately dubbed *LARGE)* contains more than 580,000 letters. There are genes containing just one coding region, or exon, whereas one gene is divided into fifty-four pieces. The gene mutated in the inherited vision disease Sorsby's fundus macular degeneration, itself more than 60,000 bases in length, resides head-to-tail within a huge gap in another gene.

The most important short-term use of this new genetic lexicon is to start matching individual genes to the dozens of disease traits that have previously been traced to chromosome 22. These include genes that are deleted in brain tumors, amplified in a neuronal cancer (neuroblastoma), and disrupted in bone cancer (Ewing's sarcoma) and forms of leukemia. Other genes are mutated in patients with mental retardation and heart disease, including individuals with DiGeorge syndrome, a relatively common congenital disorder caused by the loss of several contiguous genes.

But the most exciting gene hunt of all relates to schizophrenia. In the contentious debate about the role of genetics versus environment in complex traits, and especially psychiatric disorders, schizophrenia holds special prominence. For several decades, a host of studies have shown that relatives of schizophrenics have a tenfold increased risk of the disease, but that identical twins are 50 percent, not 100 percent, concordant in their risk. The evidence clearly points to both hereditary and environmental factors conspiring to cause a disease that affects more than 1 percent of the population. With the availability of thousands of new DNA markers in recent years, researchers have mapped putative schizophrenia genes to as many as half of the twenty-four different human chromosomes. Some of these associations probably will not hold up, but others may lead to the identification of genes that influence susceptibility to schizophrenia. The link between schizophrenia and chromosome 22 involves a disorder called velo-cardio-facial syndrome (VCFS), a rare disorder caused by deletions in a stretch of chromosome 22. Most patients have lost a critical region that spans about 500,000 bases and contains dozens of candidate genes. Curiously, there is a high incidence of schizophrenia among VCFS patients and their relatives. Moreover, some schizophrenia patients harbor similar chromosome 22 deletions to those in VCFS. Revealing the identity of the putative schizophrenia gene would be justification in itself for the chromosome 22 sequencing project.

Dunham and his coauthors concluded their landmark chromosome 22 *Nature* article with a thinly veiled jibe at Celera. "Over the course of the project, the emerging sequence of chromosome 22 has been made available in advance of its final completion through the internet sites of the consortium groups," they wrote. These data had facilitated the identification of half-a-dozen important genes long before the sequence was published, including at least one disease gene.

Completion of chromosome 22 was a moment for "unbridled exaltation, yes—but also, solemn reflection," according to the *Guardian*. That summed up Collins's reaction, even though he was not a direct participant in the chromosome 22 project. Interviewed that evening on CNN, he said, "No one knew if there would be insurmountable problems that prevented the assembly [of chromosome 22]." "[This] tells us we will be able to finish the human genome in another two or three years." Dunham was more succinct: "One down, the others to go." *Nature* published an enthusiastic commentary written by Peter Little, a leading British molecular biologist now at the University of Sydney, who seemed to have been infected with millennium fever: "As 1999 draws to a close and we approach the third millennium, a new book is being written. It will change the way we see ourselves as profoundly as did the momentous books of the first two millennia—the great books of religion and *The Origin of Species*. The first chapter of this book . . . consists of the DNA sequence of most of human chromosome 22."

Ironically, the triumph for the public genome project lifted not only Collins's fortunes, but also—in a far more literal sense—those of Venter. News of the milestone spurred a flood of investment into genomics and biotech companies, including Celera. On December 16, a popular online investment firm, the Motley Fool, announced that it would invest heavily in Celera, despite judging Venter's company to be "grossly overvalued." The news sparked a stampede, such that by the end of 1999, Celera was trading at $190 a share. Celera was now valued at close to $3 billion.

As THE COMPETITION BETWEEN Celera and the public sequencers intensified, parties on both sides grew anxious to explore the possibility of collaboration, along the lines of the highly successful Venter-Rubin collaboration on the *Drosophila* sequence, which had conclusively demonstrated that Celera's shotgun approach could work for a complex organism. With Celera ripping through 1.2 billion letters of human DNA in barely a month, Venter's chances of winning the race were looking better and better. But some scientists affiliated with Celera were worried about the long-term ramifications should Venter convincingly beat the public Human Genome Project. Richard Roberts, chairman of Celera's scientific advisory board, admitted, "It will not be good if the public effort is seen to lose. Congress would wrongly assume the public effort was wasted and might decided to cut back NIH's funding." In late 1999, urgent discussions took place between scientists on both sides, including Eric Lander for the public program and Rockefeller University president Arnold Levine, another member of Celera's advisory board.

On December 29, Francis Collins led a delegation of scientists from the public genome effort in a meeting with Venter and senior advisers. The public genome project was represented by Collins, Waterston, Harold Varmus, the outgoing NIH director, and Martin Bobrow, the head of clinical genetics at Addenbrooke's Hospital in Cambridge, England. Venter's team included PE chairman Tony White, Celera executive Paul Gilman, and Levine. The meeting was a last-ditch effort to explore the possibility of collaborating on the final derivation and publication of the human genome sequence. Collins brought to the meeting a draft statement of "Shared Principles" which he hoped to release in the event the meeting went well. The statement included eight points of general agreement:

- The public Human Genome Project and Celera Genomics each have the capacity to generate substantial coverage of the human genome.
- Humankind will be better served if we can find a viable way to join forces to produce a better product in a more timely fashion.
- The methods being used (clone-based and whole-genome shotgun) are, in fact, complementary. They provide much opportunity for cross-checking.
- A collaboration would offer the opportunity for joint optimization of experimental strategy and analytical methods.
- The current antagonism and excessive competition should be replaced with a more collaborative spirit.
- The public Human Genome Project is committed to the complete sequence of the human genome being freely available in the public domain.
- The public Human Genome Project understands that Celera Genomics will be making its consensus sequence of the human genome broadly available.
- Celera Genomics is able to see assembled data generated by the public labs and is free to use it for its proprietary databases, but a scientific publication that combines substantial data from both sources should, according to accepted scientific practice, be a joint publication involving authors from both groups.

The final point added an interesting twist into the debate. All along, Celera had enjoyed unfettered access to the public Human Genome Project's sequence data, released over the Internet every twenty-four hours. But was public release of those data equivalent to publication? It was one thing for Celera to integrate the public data into its human genome sequence. It was quite another for the company to contemplate publishing those data derived by its competitors in a peer review journal without explicit consent and without direct access to the primary sequence traces for verification purposes. (Curiously, no such complaints had been made when Venter assimilated thousands of publicly released EST sequences into his 1995 Genome Directory.) Collins's counterproposal was for Celera and the public Human Genome Project to consider a joint publication, in which scientists on both sides would share access to each other's data.

But as the face-to-face meeting wore on, serious points of disagreement remained, particularly Venter's insistence on exclusive commercial distribution rights for the joint data set for up to five years, whereas Collins considered six to twelve months appropriate (by which time the Human Genome Project would have essentially completed its sequence). Moreover, Celera insisted on the rights to other applications of the sequence, such as DNA chips, and to be the exclusive distributor of the sequence over the Internet. The meeting ended with no agreement, no release of Collins's "Shared Principles," and only a vague plan for further meetings in the New Year.

For months, Francis Collins had pointed to the spring of 2000 as the target date for the public consortium to unveil the rough draft of the human genome. But on January 10, in a crafty piece of one-upmanship, Venter held a press conference to announce he had already reached that same goal. Venter said Celera had sequenced 90 percent of the human genome—10 million DNA fragments of roughly 500 bases apiece, for a total of 5.3 billion letters of DNA. Although this amounted to less than twofold coverage of the human genome, Celera had also analyzed the freely available public sequence data. With about 50 percent of the human genome covered at that time, Celera was able to find sequence for about half of the 20 percent of the genome it had not tackled, bringing its total coverage to 90 percent. Moreover, Venter claimed that Celera had at least 97 percent of all human genes and was on course to sequence 9 billion bases of human DNA by the summer. Naturally, the sequence data were closely guarded, but Venter could not resist plugging the discovery of a novel alpha-interferon gene, a potential antiviral drug that was already attracting the interest of one of Celera's pharmaceutical clients.

Although the public consortium had sequenced twice as much

human DNA as Celera, there was no doubt which side was winning in the eyes of the media. In contrast to Celera's hare, *Time* tagged the public genome project "the proverbial tortoise." The negative press for the NIH could have been much worse had it not been for breaking news of the AOL–Time Warner merger.

THE ANNUAL CONFERENCE of the American Association for the Advancement of Science is one of the most diverse meetings on the science calendar. Subjects can range from how birds compose in sonata form to novel methods of hurricane prediction. One of the undoubted highlights of the 2000 meeting in Washington, D.C., was the first public announcement of the *Drosophila* genome sequence, a few weeks before the papers would appear in *Science*.

The meeting was the first chance for two of Celera's key scientists, Mark Adams and Gene Myers, to document Celera's shotgun sequencing approach. Adams had been a loyal lieutenant to Venter for ten years, from the early days of the EST project, and Venter was content to relax in the audience while Adams reviewed Celera's sequencing strategy. It had required more than 3 million DNA sequence reads, each averaging 500 bases in length, to produce the sequence. Adams called it "Release 1.0," because efforts were continuing to close hundreds of stubborn gaps in the genome. He was followed by Gene Myers, Celera's bioinformatics expert, who declared that the shotgun strategy had been a complete success, easily coping with the frequent stretches of highly repetitive DNA that many predicted would jeopardize the assembly of the final sequence.

The final talk was given by Gerry Rubin, the head of the academic arm of the *Drosophila* genome alliance. Although it would require years of further study, Rubin said the *Drosophila* sequence already hinted at its vast potential for understanding and modeling human disease. A comparison of the 13,600 genes in *Drosophila* with the sequences of almost 300 genes known to be mutated, deleted, or amplified in human genetic disorders revealed that more than 60 percent of the human disease and cancer genes had a related sequence in the fruit fly genome. The list of partners reads like a medical dictionary: colon cancer, leukemia, and skin cancer; Alzheimer's, Huntington's, Lou Gehrig's, and Parkinson's disease; hereditary deafness, blindness, heart disease, hypertension, and diabetes; cystic fibrosis, muscular dystrophy, and numerous congenital birth defects. Perhaps most exciting was the discovery of a previously unknown fly counterpart of *p53,* the most commonly mutated gene in human cancers. The wealth of disease-related genes in fruit flies bodes well for the continued popularity of *Drosophila* well past its first centenary.

Drosophila also offers a superb model for the study of behavior and complex traits. Flies are important animals for studying circadian rhythms, aging, sexual behavior, and addiction. As an example, Rubin highlighted the research of Ulrike Heberlein from the University of California, San Francisco, on alcohol sensitivity. This involves a contraption called an inebriometer—a tall glass cylinder that houses a series of funnels stacked on top of each other. Ethanol vapor is passed through the inebriometer at the same time that flies are released at the top. Flies that are more sensitive to alcohol have trouble clinging to the walls of the tower and progressively drop from level to level until they reach the bottom. The most seriously inebriated flies had an alcohol level of 0.2 percent, well over the legal driving limit in humans. And yet breeding for twenty generations can produce flies with threefold greater alcohol resistance (that is, flies take more than thirty minutes to fall through the inebriometer, compared to the average twelve minutes). The sensitive flies harbor mutations in proteins involved in a well-known signaling pathway triggered by alcohol at the cell surface. One of these genes is amusingly known as *cheapdate.*

Rubin ended his talk on a surprisingly candid note by addressing the heated criticism he had faced for collaborating with Venter. "Working with Celera," he said, "has been one of the most pleasurable scientific experiences I have had in my 30-year career. It's been a ball—it's been great! Seldom have I encountered a group of individuals who were so dedicated and so hard working. . . . They have exceeded all the commitments they made to me in this collaboration. They have behaved with the highest standards of personal integrity and scientific rigor."

One month later, on March 24, *Science* devoted almost half its regular issue to the *Drosophila* genome sequence. The cover reproduced two beautiful color drawings of *Drosophila* from the 1930s by Edith Wallace, Thomas Hunt Morgan's long-time assistant, crawling over row upon row of DNA script. The special issue contained a series of articles and reviews marking the completion of the *Drosophila* genome project. One could forgive the Drosophilists their enthusiasm. Mark Krasnow and Thomas Kornberg put it like this: "The *Drosophila* sequence is a critical resource that ensures that this tiny dew-lover will continue to lead the way to new biological pathways and principles. If *Drosophila* has been difficult for

workers in other fields because of an arcane nomenclature and idiosyn-
cratic husbandry, the sequence now provides access through a universal
language—the DNA sequence."

In a cute public relations ploy, Celera left a copy of the fruit fly sequence on a CD-ROM on the chair of every scientist attending a conference a few days before the *Science* publication. Venter said he still wanted
to print out one copy of the *Drosophila* sequence, although even using 8-
point type filling every square inch of paper, it would require 27,000
pages. Scientists around the world immediately began searching the
Drosophila sequence and for genes of particular interest. However, it
turned out that 0.1 percent (about 150,000 bases) of the fruit fly sequence
was actually contaminating human DNA (although it had not been incorporated into the assembled genome sequence). Seizing on the error,
some scientists accused Celera of rushing to publish unverified data.
"People who live in glass houses shouldn't throw stones," said Richard
Wilson, a close colleague of Bob Waterston at Washington University.
Rubin retorted that the contamination was "absolutely trivial," the
equivalent of a typographical error, and said he was offended by the politi-
cization of the sequence. Although the episode was momentarily embarrassing, there was no lasting damage. Given the unprecedented scale of the
sequence assembly, the accidental deposition of a small amount of foreign
DNA was not entirely unexpected.

The English poet Samuel Johnson wrote, "A fly, Sir, may sting a
stately horse and make him wince; but one is but an insect, and the other
is a horse still." But in the postgenomic era, with the *Drosophila* sequence
poised to reveal even more truths about human biology, it is the lines of
Johnson's contemporary, William Blake, that come to mind: "Am not I a
fly like thee? Or art not thou a man like me?"

IN STARK CONTRAST to the harmonious collaboration between academia and Celera on the *Drosophila* genome, negotiations on the human
genome were going nowhere. In the weeks that had passed since the
meeting between Collins, Venter, and their senior aides, Collins had
grown increasingly frustrated by what he perceived to be stonewalling. A
phone conversation with PE chairman Tony White had merely rein-
forced Celera's position, and Venter declined to return Collins's calls and
e-mails. Finally, on February 28, Collins faxed a "confidential" letter
addressed to Venter, White, Levine, and Gilman and signed by Collins,

Varmus, Waterston, and Bobrow, reiterating the major disagreements between the Human Genome Project and Celera. "While establishing a monopoly on commercial uses of the human genome sequence may be in Celera's business interest, it is not in the best interests of science or the general public." Concluding that "there is no real interest on the part of Celera in continuing to pursue this particular collaborative model," Collins offered Venter one week to resume negotiations. Failing that, "we will conclude that the initial proposal whereby the data from the public Human Genome Project and Celera are collaboratively merged is no longer workable."

On the eve of Collins's March 6 deadline, the Wellcome Trust released the letter to the media, presumably to apply pressure on Celera. The decision backfired, allowing Celera to denounce its competitors' underhand tactics. White angrily attacked the British charity's move as "slimy" and "dumb." Venter, by contrast, portrayed himself as the innocent victim, saying, "All I want to do is go home and take a shower." Even Collins, relaxing at his parents' home when he heard about the leaked letter, said he could understand Celera's feeling a little upset. However, he dismissed as "pretty fanciful" allegations that the release of the letter was timed to sabotage negotiations with Celera so that the Human Genome Project could support plans for a new public database along the lines of Celera's. Justin Gillis, writing in the *Washington Post,* summed up the ugly feud: "The Human Genome Project, supposedly one of mankind's noblest undertakings, is resembling a mud-wrestling match." Two days after the revelation of Collins's ultimatum, Celera publicly responded, stating that "we continue to be interested in pursuing good-faith discussions toward collaboration," provided that the company's commercial interests were protected. "For pure research applications, we foresee information being released at little or no cost to the end user. . . . Researchers would be free to use the published data in their research at no cost."

On March 9, as for hundreds of days before, yet another batch of human DNA sequence was deposited into GenBank, the public genome database. The significance of this particular date became clear only a week or so later, when Greg Schuler, of the NIH, ran a new piece of software he had written to check the sequence for redundancies. The subsequent calculations showed that March 9 was the date that the 2 billionth letter of DNA—for the record, a "T"—was sequenced. Most impressively, it had taken just four months for the public consortium to sequence the second billion letters of DNA, compared to four years to reach the 1 billion mark.

The long-awaited "rough draft" of the human genome was on course for a June 2000 release.

ON MARCH 14, 2000, President Bill Clinton and Prime Minister Tony Blair issued what at first glance appeared to be a rather innocuous 200-word statement about the Human Genome Project. The second half of the statement read as follows:

> To realize the full promise of this research, raw fundamental data on the human genome, including the human DNA sequence and its variations, should be made freely available to scientists everywhere. Unencumbered access to this information will promote discoveries that will reduce the burden of disease, improve health around the world, and enhance the quality of life for all humankind. Intellectual property protection for gene-based inventions will also play an important role in stimulating the development of important new health care products.
>
> We applaud the decision by scientists working on the Human Genome Project to release raw fundamental information about the human DNA sequence and its variants rapidly into the public domain, and we commend other scientists around the world to adopt this policy.

Although the statement did not specify any companies by name, it was clearly aimed at Celera and other companies, chiefly Human Genome Sciences and Incyte, which had been aggressively applying for human gene patents. But despite Clinton and Blair's reiteration of the rights of biotech and pharmaceutical companies to patent discoveries stemming from genomics research, the markets panicked, sending the technology sector of the U.S. stock market into a tailspin. Celera's share price plummeted from a previous high of $290 a share to $100, dragging the rest of the biotech sector down with it.

The timing of the statement, just one week after the leaked reports of the breakdown in negotiations between Celera and the NIH, was curious to say the least. Some White House advisers argued in favor of postponing the announcement, to no avail. In fact, the statement had been in the works for at least six months. In September 1999, Tony Blair had initiated discussions with the White House to ensure that all of the public genome project data be deposited in the public database within twenty-four hours. There had been deep concern, particularly at the Wellcome Trust, that the U.S. Department of Energy might strike a deal with Celera to assist the sequencing of the three human chromosomes it was responsible for. The British science minister, Lord Sainsbury, held talks with Neal Lane, the White House scientific adviser, to turn the Bermuda Accord on the prompt release of DNA sequence data into a formal international agreement. Lane acknowledged that "although the [Department of Energy] did have an earlier agreement with Celera, they have since withdrawn it and are working with the NIH and the Wellcome Trust as a group on any future industry agreements."

Celera and other biotech chiefs tried in vain to stem the damage. Venter penned an op-ed article for the *Wall Street Journal* under the heading "Clinton and Blair Shouldn't Destroy Our Research." Once again, he vowed to publish the human genome sequence when it was complete. "As an information company," Venter went on, "we aim to disseminate scientific information broadly in a user-friendly format [using] state-of-the-art software tools," comparing Celera to Bloomberg and Lexis-Nexis, two highly successful database companies. Bill Haseltine also weighed into the debate in the *Washington Post*. "We are not in the business of patenting humanity but rather of patenting genetic drugs," he wrote, contrasting unpatentable raw genome sequence data with the need for patent protection for gene-based inventions validated by experimentation.

Ironically, Haseltine had been thrust into the center of an ugly gene patent row just a few weeks earlier, after Human Genome Sciences was awarded a controversial patent for a gene called *CCR5*. In 1995, the company had filed patent applications on a family of 150 related genes (identified as ESTs) that encode cell surface receptors found on cells of the immune system. Although Haseltine suspected that these receptors might provide docking sites for various viruses, the applications were short on specifics, and no supporting data had been published. One of the applications was for a gene called *CCR5*.

A few months later, unknown to Haseltine, a Belgian group filed patents on the CCR5 receptor they had discovered on the surface of T cells, the natural targets of the AIDS virus. This filing specifically raised the potential relevance of CCR5 to AIDS, a suggestion that was confirmed experimentally in June 1996, when four groups independently showed that CCR5 was the portal HIV used to enter T cells. Moreover, gay men who surprisingly tested HIV negative were found to harbor variants in the CCR5 receptor that somehow blocked the viral gateway. However, the Human Genome Sciences application claimed precedence, even though one of the co-discoverers of CCR5 claimed that it contained a number of sequence errors. Haseltine acknowledged that this sequence—and that in most other patent applications—was not perfect.

The furor provided ammunition for those who argue that the current threshold for gene patents is too low. Venter draws a sharp distinction between the type of patents filed by Human Genome Sciences and Incyte on the one hand, and Celera on the other. Not only do his rivals file thousands of patents on poorly characterized genes, Venter alleges they also take advantage of the data released each day by the public genome project. Venter's rule of thumb is simple: "If patents have the word '-like' in the title, [the companies] got it from a quick computer search . . . even though someone else discovered it."

Despite the frank rebuttals from Venter, Haseltine, and others to the Clinton-Blair statement, the markets were singularly unimpressed. In the following two weeks, close to $50 billion was wiped off the valuation of the leading American companies. The British press endorsed the Anglo-U.S. statement, but cautioned that it would not be enough. "What lies ahead is a battle which will help define the 21st century," said the *Guardian* as it called for a "distinction between the discovery of genetic material and the inventions which arise from it. . . . Governments must ensure that the former belongs to the human race." Two leading academics promptly issued another Anglo-American broadside against rampant gene patenting by the biotech industry. Bruce Alberts, president of the National Academy of Sciences, and Sir Aaron Klug, president of the Royal Society of London, criticized the promiscuous patenting of gene sequences without a full understanding of their function and argued that the completion of the genome sequence was just the first step in understanding how thousands of genes act in concert to orchestrate the human body. "It is vital that all researchers have access to the full genome without charge or other impediment. The human genome itself must be freely available to all humankind."

ON MARCH 30, one week after the publication of the *Science* issue heralding the complete *Drosophila* genome, Venter hosted an open house at Celera for over 150 journalists, organized by the D.C. Science Writers

Association. Reporters were escorted around the facility, peering into sequencing labs and computer rooms. The laboratories where the DNA was purified were eerily quiet, with just a few young scientists in lab coats surveying equipment and tending to large trays of DNA samples. By contrast, around every corner there was a security guard discouraging inquisitive reporters from straying too far from the group.

In a makeshift conference room next to the cafeteria (Venter joked that his poor little start-up company could not afford an auditorium), Venter spent over an hour fielding questions. His first task was to clarify Celera's business model. "People think this is a biotech company," he told the assembled journalists. "It's not. It's an information company. We're in the same business most of you are in, and like you, we're still trying to get paid for it!" That revenue would not, he insisted, come from gene patents, although he acknowledged that Celera had filed 6,500 provisional patent applications. These are placeholders of sorts, affording the company twelve months from the date of the provisional application for it (or one of its clients) to file a patent. But Celera would file for gene patents only under stringent conditions, explained Venter. "We've set the bar with our customers that, unless it's something that they're putting into a drug development pathway, we're not going to file a patent on it."

Venter, relaxed and enjoying himself, even poked fun at his notoriously bad public reputation: "If you hunt the Internet on my name, you'll see that Celera supposedly has an agreement with the Department of Defence to make ethnic weapons against every group in the world—unless you're an American! There's a [web] site saying we have a secret pact with the Israelis to make anti-Arab weapons, and then there's a site saying we have a secret deal with the Arabs. . . . People have been asking us how we're going to make money. I guess it's through all these secret deals!"

The convivial mood was broken when a reporter from *Nature* asked Venter why he wanted to prevent academic scientists from annotating genes from Celera's database and sharing that annotation with other academics. Venter adroitly sidestepped the question by focusing on the *Drosophila* collaboration with the Berkeley genome center, but could not resist a sarcastic crack: "Maybe because it was published in *Science* and not *Nature*, you didn't read about the annotation jamboree that was here."

The most stunning sound byte of the evening came not from Venter but from a young Celera researcher outside the laboratory where the human DNA was sheared into millions of fragments prior to sequencing. He admitted that Celera had just finished sequencing the total genome of one individual, determining the order of letters an average of three times each. The human genome was history.

ON APRIL 6, VENTER ONCE AGAIN TESTIFIED before the House of Representatives Science Subcommittee on Energy and Environment, along with Neal Lane, Rubin, and Washington University's Bob Waterston. He had good reason to be in high spirits: one day earlier, President Clinton had clarified his previous statement on gene patenting. "Tony Blair and I crashed the markets for a day or two and I didn't mean to," said a humble Clinton during a White House conference on the new economy. "If someone discovers something that has a specific commercial application, they ought to be able to get a patent on it." Clinton was merely reaffirming what every analyst knew to be the case under American law, but somehow had been lost in the panic instilled by the Clinton-Blair announcement. Shares of Celera and the other genomics companies immediately jumped in relief.

With impeccable timing, just hours before Venter's testimony, Celera officially announced that it had sequenced the complete genome of an anonymous male individual, spelling the letters of roughly 30 million fragments an average of three times apiece. Within four to six weeks, the company predicted it would assemble these millions of fragments into the complete sequence of all twenty-four different chromosomes. Sequencing would also be completed for a female DNA sample.

The Celera chief's testimony that morning was vintage Venter—forceful, assertive, almost painfully direct. Recalling Maynard Olson's derisory challenge two years earlier before the same committee to "show me the data!" Venter declared, "He was wrong, and I am happy to again show the Subcommittee and the world the data." He reassured the committee that he would publish "the assembled, accurate, annotated [human] sequence, just as we did with *Drosophila.*" He testified: "One of Celera's founding principles is that we will release the entire consensus human genome sequence freely to researchers on Celera's Internet site when it is completed." There would be no restrictions on publication or patenting by third parties of these data, but Celera would not deposit all of its data in the public database GenBank, and risk allowing other companies to resell the complete contents.

By contrast, he raised concerns about the quality of the public consortium's sequence data: "I find myself in the peculiar position of warning you that in the race to complete a draft human sequence, the publicly funded human genome program may be at a stage where quality and scientific standards are sacrificed for credit." As for concerns that Celera would somehow monopolize its genomic treasury, Venter smoothly replied, "We don't have enough researchers in the world to analyze all the information that's just been generated in the past week, let alone the next decade." Venter also alluded to the patent controversy that had hidden his company's (and his personal) value so heavily. "We have to do a better job in communicating our business objectives to you and the public," he admitted. "As an information company, Celera is designed to assist researchers rather than focus on the development of new pharmaceuticals." This may change, however, as questions about the long-term profitability of this model continue to dog Celera.

Without Collins present to defend the NIH, the proceedings cast unnamed parties at the NIH in a grim light. One questioner raised the allegation that the NIH had been discouraging public universities from signing subscription agreements to Celera's databases, with possible consequences for future grant applications. "Several universities are concerned about signing up," Venter replied without naming names, "because there might be retaliation and loss of government funding." There was also controversy over plans for Celera to collaborate with the overshadowed Department of Energy's genome program. "We had a memorandum of understanding to work together to complete the three [Department of Energy] chromosomes," said Venter, but negotiations collapsed at the last minute. "I understand it was pressure from the NIH and the Wellcome Trust—they didn't want parts of the Government doing separate deals, so the memorandum of understanding did not get signed."

The House session was adjourned with California congressman Ken Calvert offering a tacit endorsement of Celera and the biotech industry in general: "Dr. Venter and others are responsible for speeding up the sequencing of the human genome by five years. For this reason at least, I would rather have the problems of private-sector involvement in the human genome field than not. Some problems are good to have, and I think this is one of them."

Interviewed on public television that evening, Venter tried again to clarify Celera's business model. "It's extraordinary," he said. "We're sequencing the genome for the world for free, not at taxpayer expense, and we're giving it to the world, because it's going to drive discovery so that people need our databases and our software and our computer capacity much more than they would now." Just as Galileo's telescope that "changed science and our view of the world," Venter said the new DNA sequencers developed by Michael Hunkapiller had given Celera a new tool set to understand the human genome.

Not surprisingly, leading figures in the public genome project poured scorn on Venter's unsubstantiated claims, complaining that they were doing a disservice to the whole field of genomics by hyping a minor milestone. Eric Lander said that a naive listener could be fooled into thinking that Celera had just sequenced the human genome, when in fact it had produced fewer data than the public sequencing consortium. Waterston estimated that Celera's modest coverage could leave as many as 40,000 gaps in the DNA sequence. Francis Collins also took issue with Celera's definition of the finish line, which had been revised downward from 10X (at the company's launch) to 4X (at the beginning of 2000) and possibly lower. During a speech at the Human Genome Organization conference in Vancouver a few days later, he warned, "You should not take at face value any claim by any group for at least two years that says we have finished sequencing a human genome sequence. It will not be true."

Collins's remarks had a chilling effect on Celera's vacillating valuation, deflating the stock price just as quickly as it had risen the previous week. It was as if Alan Greenspan himself had spoken.

WHILE CELERA FRANTICALLY SET ABOUT ASSEMBLING its human sequence, scientists with the public Human Genome Project announced a series of further sequencing milestones in advance of the "rough draft." In April, the Department of Energy announced that its Joint Genome Institute had completed the "rough draft" of the three human chromosomes it had been assigned: chromosomes 5, 16, and 19. It was a sweet moment for the agency that had started the Human Genome Project fifteen years earlier but had long since been overshadowed by the NIH program. Using automated sequencers produced by Pharmacia (rather than the PE model favored by most biotech and academic scientists), the Joint Genome Institute had sequenced about 300 million bases (compared to 33 million for chromosome 22 published four months earlier)—roughly 11 percent of the human genome—and identified between 10,000 and 15,000 genes.

Among them were likely to be genes that, when mutated, predispose to kidney disease, prostate and colon cancer, hypertension, and atherosclerosis. Several gaps remained in the sequence and would be filled in later.

Just two days before the May Cold Spring Harbor meeting on DNA sequencing, *Nature* released an electronic version of another milestone: the sequencing of 33,546,361 bases of the smallest human chromosome: chromosome 21. The sequence had been compiled primarily by a consortium of Japanese and German researchers. For all of James Watson's attacks on the cool Japanese reaction during the launch of the genome program, Japan had played a major role in the sequencing of the first two human chromosomes—this despite the fact that Shimizu's basement laboratory at Keio University in Tokyo, where much of the data were generated, was reminiscent of a Third World country.

News of the Japanese contribution to chromosome 21 was marked by widespread media coverage in Japan, but paled beside the combative tone of the Germans. The colorful language of the press announcement issued by the Deutsches Humangenomprojekt fanned the flames of tension between the rival parties:

Dear Ladies and Gentlemen,

During the last weeks, Mr. Craig J. Venter [sic] and his company Celera kept the public in suspense; fears came up, that one single person could secure the rights on man. But the assertion, that he had deciphered the human genome as a whole, has been wrong. What he holds in his hands, are millions of pieces, which still have to [be] put together by computer.

Therefore, the race for the deciphering of man has not yet been finished!

The public sequencing project, laughed at in this connection, is answering promptly. The second complete sequence being known after chromosome 22 is that of chromosome 21. . . . The four German Project Leaders participating in the consortium, Prof. Helmut Blöcker, Braunschweig, Prof. Hans Lehrach, Berlin, Prof. André Rosenthal, Jena and Dr. Marie-Laure Yaspo, Berlin, will present you chromosome 21.

The sequencing of chromosome 21 was a five-year marathon, beginning with a detailed physical map of the chromosome, ordering DNA fragments before they were sequenced. The result was the most accurate and extensive human DNA sequence released so far. Each DNA clone was sequenced at least eight times, for a final accuracy greater than 99.995 percent. Only three small gaps totaling about 100,000 bases remained, meaning coverage exceeded 99.7 percent. The DNA sequence from unrelated sources varied once per 700 letters on average.

Most scientists had predicted that chromosome 21 would contain about 1,000 genes. Incredibly, the final analysis revealed just one-quarter that number. Even including genes predicted *in silico* using various computer programs, the final catalogue contained just 225 genes, made up of 127 previously known genes and 98 predicted novel genes. Although a few genes have doubtless been missed—say, 10 percent—the paucity of genes, particularly near the pinched centromere of the chromosome, is striking. In one barren stretch dubbed the "black hole," there is a run of 7 million bases containing merely 7 genes. As one travels the length of the chromosome, the density of genes picks up. Immediately beyond the black hole is the chromosome's most famous single member. *APP* codes for the amyloid precursor protein, a large protein that is misprocessed in patients with Alzheimer's disease, resulting in the deposition of a small, internal fragment of *APP* in dense plaques in the brain, the pathological hallmark of senile dementia. In a few families, inherited *APP* mutations lead to an early-onset form of Alzheimer's disease. Farther down the chromosome are genes that, when mutated, cause amyotrophic lateral sclerosis (Lou Gehrig's disease), epilepsy, deafness, autoimmune disease, birth defects, and manic depression.

The biggest hope is that this new catalogue will provide vital new clues in understanding and potentially treating Down syndrome, the most common genetic form of mental retardation. Down syndrome, which affects 1 in 700 live births, is named after the English physician John Langdon Down, who recorded his observations of the mentally retarded in an asylum in an 1866 paper with the unfortunate title, "Observations on an Ethnic Classification of Idiots." Down believed that the disease somehow broke down barriers between races and proposed that it be classified into four subdivisions—caucasian, ethiopian, malaysian, and mongolian—and that somehow this provided evidence for the unity of the human species. Of these, mongolism was recognized as a distinct subgroup, although Down's son led the call for a more appropriate term for the disease.

In 1958, just two years after scientists finally determined that humans possess twenty-three pairs of chromosomes, Jérôme Lejeune showed that cells from children with Down syndrome contained a forty-seventh chromosome. (This chromosome was erroneously thought to be the next to smallest, hence was termed chromosome 21 rather than 22.) The presence of an extra chromosome, called trisomy, accounts for about 10 percent of spontaneous abortions. A third copy of a chromosome results in the serious overproduction of hundreds or thousands of proteins, destroying the intricate biochemical balance of the cell. The most common condition is trisomy 21, the risk of which increases dramatically with maternal age (particularly after age forty). Trisomies of chromosomes 13 and 18 are much less common, and affected babies usually die within a few months of birth. The relative health of individuals with trisomy 21 is likely due to the paucity of genes on the smallest human chromosome.

In addition to mental retardation, Down syndrome children suffer congenital heart disease, increased risk of leukemia, early-onset Alzheimer's disease, and scores of other developmental problems. It is not unreasonable to suspect that each of these traits might be attributable to the extra copy of a particular gene on chromosome 21. For example, an extra dose of a gene called *minibrain* (named after the corresponding fruit fly gene) might contribute to cognitive aspects of Down syndrome, whereas a third copy of the *APP* gene is suspected to be involved in the Alzheimer's symptoms.

Scientists have already created a mouse model of Down syndrome by breeding mice with an extra segment of mouse chromosome 16 (one of three mouse chromosomes that harbors homologues of the genes on human 21). With the full gene inventory of chromosome 21, scientists will be able to refine these animal models and begin attributing particular elements of Down syndrome to individual genes. The optimism of Down syndrome experts such as Julie Korenberg is understandable: "How many IQ points would it take to give large numbers of children with Down syndrome the ability to function in normal society? Not many."

WHILE VENTER DECIDED TO POSTPONE the announcement of Celera's first complete assembly of the human genome sequence, Celera faced some unwelcome distractions. A group of law firms filed a series of class action lawsuits against the company, alleging that customers who bought stock during Celera's secondary offering of more than 4 million shares in February 2000 at $225 each should have been informed about Celera's secret negotiations with public officials. From its peak just before the in-famous Clinton-Blair patent statement, Celera's stock price plummeted to a low of about $70 a share. But in April there was some good news, when Vanderbilt University became the first academic subscriber to Celera's human genome database. Each Vanderbilt researcher would pay $5,000 for universal access to Celera's data, or $1,000 for a three-hour session. One faculty member called it a "bold experiment," but doubted that Vanderbilt would benefit from its short-lived monopoly. The cost was too high and the information difficult to access or already in the public domain. Indeed, competition to Celera appeared from DoubleTwist, with backing from Sun Microsystems, offering a suite of software tools for researchers to navigate the public genome sequence data, identify genes, examine protein-protein interactions, and more. Lander called it "a real validation of the public sequencing project." DoubleTwist claimed to have found 65,000 genes and predicted a further 40,000, pushing the total beyond the 100,000 mark.

But what of the race itself? Even as the public Human Genome Project frantically prepared for the release of its rough draft, key members of the alliance sought to temper expectations by playing down the "race" angle. "The whole finish-line mentality is silly," said Lander, pointing out quite fairly that Celera would always have the advantage because it could access the publicly released sequence data. Besides, well over three-quarters of the human sequence was in GenBank, and researchers were already devising experiments to analyze the information without waiting for the full genome sequence. A better metaphor than a two-dimensional road race, said Harold Varmus, was that of charting a new continent.

The simmering tension and public bickering between the two parties threatened to erupt at any moment. *Nature* urged calm: "So far, good sense and manners have only occasionally been overtaken by bouts of claim and counter-claim, hype and counter-hype." But when that happened, the sequencing stakes threatened to degenerate into a farcical episode of the classic cartoon series *Wacky Races*—with no prizes for guessing who played Dick Dastardly. "Much is at stake," cautioned *Nature,* "not least the reputation of science as one of the few domains whose ultimate goal, for all its internal intrigues and personal competitiveness, remains the common good."

Similar sentiments were echoing around the corridors of the West Wing, where the chain of events that would ultimately restore some dignity to the Human Genome Project bore the presidential seal.

CHAPTER TEN

The Eighth Day

I not only think that we will tamper with Mother Nature,
I think Mother wants us to.
　　　　　　　　　　　—WILLARD GAYLIN

I T MIGHT BE EVIDENT by now that Craig Venter and Francis Collins do not see eye to eye on many things. But even before the human genome sequence was completed, the two men firmly agreed on one thing: the utterly profound impact the sequence will have on the practice of medicine.

> *Craig Venter:* Within 10 years, every baby born in a hospital in this country will have its complete genome repertoire determined. Their parents will have it on a DVD disk—or whatever the new media is at the time—before they leave the hospital. I think it will be fantastic to have my own genetic code, because every week, there are new articles in *Science* and *Nature* describing links between changes in the genetic code and disease. I would be logging onto the Celera database to try and understand those.

> *Francis Collins:* In 10 years, we should be able to make predictions for you and me for what conditions we're most likely to be at risk for, and that in itself would allow us to practice some preventive medicine strategies based on our own individualized risks. Give us 20 years, and I think you won't recognize medicine in the way the therapies are developed and applied.

For all the debates about whether the sequence of the human genome is more important than the splitting of the atom or landing on the moon, no previous technological accomplishment will have as profound an effect on human life. The explosion in genomic information fueled by the sequence will revolutionize the diagnosis and treatment of countless diseases. We are on the brink of a new era in which individuals will be diagnosed and treated based on their genome, not their symptoms. We already know of numerous examples of genes that influence our risk of developing cancer, heart disease, diabetes, and Alzheimer's disease. By 2010, if not sooner, this list will have grown dramatically to include asthma, migraine, blood pressure, and so on. As Collins states with total conviction, "There is no disease, except some cases of trauma, that don't have hereditary contributions. There's not one example I know of."

The genetic revolution hurtling toward us is the inevitable outcome of the Human Genome Project, which will deliver the sequence of every human gene. But it also relies on powerful new technologies such as DNA chips, which will enable the rapid screening of thousands of disease-related genes, perhaps, as Venter suggests, from the first week of life. This information will not only provide a comprehensive analysis of disease risks and life expectancy, but also enable doctors to customize drug treatment based on an individual's unique genetic profile, an exciting focus of the pharmaceutical industry called pharmacogenomics. In principle, these technologies could also be extended to the unborn using the technology of preimplantation genetic diagnosis, which allows doctors to screen the genetic makeup of embryos three days after in vitro fertilization.

The rapid advances in decoding the human genome and the tools for interpreting that information will forever change the practice of medicine. It will affect anyone on the planet with access to modern health care. In this chapter, we explore the brave new world that lies just beyond the genome sequence and the extraordinary consequences of this technology.

"WHEN THE WHOLE HUMAN GENOME is available on CD-ROMs," predicts Eric Lander, "we can begin to figure out how the 100,000 genes interact and talk to each other, how they respond to different diseases and developmental pressures." The systematic dissection of the human genome sequence in the next few years will rely heavily on new tools that allow the simultaneous analysis of tens of thousands of genes—a far cry from just five years ago, when the function of genes had to be analyzed individually. Advances in understanding the function of gene products will shed light on the chemical pathways that govern cell behavior and communication, as well as point to the ways these genes go awry in disease. The key to these revelations will be a device no bigger than a postage stamp: the powerful technology of DNA microarrays, popularly known as DNA chips.

The core of this revolution might be considered a nondescript building just off Route 1, south of San Francisco, where Stephen Fodor, the president of Affymetrix, donned in laboratory coat, mask, and gloves, inspects a black plastic container about the size of a microcassette, bearing the company logo. Sandwiched inside this container is the key to the future of genetic diagnosis and molecular biology research. It is a virgin microchip, no bigger than Fodor's thumbnail, but instead of silicon circuits, it contains millions of minuscule, custom-designed strands of DNA. These DNA chips have generated the biggest buzz in molecular biology circles since the advent of the polymerase chain reaction—the technique invented by Kary Mullis that amplifies minute traces of DNA—some fifteen years ago.

The Affymetrix chips resemble a microscopic chessboard containing not sixty-four but hundreds of thousands of squares. Each square contains a forest of specially designed DNA probes, just twenty or thirty letters in length, attached at one end to the slide. The sequence at each position of the grid is engineered to match part of a specific gene; each probe is synthesized one letter at a time using the same technique (photolithography) used in making semiconductors. A typical chip might contain 400,000 different DNA probes, with a row of perhaps 10 probes matching different regions of a single gene. Thus, one chip can interrogate the activity and sequence of about 40,000 genes simultaneously.

A second type of DNA chip involves spotting complete genes, or cDNAs, onto slides, rather than designing discrete DNA probes. This approach, pioneered by Patrick Brown and colleagues at Stanford University, is much more affordable, although there are fewer applications. Brown, who says a DNA chip is "like a microscope for watching a living genome in action," is pioneering a new era of genetic analysis in which it is possible to monitor the expression of thousands of genes simultaneously. For example, gene messages from a breast cell can be tagged with a fluorescent dye and passed over the chip. Each message sticks to the corresponding sequence on the chip; the more copies of the message, the brighter the fluorescence signal. Repeating the experiment with samples from a cancerous breast cell, tagged with a different dye, allows a comparison of the two signals and reveals whether specific genes are switched on (red signal) or shut down (green signal) in the tumor.

These molecular portraits—snapshots of the activity of thousands of genes in a tissue—are the future of genetic diagnosis and will enable doctors to make informed decisions about surgery or chemotherapy. They can help distinguish different classes of cancer, showing that each cancer is in fact a collection of several different diseases. For example, a survey of the activity of 8,000 genes in dozens of breast tumors by Brown and colleagues reveals striking differences in the expression of different clusters of genes. Thus, breast carcinomas can be classified as estrogen receptor–positive or –negative, with the latter subdivided into basal-like or Herceptin-positive tumors. Similar studies have shown that the most common form of non-Hodgkin's lymphoma is actually two different diseases. Future analyses will provide far more detailed definitions of these and other cancers.

The DNA chips made by Affymetrix and other companies can also be used to interrogate the precise sequence of specific genes to search for mutations. This is because even a single-base discrepancy between the sample and the probe on the chip will compromise binding and produce a telltale reduction in the strength of the signal. Chips can be made that contain hundreds of thousands of single nucleotide polymorphisms (SNPs), the variations at a certain base position that occur roughly once every thousand bases. There are on average four to eight SNPs in every gene, and about half of these will change the sequence, and potentially the function or activity, of that gene. Chips containing SNPs can already be used to identify thousands of critical genetic variations in an individual's DNA, paving the way for the routine testing of an individual's genetic makeup and personalized scorecards of disease risks and susceptibilities.

Twenty years from now—maybe sooner—a visit to the doctor's office will be as much about assessing your genetic constitution (genotype) as your physical constitution (phenotype). A few days before the appointment, you will mail a cheek swab that will be used to prepare your DNA. This sample will be analyzed using a DNA chip containing sequences of thousands of disease-related genes, providing a readout of your own sequence at thousands of critical regions in genes that scientists have tied to various diseases. You could learn—assuming you wished to find out—the likelihood of your developing Alzheimer's disease; diabetes; hypertension; breast, ovarian, or colon cancer; asthma; heart disease; as well as hundreds of classic genetic diseases such as cystic fibrosis, thalassemia, and Tay-Sachs. And in a decade or two, it is highly likely that we will have pinned down genetic factors that cause depression, attention-deficit disorder, addictive behavior, schizophrenia, and many other complex conditions.

The overpowering hope of the pharmaceutical industry is that this new technology will not only shed light on disease susceptibility, but will also provide clues to the best mode of treatment. This chromosomal crystal ball will indicate the best drugs for an individual based on the identity of SNPs in a cluster of key drug-responsive genes. This is called pharmacogenomics, one of the most expectant areas of current pharmaceutical research. Based on this information, your physician will be able to prescribe drugs not by evaluating your symptoms, but by scrutinizing your genetic code to determine which drug might be most efficacious for your specific condition. For example, a test for variants of the cytochrome P450 enzyme would reveal whether you are responsive to a major class of antidepressant drugs. Other tests could predict patients' responses to commonly prescribed drugs for asthma, hypertension, or migraine, and predict the risks of severe side effects, as seen a few years ago with deaths associated with the popular diet drug phen-fen.

Although Collins is not in the business of making rash predictions about people flashing DVDs with their genetic code in ten years, he does have a bold vision about the future of genetically based, individualized preventative medicine. By 2010, he predicts that tests for twenty-five common diseases will be widely available, allowing people to change their lifestyle accordingly—for example, by having regular colonoscopies should they be at risk of colorectal cancer or quitting smoking should they carry predisposing genes for lung cancer. By 2020, Collins predicts that gene-based designer drugs will be available for diabetes, hypertension, and many other common diseases. By 2030, scientists will have identified genes that control aging, and individual DNA sequencing on demand will cost less than $1,000. And by 2040, Collins foresees gene-based medicine being the norm, with most illnesses detected before symptoms appear and drug and gene therapies tailored to suit the individual.

There are hundreds of potential tests available, but we can already predict what some of the first commonly available tests will be. For heart disease, high blood pressure, diabetes, and Alzheimer's disease, there are growing lists of genes that are associated with Mendelian forms of the disease. These include familial hypercholesterolemia, Liddle's syndrome, maturity-onset diabetes of the young, and late-onset Alzheimer's disease, respectively. Common gene variations in these and many other genes will help determine a person's risk of these common illnesses. For example, 25 percent of the population carries one particular version of the *APOE* gene, *APOE4,* on chromosome 19 that increases risk of Alzheimer's disease about threefold; 1 percent of the population carries two E4 copies that increase the risk more than twenty-fold. Other tests for Parkinson's disease, venous thrombosis, osteoporosis, glaucoma, and many cancers, including breast, ovarian, colon, and lung, are available now or will be soon. In some cases, the importance of the test will vary depending on ethnic background. For example, Ashkenazi Jews have an increased likelihood of carrying certain mutated forms of the *APC* and *BRCA1* genes that predispose to colon and breast cancer. Tests for depression and schizophrenia may be available within ten to twenty years.

A decision to have a test will be influenced by the availability, if any, of a rational treatment. Some women diagnosed with a *BRCA1* mutation undergo a prophylactic mastectomy, and children harboring a faulty *APC* colorectal cancer gene often have their colons removed. A positive Parkinson's disease test might lead you to explore the possibility of a fetal tissue or xenotransplant. But a positive test for Alzheimer's or many other diseases is of limited use in the absence of an effective therapy.

This forecast of widespread testing raises major concerns about the right to genetic privacy. Health insurance companies would argue that it is unfair for healthy people to pay higher premiums to subsidize those people carrying a predisposing genetic flaw, but is it ethical to penalize people for their genetic heritage? Without effective legislation, the risk of discrimination in seeking health insurance and employment is very real. "All of us carry probably four or five really fouled-up genes and another couple of dozen that are not so great and place us at some risk for something," explains Francis Collins. "People don't get to pick their genes, so their genes shouldn't be held against them."

THE COMPLETION OF THE HUMAN GENOME SEQUENCE will provide a bonanza of opportunities for drug discovery. It may also prove to be the long-awaited stimulus for the troubled, but richly promising, field of gene therapy. No other area of research has seen such high expectations and bitter disappointments over the past decade than gene therapy. Indeed, it is bizarre that scientists should have succeeded in cloning a sheep before being able to claim a cure using gene therapy. But despite some tragic setbacks, there is a glimmer of hope on the horizon.

In September 1990, W. French Anderson, Michael Blaese, and Kenneth Culver at the NIH performed the first officially sanctioned trial for gene therapy on a four-year-old girl named Ashanthi DeSilva, who suffered from an extremely rare immunodeficiency called adenosine deaminase (ADA) deficiency—the "bubble boy" disease. (The disease is so rare scientists often quipped that there were more doctors studying it than patients.) ADA is an essential enzyme for an important component of the immune system called T cells. Unable to make this enzyme, Ashanthi was dependent from two years of age on expensive supplements of pure ADA protein. In Anderson's historic trial, deactivated viruses containing the healthy ADA gene were coaxed into about a billion purified T cells and then reinjected into the child. For the next two years, Ashanthi and a second patient, Cindy Cutshall, received continued treatment. As a precaution, Anderson's team continued to administer low doses of the ADA protein. Five years later, the NIH team concluded that although some of the transplanted cells were actively producing ADA, the therapy had not led to permanent cure for Ashanthi or any other patient. Italian researchers reached a similar conclusion based on their own trial.

Another notable trial was conducted by James Wilson, director of the University of Pennsylvania's Institute for Human Gene Therapy. Wilson devised a radical ex vivo therapy to treat patients with familial hypercholesterolemia, an inherited form of heart disease caused by mutations in the gene for the LDL receptor, which removes cholesterol from the circulation. This disease can lead to premature death due to massively increased levels of cholesterol. Wilson's trial involved resecting about 15 percent of the patient's liver, introducing the healthy LDL receptor gene into the liver cells, and injecting the treated cells into the patient's hepatic portal vein. Early indicators were promising—the first report of the procedure even dared to use the word *successful* in the title—but there was no lasting benefit.

Despite these setbacks, researchers pressed on with new genes to treat new diseases, convinced that eventually something would work. During the 1990s, some 3,000 patients received genetic therapy for cystic fibrosis, hemophilia, muscular dystrophy, AIDS, cancer, and other disorders. But time and again, the treatment offered only transient relief. Many problems were caused by natural immune reactions to the viruses used to deliver the therapeutic genes, particularly adenoviruses, which naturally cause respiratory infections. The body has no way to distinguish the cargo of a potentially helpful virus from the more pernicious variety. Inflammation, swelling, fever, and other symptoms inevitably occurred, sometimes requiring the trials to be halted.

In September 1999, tragedy struck. Jesse Gelsinger, an eighteen-year-old boy from Arizona, flew to the University of Pennsylvania Medical Center to volunteer for gene therapy for a rare genetic disorder, ornithine transcarbamylase (OTC) deficiency, characterized by the buildup of toxic ammonia in the body. Wilson's team administered millions of viral particles containing the normal OTC gene. Gelsinger developed a severe inflammatory reaction leading to multiorgan failure and died a short time later, on September 17. Wilson called the event "tragic" and "unexpected," while Gelsinger's dad mourned the loss of "a hero." Collins, an early advocate of gene therapy, spoke for many when he said, "This tragedy has rocked the field down to its toes. It is sobering to consider that this approach not only hasn't led to cures in the past 10 years, but is actually capable of harm." Several other deaths in patients participating in gene therapy trials have subsequently come to the light, although it is not possible to say definitively that the two were linked.

Just as the field of gene therapy had reached rock bottom, word arrived of the first tangible success for this heavily touted field, reported by a French research team led by Alain Fischer. The disease in question is severe combined immunodeficiency-X1, an X-linked disorder similar to that affecting Ashanthi DeSilva, characterized by a block in the development of T cells. Three infants, all less than one year of age, were treated using an ex vivo approach, in which extracted bone marrow cells are corrected with the functional gene and infused back into the patients. Ten months after treatment, the patients' immune systems were functioning normally, and the children were home, free from medication, indicating that the gene therapy had worked.

The apparent success of the French experiment not withstanding, researchers are eagerly exploring a number of alternative strategies for delivering therapeutic genes. One particularly promising vehicle is a heavily modified version of the AIDS virus championed by Inder Verma at the Salk Institute. Verma has engineered this virus so that it is capable of infecting a broad range of human cells while removing the genes that make it pathogenic. The result is a virus that "is easy to make, that delivers genes at very high efficiency, that can infect a nondividing cell and that enables its therapeutic gene to become part and parcel of the chromosome."

Other stealth delivery strategies dispense with viruses altogether, using liposomes, spheres of fatty molecules, to house the therapeutic genes, or simply naked DNA. Jeffrey Isner has used purified copies of the *VEGF* gene to stimulate the growth of new blood vessels in patients with clogged blood vessels in the legs and heart. Also on the horizon are human artificial chromosomes. These are larger custom-designed DNA fragments, designed to mimic normal human chromosomes. They are stable in the cell nucleus, can carry any desired gene, and have none of the side effects associated with viruses.

The strategies described so far use different vehicles, but all seek to introduce a healthy gene to compensate for a faulty gene. Ideally, scientists would prefer to fix the inherited flaw in the relevant gene in the nucleus directly. Kimeragen, a biotech company founded by Penn researcher Eric Kmiec, thinks it has found a way to do just that. Instead of "remodeling the whole kitchen to repair a leaky faucet," Kmiec is pursuing a technique called chimeraplasty, which involves a specially designed patch of DNA to repair a stretch of DNA sequence containing a mutation. The idea is to introduce a small segment of DNA that will align itself with the mutant gene. The patch sequence matches the gene precisely, except that it contains the normal sequence instead of the mutation. Because the two sequences differ, a distortion in the double helix results that attracts the attention of the normal DNA repair machinery, such that the correct base is stitched into the gene.

Chimeraplasty has not yet worked in humans, but studies in animal models have been encouraging. Clinical trials are being considered for Mennonite children suffering from Crigler-Najjar's syndrome, a recessive disorder in which the body fails to metabolize bilirubin, a by-product of red blood cell recycling, which eventually leads to liver failure. This standard treatment is phototherapy, which requires patients to sleep on tanning beds under bright blue lights to prevent serious jaundice. The prospect of this therapy offers a lifeline to the affected families and the chance to give something back to the Amish and Mennonite communities of Pennsylvania for decades of invaluable service to genetics research.

The concept of gene therapy is so inherently simple that it is hard to believe that it will thwart researchers much longer. Ten or twenty years from now, gene therapy may be a viable option for dozens of genetic diseases. If the technology does become successful, there will be those who advocate using gene therapy to modify genes in the germline—sperm and egg cells—so the errant gene can be prevented from being passed down to future generations. Some scientists would go even further: they harbor dreams of enhancing memory or postponing aging.

"Dare we be entrusted with improving upon the results of the several million years of Darwinian natural selection?" asks James Watson. "Are human germ cells Rubicons that geneticists may never cross?" Watson emphatically answers his own rhetorical questions in the affirmative. "We all know how imperfect we are. Why not make ourselves a little better?" One reason, warns Lander, is the dire possibility of something going awry: "The prospect of a 'product recall' from the human gene pool is too surreal to contemplate." Another reason is that we will never know what we might miss. Some of the most famous figures in history suffered serious genetic diseases: Abraham Lincoln had Marfan's syndrome, Van Gogh epilepsy, Albert Einstein dyslexia, Lou Gehrig and Stephen Hawking amyotrophic lateral sclerosis, Ray Charles glaucoma. Manic depression is a debilitating disease that is highly overrepresented among artists and writers. An appreciation of the rich diversity of human life is in order.

GERMLINE GENE THERAPY is many years away, but another genetic screening technology is already having a major impact on the makeup of the next generation. One of the first appointments named by Francis Collins to the newly formed genome center on the NIH campus in 1994 was Mark Hughes, a precocious young geneticist from the Baylor College of Medicine. Hughes had been sold on the potential of an in vitro fertilization technique from the moment he heard about the pioneering experiments of Alan Handyside and Robert Winston at London's Hammersmith Hospital in 1990, twelve years after the birth of Louise Brown, the first test tube baby. The case involved a couple with a family history of an X-linked disorder, meaning that a son would have a one-in-two chance of developing the disease. Rather than contemplate an abortion should the fetus be affected, the family turned to preimplantation genetic diagnosis, which for the first time, would afford a couple an element of control over the fate of their newborn.

Preimplantation genetic diagnosis works like this. Somewhere from ten to twenty or more eggs are harvested from ovarian follicles using ultrasound before being fertilized in a petri dish. (Intracytoplasmic sperm injection can be used to increase the chances of fertilization.) The resulting fertilized embryos are incubated for two to three days, during which time they divide three times until they are microscopic bundles of about eight cells. At this stage, a small hole is drilled into the outer zona pellucida and one or two cells, or blastomeres, are delicately teased away from each embryo with a fine glass needle. (This does not harm the embryo, because the cells have not yet begun to specialize.) DNA from the blastomere is then amplified using the polymerase chain reaction to provide enough starting material to test for the presence of the diagnostic gene. After the biopsy results, healthy embryos—usually no more than three at a time— are implanted into the mother's uterus.

In the first case in 1990, Handyside's team determined the sex of the in vitro fertilized embryo by looking for DNA from the Y chromosome. Because X-linked disorders affect only males, those embryos that tested negative (that is, were female) could be reimplanted to guarantee the couple an unaffected baby girl. Hughes began a transatlantic collaboration with Handyside to develop a range of tests for cystic fibrosis and several other recessive, dominant, and X-linked diseases. Not only could couples be assured that only healthy pregnancies would result, but they would also be spared the agonizing decision of a potential abortion.

When he joined the NIH, Hughes set up another laboratory at a private hospital so that he could continue his human embryo research without violating any federal sanctions. But two years later, Hughes ran afoul of those government restrictions by transferring NIH equipment to his hospital laboratory and allowing NIH-funded colleagues to perform testing. He was fired in October 1996 for what Harold Varmus later termed the "surreptitious pursuit of prohibited research." The NIH maintained that Hughes was told that all embryo research, including that on single cells, was not permitted; Hughes says he thought that single cell DNA analysis was allowed. Subsequent investigations revealed that in vitro fertilization tests were also being performed in Hughes's NIH lab, including one that led to the reimplantation of an embryo that had mistakenly tested negative for cystic fibrosis.

Hughes eventually took a new position at Wayne State University in Detroit, where his services are in more demand than ever before. He offers screening for more than two dozen genetic disorders and has successfully arranged the births of twenty-five babies spared the Huntington's gene. In 1999, he collaborated with researchers at New York's Cornell University to perform the first successful preimplantation diagnosis for sickle cell anemia. His team harvested sixteen eggs and fertilized eight, seven of which began cell division. Four of the blastomeres tested negative for the sickle cell mutation, but only two embryos were sufficiently healthy for implantation. A third embryo carrying one copy of the sickle cell gene was also implanted, because the mother was willing to accept a healthy carrier, but this embryo did not implant. In the spring of 1998, healthy fraternal twin girls were born.

Despite a $20,000 price tag in the United States, preimplantation genetic diagnosis is rapidly growing in popularity, resulting in several thousand births worldwide over the past decade. But as Hughes and other researchers offer hope to more and more couples, he increasingly hears the charge that he is helping give rise to "designer babies." For the past few years, in addition to selecting embryos free from a disease gene, Hughes has added another test to his repertoire: one that offers a lifeline for diseased children already born. For couples who already have a child with a serious genetic illness, preimplantation genetic diagnosis offers the possibility of sparing future children the same fate. But while selecting embryos that lack the disease genes, Hughes can also pick an embryo with a compatible tissue type, so as to provide a potentially life-saving bone marrow transplant for the affected sibling. The decision to have a baby in order to save an existing child's life is highly controversial. But persuaded by the desperation of the parents he meets, Hughes has made this service available for diseases such as severe combined immunodeficiency syndrome and beta-thalassemia, on condition that the parents genuinely want a second child and that the tissue typing test is secondary to the principal genetic screen.

As preimplantation genetic diagnosis becomes more widely available and all human genes are identified, the demand for testing inevitably will extend beyond genes that cause simple inherited diseases. A growing number of genes contribute a high, but by no means certain, risk of disease. The Hammersmith group has conducted tests for a colon cancer gene, which confers a 90 percent chance of cancer, and other clinics offer tests for the *BRCA1* breast cancer gene, although many carriers will still have a long healthy life. The growing number of tests is widely welcomed, although Hughes cautions that "the technology has got a long, long way to go" before screening for polygenic disorders becomes a reality. This is absolutely correct. Fears that the new technology will usher in a brave new world of designer babies are wholly unwarranted for many reasons, at least for the time being. But if demand were sufficient in twenty or fifty or one hundred years from now, clinics could expand the repertoire of genetic screening services to offer wealthy couples the chance not only to select the healthiest embryo, but also one that carries the desired behavioral and/or physical characteristics.

There should be no doubt that people would clamor to take advantage of this chance to maximize their children's genetic potential if the day comes. From prayer to private education, psychoanalysis to plastic surgery, people will do almost anything to maximize the chances of success in life for themselves and their children. The concept of Nobel sperm banks and fashion models selling their eggs over the Internet may seem ludicrous, but private fertility clinics already offer sperm sex selection to aid in "family balancing." James Watson wrote, "If we could honestly promise young couples that we knew how to give them offspring with superior character, why should we assume they would decline? If scientists find ways to greatly improve human capabilities, there will be no stopping the public from happily seizing them."

A few years ago, before becoming the toast of Hollywood for writing *The Truman Show*, New Zealander Andrew Niccol conceived and directed another film set in the "not-too-distant-future" called *GATTACA* (the title is derived from the four letters of DNA). The film depicts a society in which genetically gifted individuals are created by eliminating disease genes and enhancing behavioral and physical characteristics, and DNA dispensaries offer complete sequence profiles based on hair or saliva specimens. Niccol's direction was rather pedestrian, and the casting of Uma Thurman as a character with a minor heart condition was curious ("I mean, if she's not a perfect genetic specimen, then what hope is there for the rest of us?" said Niccol). However, Niccol's vision of the "not-too-distant-future" was in many respects uncomfortably plausible.

In a scene in *GATTACA,* a young couple at a fertility clinic must select one of their four genetically typed embryos, which have been vetted for traits such as myopia and obesity. A boy would be nice to provide a playmate for their young son, but (in a scene deleted from the final film) they would also like to ensure that they can have grandchildren. "I've already taken care of that," the doctor nonchalantly replies. The couple is thus assured of having a healthy, heterosexual son, but cannot afford the optional extras—genes for heightened musical or mathematical ability. The film also raises the possibility of tampering with the human anatomy, whether by adding an extra finger for a concert pianist or enhancing other appendages ("Beautiful piece of equipment—I don't know why my folks didn't order one like that for me!") In *Remaking Eden*, Princeton

University geneticist Lee Silver goes even further, speculating that 1,000 years from now, the human race may have split into two separate species, the GenRich and the Naturals, unable to interbreed.

But let us return to the immediate future. Several technological developments would be necessary before an era of embryo selection becomes feasible, but none is beyond the realm of possibility. Methods will be developed to increase the number of mature eggs that can be obtained from the mother, thereby offering couples the potential to compare characteristics of dozens of different embryos rather than a few. DNA chips already provide the means to assay thousands of gene variations simultaneously, producing an instant profile of disease risks and life expectancy. The identity of those variations is rapidly becoming known through the analysis of the human genome sequence.

But the biggest question surrounding the future use of genetic diagnostic technology remains the classic conundrum: How much do genes shape human behavior and personality?

THE NINETIES SAW ENORMOUS PROGRESS in fleshing out examples for the "one gene, one disorder" paradigm (known in some circles as OGOD). To be sure, the trend went too far at times, giving the mistaken impression that genes can explain every quirk of human character. A good example came in 1999. A particularly provocative cover of *Time* featured an infant clutching a model of the double helix, under the teasing headline, "The I.Q. Gene?" The magazine reported that Joe Tsien, a professor at Princeton University, had engineered a strain of superintelligent mice simply by inserting a few extra copies of a single gene that codes for a protein with an important role in neurotransmission in the brain. Using the rodent equivalent of the Rorschach test—finding hidden platforms in water tanks, sniffing out unfamiliar objects, and so on—Tsien recognized that his supermice could store and recall information far better than their hapless littermates. The supermouse strain was nicknamed "Doogie," after the precocious teenage title character of the American television series *Doogie Howser, M.D.*, and was a big hit at Tsien's son's "show-and-tell" day.

Stephen Jay Gould and many others sharply criticized the notion of a single gene's controlling IQ. And so the Doogie mice became one more pawn in the fractious—not to mention tedious—debate between those who argue that behavior is genetically hardwired and the "not-in-our-genes" crowd, which insists that human cognitive development is the product of nurture, not nature. This fracas has been simmering for years, fueled by a procession of gaudy media headlines invoking genes for almost every facet of human development and behavior.

The truth is that all diseases have some genetic component. For example, our susceptibility to infectious diseases, including AIDS, tuberculosis, and malaria, is strongly influenced by the makeup of our own genes, as is our risk of developing cancer. In fact, we are so acclimated to linking cancer with genes that it is surprising to be reminded that cancer is not entirely determined by heredity. This was dramatically borne out when the results of the largest twin study of cancer on record were published in the *New England Journal of Medicine*. The results, surprising to some, showed that environmental factors were more important overall than genetic factors. The study, conducted at the Karolinska Institute in Stockholm, examined government records of the incidence of twenty-eight different cancers in almost 90,000 twins. Some of the twins were identical, possessing the exact same DNA, whereas others were nonidentical, sharing 50 percent of their DNA and thus no more similar to their twin as any other sibling. The results were striking: environmental factors such as cigarettes, pollution, diet, and lifestyle caused twice as much cancer as hereditary factors. The most strongly inherited cancers were those of the prostate (42 percent), colon (35 percent), and breast (27 percent). Perhaps the most remarkable finding was that if one identical twin developed cancer, the other one remained disease free 90 percent of the time.

In response, Collins and others reiterated that cancer is a genetic disease—a disease caused by damage to genes—even if inherited predisposition to cancer accounts for only a minority of cases. Identifying the key genes that safeguard the stately growth and division of cells will be the route to curing cancer. Indeed, examination of genes that have been tied to various cancers over the past twenty years suggests that they can be classified into six classes, neatly defining the key stages involved in tumorigenesis. They are (not necessarily in this order): the activation of an oncogene, the loss of a tumor suppressor gene, metastasis and tissue invasion, cell immortality, the growth of new blood vessels, and the evasion of programmed cell death.

Assessing the relative role of genes and environment—whether home, school, or the womb—in sculpting human behavior and personality is extraordinarily difficult and controversial. Twin studies routinely point to a roughly equal contribution between genes and environment for numerous traits, but a recent case strikingly showed that environment alone cannot shape human behavior. In *As Nature Made Him*, John Colapinto describes the tragic case of a young boy who, after a botched circumcision, was raised as a girl. The boy endured a troubled childhood until he was finally told the truth as a teenager and could begin to live a natural life as a male.

Most geneticists recognize the absurdity of overemphasizing the role of specific genes in shaping human nature. One of them is Jane Gitschier, a charming geneticist at the University of California, San Francisco. Recently, while searching for the gene that causes a disease involving the transport of metal ions in the body, she built a shrine in her laboratory dedicated to Hephaestus, the Greek god of metallurgy and mining. Entirely by coincidence, her search hit pay dirt a few weeks later, and the gene was appropriately named Hephaestin. An earlier example of her handiwork adorned the walls of many geneticists' offices, including that of Francis Collins. She drew up a cartoon map of the Y chromosome, depicting the locations of various genes that control male behavioral traits. The list included instantly recognizable behaviors such as "channel surfing," "total lack of recall for dates," "inability to ask for directions," and "air guitar" (in older men this is usually manifested as "air violin").

Lately, however, Gitschier's passing interest in behavior has influenced her own research. A talented vocalist, Gitschier is fascinated by the precious gift possessed by some musicians: perfect pitch (also known as absolute pitch), that is, the ability to identify the pitch of a note without any reference. Using a rigorous test she has devised to assess whether individuals possess perfect pitch, she has demonstrated that perfect pitch is a highly hereditary trait—provided that individuals who inherit the putative gene receive musical training before the age of six. She is currently gathering families in the hope of mapping and eventually identifying the putative gene.

The notion that genes can influence human behavior and personality is a powerfully seductive proposition. In the past five years or so, there has been a steady rise in evidence that many human behavioral traits are at least partially influenced by variations in our DNA, with several provocative studies suggesting that complex human behavior can be shaped by alterations in a single gene. The most vivid example came a few years ago, when the behavior of a highly antisocial Dutch family, in which men had regularly committed rape and arson, was attributed to a mutation in the gene for an enzyme in the brain called monoamine oxidase. The mutation is extremely rare, but its significance has been supported by studies of mice lacking the corresponding gene.

But can genes shape more common, intangible variations in human behavior? Witness the media blitz that greeted a report by the respected University of Minnesota researchers David Lykken and Auke Tellegen, part of the twenty-year-old Minnesota Twin study, in an obscure psychology journal. The authors set out the premise as follows: "Are those people who go to work in suits happier and more fulfilled than those who go in overalls? Do people higher on the socioeconomic ladder enjoy life more than those lower down? Can money buy happiness? As a consequence of racism and relative poverty, are black Americans less contented on average than white Americans?"

Lykken and Tellegen asked more than 1,300 pairs of twins enrolled in the Minnesota Twin Registry to complete a questionnaire assessing their contentment and well-being in life and compared the results against various parameters, including income, marital status, and education level. Identical twins consistently demonstrated a higher correlation for happiness than fraternal twins, whereas variables such as religion, age, and education have a negligible effect. The conclusion was that just as body weight is dictated by a metabolic "set point," one's natural sense of contentment may be controlled by a heritable set point for happiness. According to Lykken, "About half of your sense of well-being is determined by your set point, which is from the genetic lottery, and the other half from the sorrows and pleasures of the last hours, days or weeks." The authors also found evidence for a significant genetic contribution to well-being over five or ten years. National Cancer Institute geneticist Dean Hamer suggests, "How you feel right now is about equally genetic and circumstantial, but how you will feel on average over the next ten years is fully 80 percent because of your genes."

The Minnesota authors drolly conclude, "It may be that trying to be happier is as futile as trying to be taller and therefore is counterproductive." The intense emotions associated with losing a job or winning the lottery will inevitably wane within a short time as your "happiness meter" reverts to its set point. Lykken tries to nudge his own set point higher by savoring life's simple pleasures, such as gardening or a fine glass of wine, contentedly following the advice of the English poet William Cowper, who wrote: "Happiness depends, as Nature shows, less on exterior things than most suppose."

If we indulge the happiness gene concept for a while longer, then one of the prime candidates to mediate such an effect is the neurotransmitter dopamine, the brain's "pleasure" chemical that is released after a pleasurable meal or experience. The release and turnover of dopamine and its fellow neurotransmitter serotonin are prime candidates to mediate many potential genetic influences on behavior, at least according to findings from Dean Hamer and others. Hamer made international headlines in 1993 when his study of homosexual brothers found evidence for a gene on the long arm of the X chromosome that helps determine sexual orientation. Gay activists paraded in "Thanks, Mum" T shirts, paying tribute to the importance of their maternal X chromosome in determining their sexual orientation. Although some studies have subsequently diminished the case for a "gay gene," Hamer's research lent a measure of credibility and excitement to the previously fallow field of psychiatric genetics.

In 1996, Hamer's group confirmed findings of an Israeli study that correlated variation in the length of the gene for the dopamine D4 receptor with a well-characterized psychological trait called novelty seeking—a measure of extrovert, thrill-seeking behavior. But with the ensuing hype attached to these results, the small print in Hamer's *Nature Genetics* study was unfortunately overlooked. This was far from the gene for bungee jumping, as some newspapers reported; the D4 receptor variation could explain at best only a small percentage of the variation in novelty-seeking behavior, which was itself only 50 percent hereditary. There could be ten or a hundred other genes contributing to the hereditary variation. Perhaps not surprising, such a slim effect has meant that the findings have not been widely replicated.

By contrast, serotonin is the brain's "punishment" chemical, and changes in the way this neurotransmitter is released and recycled in the brain have important effects, as seen by the effect of the popular antidepression drug Prozac, which acts on the serotonin transporter. Naturally occurring variations in a repetitive DNA sequence close to the serotonin transporter gene alter the expression of the gene, which may in turn influence serotonin levels. Individuals with the shorter form of the gene produce less of the transporter and consequently higher levels of serotonin. Interestingly, this version of the gene correlates with harm avoidance and anxiety.

In the years leading up to the completion of the genome sequence, Hamer and other psychiatric geneticists have had little choice but to study the effects of a handful of plausible candidate genes. Genes that influence dopamine and serotonin levels certainly fit into this category, but there has been a temptation to extrapolate even further. Hamer argues that one positive benefit of having the "high-anxiety" form of the serotonin transporter gene is more sex. Furthermore, he suggests that the D4 dopamine receptor "thrill-seeking" gene indirectly affects the number of sexual partners, leading to the inane proposal of a "promiscuity gene." This trend has led to other outrageous claims—the "infidelity" gene, the "Viagra" gene, even the "religiosity" gene—all of which are unlikely to prove correct.

On the other hand, there is good reason to believe that despite the furor that accompanied publication of *The Bell Curve* in 1994, general cognitive ability or "general intelligence," known as g, is a heritable behavioral trait. "g is one of the most reliable and valid measures in the behavioral domain," says Robert Plomin of the Institute of Psychiatry in London. "Its long-term stability after childhood is greater than for any other behavioural trait, and it predicts important social outcomes such as educational and occupational levels far better than any other trait." g is a measure of the core value of a diverse range of cognitive tests, including verbal ability, spatial ability, and memory. It is not synonymous with intelligence, merely one definition of it. But Plomin's group and others are conducting widespread searches for genes that influence g and believe that some of these factors will be identified in the coming years. Plomin plays down the notion that screens for g could be conducted using methods such as preimplantation genetic diagnosis, although for pragmatic reasons such as too few embryos and too many other priorities to screen. But postnatal testing might be an attractive option, he says, particularly if simple dietary measures (such as are used now for cases of phenylketonuria) were found to be beneficial.

Many will be uncomfortable with the notion that genes influence intelligence, but researchers have solid evidence that deletions of many genes cause varying levels of mental retardation, which affects about 3 percent of the population. Genes are also being sought for language development, reading ability, speech disorders, and countless other traits. Patients with Williams's syndrome are missing several genes on chromosome 5, and while they have poor reading and writing skills, they are unusually loquacious, with remarkable verbal and musical abilities. One of the genes implicated in this disorder has been tied to cognitive function. Studies of patients with Turner's syndrome, who lack a portion of the X chromosome, may have revealed the location of a gene for female intuition, perhaps explaining in part why boys are more vulnerable to developmental disorders such as autism than girls.

WHILE THERE IS GREAT INTEREST in genetic influences on behavior, genes clearly play a role in physical development. A similar list of human genes—some real, some hypothetical—could be produced for physical attributes. Genes clearly affect stature and musculature, and may play a significant role in athletic ability. Take speed, for example. The spectacular success of athletes of African ancestry in modern athletics has been well documented. West African athletes excel in sprint events, whereas East and North African runners are almost invincible in marathons and endurance events. The manager of the Arsenal soccer club in London, Arsene Wenger, who has invested heavily in players of African descent, is no expert in genetics, but his thoughts are revealing: "I think black sportsmen have a certain advantage, and in [soccer] it shows itself in explosive speed. You can't contradict that . . . and that bit is genetic. The rest is culture and education. But the genetic bit can't be added. Not yet, anyway!"

At the University of London, David Hopkinson's group is measuring the heritability of facial features, with a view to seeking genes that influence the development of facial characteristics. Perhaps one day parents will be able to put a computer-generated face to the embryos they are selecting.

The sequencing of the human genome will finally furnish researchers with the tools to find definitive answers to the perennial question: Are complex human traits governed primarily by nature or by nurture? Despite the bleak scenarios depicted in GATTACA, the movie's subtitle—"There is no gene for the human spirit"—was somehow reassuring. Whether that will remain the case as the sequence is analyzed remains to be seen.

CHAPTER ELEVEN

The Language of God

The Book of Life begins with a man and a woman in a garden. It ends with Revelations.

—OSCAR WILDE

ON MONDAY, JUNE 26, 2000, President Clinton strode into the East Room of the White House, followed closely by two proud men: Craig Venter and Francis Collins. The hastily arranged occasion was to mark the joint announcement of the completion of the rough draft of the public Human Genome Project and Celera's "first assembly." Among the invited scientists, government officials, and media in the audience, pride of place went to James Watson, who as a scientist and statesman was most responsible for instigating the Human Genome Project.

Speaking live via satellite from London, British prime minister Tony Blair hailed the breakthrough as "the first great technological triumph of the 21st century." President Clinton congratulated the leaders of the public and private genome efforts. His jaw clenching characteristically, the president opted for theological imagery: "Today we are learning the language in which God created life. We are gaining ever more awe for the complexity, the beauty, the wonder of God's most divine and sacred gift." Alluding to the 99.9 percent similarity of everyone's genetic code, Clinton added, "Modern science has confirmed what we first learned from ancient faiths. The most important fact of life on this earth is our common humanity."

Collins, following the president, warmed to the biblical theme. "It is humbling for me and awe inspiring," he said, "to realize that we have caught the first glimpse of our own instruction book, previously known only to God." For Collins, however, this triumphant moment he had dreamed of for seven years was tinged with sadness. Less than twenty-four hours earlier, he had attended the funeral of his sister-in-law, a marionette puppeteer, who had died of breast cancer. Sadly, he said, the wonders of the genome achievement could not come in time for her. As for the competition with Celera, Collins had crafted the perfect response. "I am happy that today," he concluded, "the only race we are talking about is the human race."

Unlike the previous two speakers, Venter opted for a secular rather than spiritual approach. He described the achievement as "an historic point in the 100,000-year record of humanity," and he thanked the president and the U.S. Congress for their generous support of biomedical research. Of course, he also gave credit to Celera's parent company, PE Corporation, and its $1 billion investment. Venter predicted a plethora of practical applications from the genome sequence, including the hope that cancer deaths might be reduced to zero in our lifetime. He closed on a philosophical note:

> Some have said that sequencing the human genome will diminish humanity by taking the mystery out of life. Poets have argued that genome sequencing is an example of sterilizing reductionism that will rob them of their inspiration. Nothing could be further from the truth. The complexities and wonder of how the inanimate chemicals that are our genetic code give rise to the imponderables of the human spirit should keep poets and philosophers inspired for millennia.

Press conferences were held on both sides of the Atlantic. The conference in London had a backslapping atmosphere reminiscent of an awards ceremony, but some of the British scientists still clung to the rhetoric of competition. Said the Sanger Centre's Michael Stratton, discoverer of the *BRCA2* breast cancer gene, "Today is the day that we hand over the gift of the human genome to the public. It is very fragile and beautiful and a powerful force for great good or evil."

Following the White House ceremony, the teams of senior scientists from Celera and the public Human Genome Project gathered at the nearby Capital Hilton hotel for a huge press conference. Almost all traces of the vituperative rivalry between Venter and Collins were gone. In its place, there was a relaxed camaraderie between the two men and their groups, with Venter in particular relishing the opportunity to fire a few good-natured barbs at the public group leaders sitting across the stage from his Celera coworkers. At one point, Venter apologized for the absence of his PE colleague Michael Hunkapiller, the driving force behind the automated DNA sequencer, who had contracted chicken pox. "I told him if he came he had to sit on the public side!" quipped Venter, drawing laughter from both groups.

It was fitting that the Washington press conference was chaired by Aristides Patrinos, the director of the Department of Energy's genome effort. In the spring, concern about the incessant bad-mouthing between the two sides had reached all the way to the Oval Office. President Clinton instructed Neal Lane, his chief science adviser, to "fix it . . . make these guys work together." That task fell to Patrinos, who one Sunday evening in early May lured Venter and Collins to his Rockville, Maryland, townhouse to discuss a rapprochement over pizza and beer. "I don't think I've ever seen them as tense as they were that day," said Patrinos. But the two protagonists realized the urgent need to call a cease-fire to the verbal hostilities. After several more meetings with Patrinos, the two rivals agreed in principle to the terms of a joint announcement. The first public signs that an accord was within reach came in June, when Venter and Collins appeared together without incident at a NIH cancer conference. As the final preparations were hastily laid for a White House ceremony to make the official announcement, Venter and Collins, clothed in ceremonial lab coats, posed for the cover of *Time* magazine.

In the weeks leading up to the announcement, Collins and his allies had been ardently complaining to the media about the constant references to "the race." Shortly before the White House announcement, the *New Yorker* published an article by Richard Preston that revealed Venter's former feelings about the competition with the public program. "They're trying to say it's not a race, right?" Venter asked. "But if two sailboats are sailing near each other, then by definition it's a race. If one boat wins, then the winner says, 'We smoked them' and the loser says, 'We weren't racing—we were just cruising.' " Collins's response was that if they were in a race, then the teams were on different tracks and using a different finish line. Lander was also put out by the tenor of the media coverage. "The journalistic feeding frenzy has served no one terribly well and I am just not impressed by it . . . this is a cheap and easy way for science writers to get their story on the front page."

At the packed Washington press conference, Collins returned to this theme. He made a point of reminding the media that the public consortium's data, collected at a pace of 1,000 bases per second, had been deposited in a publicly accessible database every evening, where it was available to anyone who wished to use it, including Venter. This was duly acknowledged by Venter, who also pointed out that his parent company, PE, had sold hundreds of the same DNA sequencers that Celera used to four of the five major genome centers, hardly the act of a company determined to win the "race" at all costs.

Although both parties said they were hopeful of arranging simultaneous publication in a few months, and perhaps comparing their respective sequences and sequencing methodologies, the agreement was one of mutual cooperation, not necessarily collaboration. And it would be foolish to deny that the primal competitive instincts aroused in both camps had expedited the completion of the genome sequence. "It gets the juices flowing," said Collins. "Competition is an essential part of science," Lander acknowledged. "The Scientific Method works by having this friendly exploration of alternatives." To which Venter replied sotto voce, "with friends like these . . ."!

THE UNUSUAL SATELLITE LINK-UP between the White House and Downing Street emphasized the magnitude of the genome announcement. But just what was being celebrated? According to Collins, the public consortium had sequenced more than 22 billion bases of DNA, providing a seven-fold coverage of the genome, with more than 60 percent of the sequence being collected in the past six months. The human genome was estimated to contain 3.15 billion bases. Collins said that a quarter of the sequence was "finished," almost 50 percent of the sequence was in "near-finished" form, and a further 38 percent was in draft form. Ninety-five percent of the sequence had an accuracy of 99.99 percent, but there was less certainty about the remainder, so the net accuracy of the sequence was just 99.9 percent. A cursory analysis of the public's sequence contained 38,000 confirmed genes, providing a lower boundary for the international gene lottery.

    It was a typically eloquent, dignified performance by Collins, who summed up the import of the achievement with a quotation from Antoine de Saint-Exupéry, author of Le Petit Prince: "Your goal is not to foresee the future, it is to enable it." He added, "The sequence that has been generated today . . . should be an enabler of the future understanding of human biology like no other." It was also time to pay tribute to the many men and women who had contributed to the historic event. "We are standing on many shoulders," said Collins approvingly. "The Human Genome Project did not arise like Venus out of the head of Zeus from nowhere!" This raised a few eyebrows in the audience, none more so than Ari Patrinos, the Athenian, who could not resist pointing out to his friend that it was actually Athena, not Venus, who emerged from the head of Zeus. (Collins started to apologize for confusing his Greek and Roman mythology before realizing he was just making matters worse.)

Venter followed Collins to provide some statistics on Celera's version of the sequence, which covered 99 percent of the genome and, appropriately, had had a gestation period of precisely nine months. It was based on the analysis of the genomes of five people: an African American female, a Chinese female, a Hispanic female, and two Caucasian males. There has been considerable speculation that the first genome to be sequenced belonged to Venter himself, but Venter merely says that company policy is not to divulge the identity of the donors. Venter reaffirmed that the raw sequence would be made publicly available on publication, but customers seeking instant access, annotation, and comparisons with other sequences, such as the mouse, would have to become paid subscribers.

Celera calculated that the human genome contains 3.12 billion bases, some 30 million bases fewer than the public consortium's estimate. Celera had sequenced 26.4 million DNA segments of about 550 letters apiece. The total amount of DNA sequenced was 14.5 billion letters, providing 4.6 times coverage of the genome. Venter attributed much of the success to a "secret weapon": a "paired-end" sequencing strategy devised by Hamilton Smith and Robert Holt that allowed Celera to fill in gaps in the sequence that other methods could not close. The sequence assembly required more than 20,000 central process unit hours on Celera's supercomputer and more than 500 million trillion DNA comparisons. "I asked Gene [Myers] what that really was," joked Venter, "and he said, 'I think it's something like a gazillion!' " Whatever the number, Venter claims it is the largest computational biology calculation in history.

All in all, it was one of the proudest, most satisfying days of Venter's two-year tenure at Celera. He heard that his first biotechnology partner had signed a deal while he was at the White House, and the finishing touches were being put on negotiations with a major academic subscriber—Australia. The only sour note was provided by the fickle stock market. Signs of Celera's willingness to cooperate with the public scientists apparently left some investors nervous about the long-term viability of Celera's business model. On the day that Celera made history, its shares dropped 11 percent.

JUNE 26 WAS A HISTORIC DAY, and scientists and commentators struggled to find the most appropriate metaphor to capture its significance. Curiously, the result was a list that mimicked the letters of the genetic code: the Apollo moon landings, Copernicus's astral visions, Galen's study of human anatomy, Turing's cracking of the Enigma code. Even the analogy to the invention of the wheel was sounding a bit tired, acknowledged Michael Dexter, the Wellcome Trust director who had coined it just a few months earlier. "I can well imagine technology making the wheel obsolete," said Dexter. "But this code is the essence of mankind, and as long as humans exist, this code is going to be important."

Most scientists simply ran out of superlatives, calling the declaration a revelation and one of the most significant days in human history. The evolutionary biologist Richard Dawkins waxed lyrical: "Along with Bach's music, Shakespeare's sonnets and the Apollo Space Program, the Human Genome Project is one of those achievements of the human spirit that makes me proud to be human." (One is tempted to ask, Is there any other choice?) Appearing in one of many joint television appearances with Collins after the press conference, Venter said on *Nightline* that it had been "a phenomenal day . . . a historic day for us as a species," to which the quick-witted host countered, "Well, on behalf of our species, thank you!"

But the hastily orchestrated Anglo-American announcement left some wondering if the celebration was not a little premature. After all, Celera's sequence data, although essentially complete, were securely beyond the reach of the general public, available only to subscribers—mostly in the pharmaceutical industry—to Celera's database. By contrast, the public consortium's data were freely available over the Internet but were still short of the rough draft target of 90 percent coverage of the genome. "I'm a little confused on timing," a puzzled reporter asked the Sanger Centre scientists at the press conference in London. "You've mapped 97 percent of the genome, sequenced 85 percent and finished 24 percent. So why did you choose now to make your announcement?" There was a good deal of cynicism in the United States as well. If the genome project was over, journalists joked, then Schubert's Unfinished

Symphony could be declared finished, baseball games would be called after eight innings, and incomplete passes in American football games would be ruled complete.

The reason for the carefully orchestrated announcement, according to the *Economist,* was reminiscent of the race scene in *Alice's Adventures in Wonderland,* when the judge (the dodo), after careful deliberation, declares, "Everybody has won, and all must have prizes." The prize to the public project's scientists "was the pretence that the race was actually a tie, and that a commercial upstart that is barely two years old did not really beat them to it." As for Venter, he receives "the officially sanctioned respectability that should give him a place on the invitation list . . . to receive a Nobel award for medicine from the hands of the King of Sweden." There was another more pragmatic reason for the June 26 announcement. Michael Morgan suggests it was the only time that a gap could be found in both Clinton and Blair's schedules.

The reaction from the biotechnology community was mixed. Venter's former partner, Bill Haseltine, was skeptical about Celera's prospects: "It's like a private company in 1967 announcing they are going to race NASA to the moon. And they did it! The next question is, 'What is the business plan?' " Another competitor, Millennium Pharmaceuticals' Steven Holtzman, said: "The race at this point is not for the DNA. . . . The race is in assigning to genes and to variations in genes a role in disease initiation and progression and drug response." Incyte's president and CEO, Roy Whitfield, ventured, "I would describe it as the beginning of thousands of races. If you have colon cancer, the race is about curing colon cancer. If you have arthritis, it's a race to cure arthritis. It's the start of a really long race to have a tremendous impact on human health."

The considered opinion of the incomparable Sydney Brenner is that there is still much more to do. "The idea that this is a tremendous scientific accomplishment is simply ridiculous," says Brenner. "It is an entrepreneurial accomplishment, a great managerial achievement, but there isn't any new science in it. It's exactly like sending a man to the moon. Sending him there is easy, it's getting him back that's the problem!"

Some of the media coverage was not quite so erudite. On *Good Morning America,* Diane Sawyer succinctly summed up the conquest of the human genome as a story about "personality, profit, passion—and pizza!" She went on to ask the ABC science correspondent, "So, does this mean we can create a human being in a petri dish?" On CNN, Larry King succeeded in embarrassing the Dalai Lama, who had clearly not heard about the genome announcement and confessed he did not know much about DNA. And the host of one late-night television show jibed: "Now scientists are concentrating on their next big project—getting laid!"

There was naturally speculation about a genome project movie. John Hodgson, editor-at-large of *Nature Biotechnology,* took on the job of casting director: John Malkovich would star as Venter, Tom Hanks could play Collins, Billy Connolly would be perfect as John Sulston, and President Clinton would play himself.

IN THE WEEKS and months following the White House declaration, Celera steadily expanded its list of university subscribers to its DNA databases, including Harvard University and the University of Texas Southwestern Medical Center. The biggest coup was the agreement signed in September 2000 with the Howard Hughes Medical Institute, the largest U.S. medical philanthropy, which supports more than 350 of the leading biomedical researchers around the country. Institute investigators could subscribe to the Celera database for about $15,000 per year.

It was predictable that more and more leading universities would eventually strike deals with Celera. Despite questions about the advantage offered by the Celera database over the publicly deposited sequence data in GenBank, few researchers could stand to think that their rivals at another institution were gaining an advantage by having access to Celera's human sequence. In addition, subscribers can access mouse genome data, allowing researchers to learn more about the role of putative human genes by comparing them with their mouse counterparts. Celera also began offering access to its own polymorphism catalogue, which includes some 2.5 million SNPs.

But while many researchers were clamoring to gain access to Celera's genome database, others were growing increasingly nervous about it. Five years ago, several start-up biotechnology companies decided to keep quiet about medically significant gene discoveries, sparking angry protests from academics, who charged that the industry was putting profits before patients. Now the tables are turning. With virtually every gene accessible over the Internet, some university researchers are guarding their leads for fear that well-funded biotechnology companies might exploit their findings. Angela Christiano, a dermatologist at Columbia University, made headlines in 1997 when she discovered the gene mutated in a rare inherited form of alopecia, causing a complete loss of body hair. The potential market for drugs based on baldness genes is immense, but as Christiano continues her search for other genes related to hair growth, she is reluctant to reveal any tentative leads about the chromosomal locations of other baldness genes until she has characterized the gene. Publishing the approximate location of a baldness gene could allow a biotech company to seize on the map location and identify and patent the gene first.

Even before Venter proved his sequencing savvy on larger genomes with the fruit fly genome, he was being inundated with requests to sequence the genomes of different organisms. The mouse sequence, which Celera will have essentially completed by the end of 2000, will add immense value to Celera's genome database. By comparing the sequence of three inbred strains of mice with the human genome and comparing the conservation of different sequences between the two species, scientists will be able to predict which segments are likely to constitute genes. In 1999, Venter caused a major stir by announcing that Celera could sequence the entire genome of rice in just six weeks. But in April 2000, Monsanto announced that it had completed a rough draft of the rice genome, largely the work of Leroy Hood's group in Seattle and which is now publicly available. Venter is also weighing the merits of sequencing the genomes of cow, camel, various insects, and the chimpanzee. The dog genome is also an attractive potential target, since a genomic analysis of the remarkable variety of dog breeds in existence could have major medical benefits, as glimpsed by the recent identification of a narcolepsy gene in dobermans. During an informal lunch at Celera in 1999, Venter casually wondered how many chromosomes there are in an apple, having been approached about the feasibility of sequencing the apple genome while visiting Australia.

Over the past ten years, Venter's restless ambition and single-minded opportunism have alienated him from many of his fellow scientists. But for a man worth hundreds of millions of dollars, a man so wealthy he can afford to donate half of his Celera stock options to his former institute, Venter still craves the respect and admiration of his peers. He talks openly about winning the Nobel Prize and says his accomplishments compare favorably with those of many past winners. He once joked with his close colleague Hamilton Smith that they might one day share the Nobel Prize. Smith, who won the prize in 1978 and subsequently endured many personal and professional difficulties, replied simply, "You can have it." Many observers believe Venter's trailblazing accomplishments in DNA sequencing over the past decade—pioneering the EST approach, sequenc-

ing the first genomes of living organisms, and catalyzing the human sequence—justify a Nobel Prize. But regardless of whether he receives the
call from Stockholm, Venter has plenty more to do.

While Celera is laying plans to keep its army of DNA sequencers
gainfully employed in the coming years, it is looking toward the next big
target of the biotechnology industry: the proteome project. One month
after the human genome announcement, Venter told delegates at an international biochemistry meeting, "It's only by understanding protein
function that we can truly understand and predict medical outcomes."
Celera's sister company, Applied Biosystems, is producing a new generation of mass spectrometers to speed up the sequencing of proteins. This
technique could be used to detect abnormal proteins on the surface of
cancerous cells, for example, including proteins that have been chemically
modified by the addition of sugar, sulfur, or phosphate groups. These
modifications can have a drastic impact on protein function but are not
detectable by genome analysis. Celera has even hinted that it might diver-
sify from its original mission as an information company to explore new
avenues, such as patient-specific vaccines and other forms of drug development.

LONG BEFORE THE HUMAN GENOME SEQUENCE was in the books, scientists were relishing the immense challenge of sorting through the genetic parts list and figuring out how they all work together. The task
ahead cannot be underestimated. "I am certain that a century from now,
scientists will still be making major, insightful discoveries on the genetic
sequence that's going to get determined [in 2000]," predicts Venter.

Scientists in academia and industry are assembling the next generation of tools to ask the key questions: What is the function of the proteins
encoded by the roughly 60,000 human genes? How can those functions
be massaged to treat disease? The answer is encapsulated in the new buzzwords of the postgenome era—terms such as proteomics, functional genomics, and structural genomics. The emphasis is already shifting from
merely determining the identities of all of the human genes to understanding the properties of the hundreds of thousands of proteins they encode (proteomics). Major efforts are also underway to analyze the
functions of genes (functional genomics) and the structures of the corresponding proteins (structural genomics). These terms sound enticing to
industry analysts and venture capitalists, but do not be deceived. "The dif-

ference between physiology and functional genomics," says University of Wisconsin geneticist Howard Jacob, "is marketing."

In the postgenome era, scientists are no longer constrained to study one gene at a time or one protein at a time. A new collection of tools is allowing them to determine properties of the entire cellular repertoire of genes and proteins. Just as DNA chips have tremendous potential to monitor the expression of genes in healthy and diseased cells, researchers are developing similar methods to assess the properties of proteins. For example, Stuart Schreiber and colleagues at Harvard University have developed a prototype protein microarray consisting of 10,000 protein spots robotically attached onto a common microscope slide, which can then be used to screen the binding of small-molecule drug candidates and take protein snapshots of cells.

Another powerful approach to characterizing proteins is to determine which proteins physically interact with each other, providing telling clues to the pathways and functions of otherwise poorly characterized proteins. A clever technique called two-hybrid analysis, developed several years ago by University of Washington geneticist Stanley Fields and coworkers, uses one protein as a bait to detect binding of another. Fields has extrapolated this method to monitor the protein-binding properties of all 6,000 gene products in the yeast genome. The first large-scale results, performed in collaboration with CuraGen, identified partners for more than 1,000 yeast proteins, providing valuable clues as to the function of hundreds of unknown gene products. This approach is now being scaled up to analyze the proteins encoded by the completely sequenced fruit fly genome. Together with other biochemical, computational, and genetic approaches, researchers will assemble a new global picture of protein interactions inside the cell.

In conjunction with techniques to determine protein function on a massive scale (functional genomics), another priority is to determine the three-dimensional structure of proteins on an industrial scale, or structural genomics. "The first look at a new protein structure provides a thrill that could be compared to that felt by explorers in the late 19th century upon discovery of a new biological species," marvels John Kuriyan, a leading structural biologist at the Rockefeller University. However, this nostalgic romanticism is threatened by the creation of large consortia and companies to do for protein structure what the Human Genome Project did for DNA sequencing.

Kuriyan's colleague, Stephen Burley, is leading the call for a structural genomics initiative to speed up the determination of protein crystal structures. Although there are hundreds of thousands of different proteins in the human body, they employ a surprisingly small repertoire of just a few thousand discrete protein folds. These domains, some with catchy names such as "zinc fingers" or "leucine zippers," are pressed into service repeatedly to help a diverse range of proteins bind to DNA or other proteins. The NIH plans to spend about $100 million over the next five years to fund a few select groups to catalogue the most important protein shapes or folds. Meanwhile, the Wellcome Trust is contemplating the creation of an international structural genomics consortium between academic centers and industry, to speed up the public release of protein structures.

Impressive as these initiatives are, a host of new companies are gearing up to generate these structures even faster. Tim Harris, the mercurial British CEO of the imaginatively named Structural GenomiX, a new company based in San Diego, is gambling that his firm's industrialized approach to X-ray crystallography can determine the three-dimensional structures of 5,000 proteins in five years. These structures can then be sold to pharmaceutical companies looking for new and improved drug targets.

Peter Schultz, director of the Genomics Institute of the Novartis Research Foundation, also in San Diego, says that structural genomics will be fundamental to the future of drug discovery: "If you can input 200,000 or 400,000 compounds in a computer, and actually dock *in silico* this entire library of molecules against a particular protein structure, one could, in theory, virtually identify leads for new drugs." Schultz is launching a new company, Syrrx, to find out.

There are some striking parallels between the launch of these and other structural genomics ventures and the way that Celera audaciously challenged the public consortium running the Human Genome Project, improving the efficiency of sequencing technology rather than reinventing it. Just as happened fifteen years ago, when the first glimmerings of the Human Genome Project began to gain public attention, some prominent protein chemists are not happy about a blind, big science approach to solving protein structures. Similar criticisms were muted as the value of the genome sequence became apparent, but just as the linear DNA code does not automatically reveal the function of genes, the three-dimensional shape of a protein does not promise to elucidate the role of the protein.

Both the public and private structural genomics efforts face an enormous challenge. There are many more types of protein than there are genes, because most genes are interpreted in slightly different ways. Moreover, many of the most medically important proteins—those lodged in the cell membrane that recognize signals from hormones or transmitters or other cells—are the most difficult to crystallize. When Rockefeller University's Rod MacKinnon managed to determine the first crystal structure of a membrane channel in 1998, his announcement at a conference in Utah was greeted with a rare standing ovation.

Some scientists hope that by gaining a better understanding of the rules that govern the way a protein's one-dimensional chain of amino acids governs its folding into a complex three-dimensional shape, the problem of solving protein structures will move from the experimental to the computational. A key weapon in this quest, which is already being hailed as biology's latest holy grail, is Blue Gene, a new supercomputer being developed by IBM. Blue Gene is 1,000 times more powerful than Deep Blue (the machine that famously beat world chess champion Garry Kasparov a few years ago) and 2 million times more powerful than a competent personal computer. When completed in a few years' time, Blue Gene, based at the Watson laboratories in New York, will have 1 million microprocessors and an optimal speed of a petaflop—1 quadrillion operations per second. Blue Gene will then be pressed into service modeling the folding of a single protein from a linear chain of amino acids into its precise three-dimensional shape. This initiative will be of medical relevance for a number of diseases that are caused by problems in protein folding, such as Alzheimer's disease, cystic fibrosis, and prion diseases such as mad cow disease.

Sydney Brenner, for one, is looking forward to the postgenomic era: "When all of this [genome project] mania dies down, we'll get back to hypothesis-driven normal science. What people are forgetting is that there are several tens of thousands of biochemists who can now get stuck in and study protein function. What you've done is provided them with a tool, and that's fine."

WHILE SCIENTISTS ANTICIPATE divining the secrets of the human genome and the thousands of proteins it encodes, a few brave souls are contemplating something far more radical—the warp-speed evolution of the universal genetic code to create new forms of life. The issue is encapsulated neatly by Peter Schultz: "Why are there only four bases in our

DNA and only 20 amino acids in our proteins? What would life lo
if God had worked on the seventh day and made a few more?"

The genetic code contains 64 possible triplets (the number of
mutations of 4 letters in a triplet pattern is 4 × 4 × 4), yet there are only
20 different protein building blocks, meaning that many changes in the
DNA sequence do not alter the encoded protein. Schultz hopes to take
advantage of the degeneracy of the code to expand on it a little. To incorporate an unnatural amino acid into proteins, for example, Schultz must
engineer both ends of a molecule called transfer RNA, the adapter RNA
that carries the appropriate amino acid to the site of protein synthesis after
recognizing a specific triplet in the RNA message. First, he modifies the
transfer RNA anticodon so that it recognizes a different triplet—for example, UAG, which is normally the "stop" codon. Next, he switches the
amino acid attached to the other end of the adapter RNA. By screening
special strains of bacteria, he can engineer new proteins that incorporate
these foreign amino acids. Indeed, such unnatural bacteria already exist in
suspended animation, safely stored in a deep freeze in the laboratory of
Jeffrey Tze-fei Wong at Hong Kong University of Science and Technology. Wong has successfully incorporated a synthetic amino acid, fluo-
rotryptophan, into a strain of *Bacillus subtilis*.

But an even more remarkable attempt to go nature one better is in
the cards. Researchers are exploring ways to expand the genetic lexicon
itself. Japanese researchers have succeeded in using a four-base code to introduce a pair of unnatural amino acids into proteins. Meanwhile, other
groups are incorporating unnatural bases into DNA that are able to pair
just as efficiently as the A–T and C–G base pairs of the double helix. They
are even making progress in identifying different enzymes that can copy
DNA strands containing these artificial bases, paving the way for the creation of bacteria that can make totally unnatural proteins. This research
certainly has potential practical applications, such as the creation of organisms to clean up toxic waste, but this seems so mundane compared to
the philosophical implications: to understand why the universal genetic
code evolved the way it did and to see just how far we can modulate it.

The prospect of exercising control over the human genome is
daunting. The brilliant physicist Stephen Hawking says, "There haven't
been any significant changes in human DNA in the past 10,000 years. But
soon we will be able to increase the complexity of our internal record, our
DNA, without having to wait for the slow process of biological evolution
. . . by increasing our brain size, for example." Hawking is in no doubt

that no matter how stringent the legislative regulations are on human genetic engineering, "someone will improve humans somewhere."

Many scientists are understandably uncomfortable with this prospect, but as James Watson points out, "Genetics *per se* can never be evil. It is only when we use or misuse it that morality comes in." However, in the wake of enormous controversy over the mad cow disease debacle in the United Kingdom and the perceived dangers of genetically modified food, the public will take much more convincing than self-assured statements from a Nobel laureate. Polls conducted after the June 2000 genome announcement showed that a majority of those questioned were morally concerned about genome research. Effective legislation against the specter of genetic discrimination would be one way of restoring the public's trust.

ON MAY 8, 1900, the British zoologist William Bateson settled into his seat on the Great Eastern Railway for the journey from Cambridge to London. Two weeks earlier, he had stumbled across a reference to some extraordinary breeding experiments carried out by a Bohemian abbot thirty-five years earlier. Bateson had tracked down the original report, authored by Gregor Mendel, and was entranced. As the whistle of the locomotive pierced the calm of the Cambridgeshire countryside, the first half of genetics was officially underway.

A few hours later, during a lecture delivered at the Royal Horticultural Society, Bateson fired the first shot. "An exact determination of the laws of heredity," he predicted, "will probably work more change in man's outlook on the world, and in his power over nature, than any other advance in natural knowledge that can be foreseen." Bateson became an ardent advocate of Mendel's work and coined the term *genetics.*

A century later, the game has changed almost beyond recognition. Genetics has been transformed from a somewhat esoteric debate of the practicalities of plant breeding to a subject of breathtaking importance and relevance to everyday life. The world of biomedical research will be divided into the two halves: the gene-by-gene, clone-by-clone approach that marked the first half, and the global, population-based, genome-centered analysis that will dominate the second.

In addition to the major impact genome sequence surveying will have in gene therapy and pharmacogenomics, the genomic revolution will lead to other major breakthroughs. One area with excellent potential involves the use of stem cells, primordial undifferentiated cells that retain the capacity to develop into almost any kind of tissue. The isolation of human embryonic stem cells by two groups in 1998 provoked enormous excitement—and an ethical controversy. The excitement arises from the potential to bathe stem cells in a cocktail of factors that could coax the production of insulin-secreting pancreatic tissue from the stem cells of a diabetic patient or replenish wasted muscle in patients with muscular dystrophy. The controversy centers on ethical concerns that the best source for these cells are human embryos, although adult stem cells will have useful applications. An even more impressive feat in genetic reprogramming occurred in 1996 with the birth of Dolly the sheep, cloned from a bladder cell from a six-year-old sheep. Since then, researchers have reported the successful cloning of mice, cows, pigs, and more sheep, raising the possibility of therapeutic cloning in humans.

THE GENERATION of the complete genome sequence has been the greatest adventure in modern science, one that has been propelled by the commitment of many remarkable men and women. Craig Venter remains as controversial as he was nine years ago, when he burst onto the scientific stage, but even his detractors would acknowledge that his vision has brought the Human Genome Project to completion five years ahead of schedule. He did not do it in isolation: Francis Collins and the leaders of the public genome consortium toiled for years before Venter's dramatic entrance to place bookmarks throughout the human genome. They have redeemed an unwieldy multinational project to produce a working draft of the sequence. In the next few years, this will be polished to produce the gold-standard human genome sequence, a reference for the rest of time.

In his classic novel 1984, George Orwell wrote, "All history was a palimpsest, scraped clean and re-inscribed exactly as often as necessary." (A palimpsest is a text that has been erased and overwritten.) The genome sequence can be considered a genetic palimpsest, a text that has been overwritten time and again by evolution. Buried in this sequence, once scientists have developed sufficiently sophisticated tools and computer programs to unearth them, are the answers to the origins of life, the evolution of humanity, and the future of medicine.

Genomania!

> The skeptics were right when they said Craig Venter
> would never crack the human genetic code on schedule.
> He was two years early.
>
> COMPAQ ADVERTISEMENT

> I've seen a lot of exciting biology emerge over 40 years.
> But chills still ran down my spine when I first read the
> paper that describes the outline of our genome.
>
> DAVID BALTIMORE, Nobel laureate

SUNDAY, FEBRUARY 18, 2001: The atmosphere outside the
Continental Ballroom in the San Francisco Hilton and Towers Hotel
was more reminiscent of a rock concert than a staid scientific convention. When the doors finally opened, thousands of people rushed
inside to claim a seat for the most anticipated lecture of the American
Association for the Advancement of Science annual convention. It
was just one week since newspapers around the world had trumpeted
the completion of the human genome sequence. That same week,
blockbuster issues of *Nature* and *Science* had featured dozens of reports
on the historic achievement.

As the delegates settled into their seats, a crowd began to gather
around a familiar figure at the front of the stage. It was Craig Venter,
resplendent in formal black tie, grinning broadly as he autographed
copies of the latest issue of *Science,* featuring the report of the Celera
genome sequence. Two months earlier, *Time* magazine had named

him 'Scientist of the Year' for 2000 (he might even have earned the accolade 'Person of the Year' had it not been for the bizarre outcome of the 2000 US presidential election). The formal attire signaled the magnitude of the talk he was about to give—and might even be a dress rehearsal for the ultimate scientific award.

VENTER'S LECTURE CAPPED a frantic week marked by the publication of two simply extraordinary issues of *Nature* and *Science,* featuring the detailed genome reports of the public consortium and Celera, respectively. *Science* magazine was the logical place for Venter to publish his account of the human sequence. After all, Venter had published his three most famous reports in *Science*—expressed sequence tags (ESTs) in 1991, the first microbial genome sequence in 1995, and the *Drosophila* genome sequence in 2000.

But the decision to consider the Celera report was problematic for the editors of *Science.* Unlike the fruit fly sequence, which was released without any restrictions in part to validate its DNA sequencing strategy, Celera argued that its human sequence was far too valuable simply to be deposited in GenBank, the traditional public repository for DNA sequence data administered by the National Institutes of Health. To do so would make the Celera sequence accessible to anyone—including rival commercial firms—without any company-imposed restrictions or safeguards. Having spent hundreds of millions of dollars to obtain the sequence, Celera was not about to help its competitors by giving it away for free.

After months of negotiation, the *Science* editors reached a controversial deal with Venter for the right to publish the Celera paper. Setting a major precedent, *Science* agreed to allow Celera to monitor and control downloads of its sequence data from its own website rather than from GenBank. Celera said academic users could download segments of one million bases per week without restriction, but commercial users would have to sign a license agreement before they could see the sequence. *Science* would hold a copy of Celera's database in escrow, as security against possible changes in company policy.

The agreement between *Science* and Celera provoked a furious outcry from leading scientists, who argued that it was unethical to discriminate in the release of published scientific results, and lobbied editor-in-chief Donald Kennedy, the former president of Stanford University, to reverse his decision. Eric Lander, director of the MIT genome center, said *Science* was "confused about the purpose of scientific publication," arguing that if authors could restrict access to certain users, "the pace of discovery will be slowed and the public will lose." Michael Ashburner, joint head of the European Bioinformatics Institute and a former member of *Science*'s editorial board, charged that "the only intellectually honest position would be to not publish the [Celera] paper. If it came to me, I would simply chuck it in the bin without even opening the FedEx package."

But some senior academics took a more even-handed view. One Nobel laureate, David Baltimore, said the critics of *Science* were being "short-sighted." Lobbying to prevent publication of the Celera data "is basically cutting off your nose to spite your face," said Baltimore. *Science* ignored the furore, and proceeded with its review of Venter's paper, which was formally submitted to the journal on December 6, 2000.

Science's consideration of the Celera paper quashed hopes of simultaneous publication of both papers in the same journal. Although the *Nature* editors had also discussed publication terms with Celera, they were uncomfortable with the prospect of access restrictions to the sequence. Meanwhile, senior staffers were conducting a subtle solicitation of the public genome paper, for example, sponsoring a talk by Lander at the Hay-on-Wye literary festival in Wales. The day after Celera submitted to *Science,* the public consortium declared they did not feel comfortable submitting its paper alongside the Celera manuscript and plumped for *Nature* instead. The genome countdown had begun.

ROBIN MCKIE, THE VETERAN SCIENCE reporter for the *Observer,* was in a bind. Press conferences had been scheduled for Monday, February 12 in London and Washington, D.C. to mark the official online publication of the two sets of genome reports in *Nature* and *Science*. Both journals had distributed advance copies of the reports to select members of the media, allowing reporters a few days' notice to sift through the dozens of articles, interview experts, and prepare their stories. In return, the press agreed not to print or broadcast any story until the journals officially released the news by lifting their embargo that Monday.

For McKie, the timing could not be worse. If he abided by the terms of the journals' embargo policy, he would have to wait almost a week to run a story—an eternity in the cutthroat world of journalism. McKie's solution was simple: relying on alternative sources and ignoring the journals' press releases, he had occasionally scooped some major scientific stories. His biggest coup by far came in February 1997, when he broke the sensational story of Dolly, the famous cloned sheep, four days before the news was published in *Nature*. (McKie says he obtained the news from a television producer.)

A few days before the London press conference, McKie, along with many newspaper science reporters, was in Lyons, France, to cover the BioVision 2001 international science conference, where both Collins and Venter were giving talks. McKie was looking for a hook to scoop the world's press, and Venter duly obliged. During Venter's address to about a thousand delegates on Friday, February 9, he previewed his upcoming *Science* paper, displaying the magazine cover, and unfolding a huge poster portraying the Celera genome map. He also alluded to the surprisingly low total number of genes his team had found (confirming some previously published reports), which he said invalidated the simple notion of "one gene—one function."

The next day, McKie drafted his newspaper's lead story. Under the title "Revealed: the secret of human behaviour," McKie wrote that the surprising paucity of genes meant that humans were much more products of the environment than previously thought, and quoted Venter extensively. As word escaped that the *Observer* was leading with the human genome exclusive, editors at *Nature* and *Science* bowed to the inevitable and officially lifted their journals' embargoes. The late editions of some American Sunday newspapers also ran related stories, and by Monday, the genome was front-page news, even before the press conferences had taken place.

Much of the ensuing press coverage took its cue from McKie's declaration that nurture had overcome nature, based on the surprisingly low gene total. The logic of his piece was dubious to say the least: presumably if the gene tally had actually been higher than expected, geneticists could have triumphantly declared humans to be exclusive products of nature! Cambridge University geneticist Martin Bobrow poured scorn on such naïve conclusions. Humans may only have twice as many genes as the fruit fly, he said, "but flies have two pairs of wings and can fly, which is more than we can do."

★

DESPITE THE CAREFULLY ORCHESTRATED press conferences in London and Washington, D.C., any hopes that the two groups would cast aside their differences were soon dashed. The Wellcome Trust press release announcing the London press conference declared gruffly that copies of *Science* would not be made available. Sir John Sulston (who was knighted in the Queen's New Year's Honours List) wasted no time in chiding Celera for its reliance on the public sequence data. When Sulston's team had pored over advance copies of the Celera report, they were stunned to discover that Celera had made extensive use of public mapping data to help assemble the final genome sequence. Consortium scientists dashed off a four-page press release, arguing that Celera's original shotgun strategy had failed. "Celera's assembly is remarkably similar to the HGP's assembly," the release deadpanned. "This is perhaps not surprising, given the dependence of the Celera assembly on the HGP map and sequence."

Tensions at the Washington, D.C. press conference were, if anything, higher in the audience than on stage, as rival journal editors squabbled over the seating arrangements. Venter and Collins amicably shared the stage, with Venter taking the opportunity to announce Celera's rapid progress in sequencing the mouse genome. But offstage, the arguments over the relative quality of the alternative genome assemblies continued unabated. "It didn't work," Lander insisted, suggesting that the Celera shotgun method should be reserved for relatively unimportant species. "He's playing with half-truths and innuendos," Venter retorted. "I'm getting so I really don't care what his opinion is."

Venter insisted that, while Celera had opted to use more public data than expected to save time and money, his company's whole genome shotgun sequencing strategy was still viable. Moreover, Venter claimed that the Celera genome assembly had greater accuracy, with more of the sequence pieced together into longer stretches. While the public team acknowledged Celera's superior coverage in some respects, they felt they had won a significant moral victory. The lingering disagreements even affected the celebratory party thrown by *Nature* at the National Building Museum in Washington, D.C., as a last-minute approach by Celera's sister company, Applied Biosystems, to co-sponsor the event was rebuffed.

What should have been one of the most dignified moments in scientific history degenerated into a transatlantic spat for bragging rights. While the British press lavished praise upon Sulston and his campaign to ensure that the sequence was delivered freely to the public, the *Wall Street Journal* published testimonials from pharmaceutical clients praising the quality of the Celera database. Under the striking headline "Private Sector Wins Genetic-Code Race," the Toronto *Globe and Mail* quoted Steve Scherer, a senior scientist at Toronto's Hospital for Sick Children, as saying that the Celera database was "twice as good" as the public data, and had enabled his group to identify a gene important in brain development. "We are the envy of other scientists," Scherer admitted, because the Celera database was too expensive for most Canadian academic institutions. But other scientists resented Celera taking public data and trying to resell it. "This undoubtedly was—and is—a large step for mankind," summed up an editorial in *Nature Biotechnology*, "but one obscured by the small-mindedness of men."

WHATEVER THE MERITS OF THE rival genome sequences, their official publication was marked by two spectacular issues of *Nature* and *Science*, organized primarily by the journals' respective genetics editors, Carina Dennis and Barbara Jasny. The 62-page *Nature* report prepared by the public consortium is undoubtedly one of the most enthralling scientific reports ever published. The hundreds of co-authors were billed as the International Human Genome Sequencing Consortium, but the principal annotator and lead author was Eric Lander. The MIT scientist confessed that his intense devotion to the manuscript for months after the June 2000 White House ceremony had obliged him "to explain periodically to my three wonderful children why Dad is holed up working on the weekend." Lander even devised the *Nature* cover—a stunning photomosaic of a double helix, composed of more than 500 miniature photographs of humans of every creed and colour. Closer scrutiny revealed one especially familiar picture of two famous Caucasians: Francis Crick and James Watson, admiring their original model of the structure of DNA.

Lander also paid tribute to the doyens of DNA in the text of the article. "The sequence of the human genome is of interest in several respects," penned Lander, in perhaps the most brazen understatement in scientific literature since the famous opening lines of Watson and Crick's letter to *Nature* almost half a century before. The public group predicted that there are between 30,000 to 40,000 protein coding genes. These coding sequences, lined end-to-end, would span only about 1.5 percent of the genome, although as most genes are interrupted by "junk" sequences (introns), genes actually make up about a quarter of the genome overall. The report also shed light on the bizarre world of junk DNA, families of viral- and bacteria-derived sequences variously characterized as freeloaders and parasites. Surprisingly, some of these DNA elements are not distributed randomly across the genome but tend to congregate near genes, suggesting they may have some unknown function.

Aside from the paltry total number of genes, the biggest surprise in the consortium report was the observation that more than a hundred genes in the human genome appeared to be bacterial in origin. This conclusion stemmed from the fact that the sequences of these genes bear a closer resemblance to bacterial genes than any other organism. Lander postulated that they might have become lodged in the human genome by a process known as horizontal gene transfer, by which numerous viral genes have taken up permanent residence within the human genome. (Subsequent studies, including one from researchers at TIGR, Venter's former institute, discounted this idea, arguing that other organisms might simply have lost these genes during evolution. Six months after publishing the original report, the consortium published a correction essentially conceding the point.)

It was left to Francis Collins to author the final few pages of the *Nature* article, highlighting the medical impact of the genome sequence. "We find it humbling to gaze upon the human sequence as it comes into focus," he wrote, and concluded the text with a passage from T. S. Eliot: "We shall not cease from exploration. And the end of all our exploring will be to arrive where we started, and know the place for the first time."

The cover of *Science* featuring the Celera paper depicted five adults of varying ethnicity, corresponding to the backgrounds of the five DNA donors selected by Celera for its sequence analysis. Also included was a newborn baby, which Venter quipped was to represent the nine-month gestation of the Celera sequence. The Celera article was co-authored by 274 scientists, with Venter unabashedly listed first. Celera identified 26,588 genes, and predicted the existence of a further 12,000—in reasonable agreement with the public consortium's estimates. Venter also pointed out ways in which the Celera genome assembly was superior to the public effort, such as in the greater average size of each contiguous DNA sequence. The Celera team compared regions of the genome for similarities, finding copious signs of genes that had been copied during evolution. For example, half of chromosome 20 has been duplicated onto chromosome 18, and many genes have been duplicated onto the gene-rich chromosome 19. The centerpiece of the article was a poster, five feet tall, depicting the location of every known gene in the human genome.

Venter closed the article by stressing the need to guard against genetic determinism—the notion that humans are "hard-wired"— and reductionism—the belief that cataloguing every gene will lead to a complete understanding of human variation. "The real challenge of human biology," Venter concluded, "beyond the task of finding out how genes orchestrate the construction and maintenance of the miraculous mechanism of our bodies, will lie ahead as we seek to explain how our minds have come to organize thoughts sufficiently well to investigate our own existence."

FIVE DAYS AFTER THEIR joint press conference in Washington, D.C., Venter and Collins were in the spotlight again, this time delivering keynote lectures on consecutive evenings before packed houses in San Francisco at the AAAS convention. Collins likened the consortium's *Nature* publication to a grand book report, and with apologies to US television talk-show host David Letterman, offered a "Top Ten" list of genome surprises, summarised as follows:

10. The genome is very lumpy—some chromosomes, such as 17 and 19, have a much higher density of genes than others, such as 18.
9. The number of human genes is much lower than expected.
8. Human genes produce a larger number of proteins than simpler organisms.
7. Human proteins are more elaborately constructed than in other organisms, by the accrual of additional domains during evolution.
6. Scores of human genes appear to have closer evolutionary counterparts in bacteria than higher organisms, including monoamine oxidase, a protein linked to depression. Collins said this discovery "puts a new face on recombinant DNA" and casts a new light on the genetically modified food

debate (although subsequent studies have led to more prosaic interpretation).

5. Repetitive DNA provides a "fossil record" that looks back some 800 million years.

4. "Junk DNA" actually has an important function. For example, there are more than one million Alu repeats, but the oldest Alu elements have been selectively retained in gene-rich regions.

3. The mutation rate in males is twice that in females. While males bear responsibility for introducing the majority of mutations into the gene pool, they also deserve credit for being the prime engines of evolutionary progress.

2. Humans are 99.9 percent identical. Most of our genetic differences are shared among different ethnicities and races. There is no scientific basis for precise racial categories.

1. These revelations underline the importance of free, unfettered access to the genome sequence.

Concluding his lecture with the same Eliot quotation in the *Nature* article, Collins was greeted with a standing ovation.

TWENTY-FOUR HOURS LATER, before a rapt audience of thousands, Venter delivered a much more self-centred account of the quest for the sequence. He spoke in deeply personal terms of his decade-long battles with the scientific establishment. From his life-changing experience in Vietnam to his feuds with James Watson and the NIH bureaucracy, he chronicled his fascination with genes and genomes to become the first person to sequence the genome of a free-living organism. His favorite microbe was *Deinococcus radiodurans,* the invincible radiation-resistant bacterium that is so popular among supporters of the panspermia theory of the origin of life. Confirming that theory would be difficult, Venter joked, because every time the cosmonauts on the Mir space station flushed the commode, they released billions of bacteria into space. "If it wasn't there before, it's there now!" Venter also plugged his company's new collaboration with Compaq to build a 100-teraflop computer (two orders of magnitude bigger than the existing Celera computer) and remarked wistfully about what he might have discovered had such technology been available a decade earlier.

Venter noted that Celera had read 27 million DNA sequences to assemble the genome sequence, but omitted any acknowledgement of the public data that contributed to the Celera assembly. About 90 percent of the human genome sequence had been assembled into segments of 100,000 bases of DNA or longer. Comparisons with *Drosophila* revealed that humans contain more genes with roles in the immune system, the blood system, the nervous system, and gene expression. Alignment with the new mouse genome sequence helped identify the key regulatory regions in the human sequence (because stretches of DNA that are functionally significant are more likely to be conserved during evolution). Ironically, Venter said the chimp genome was just too similar to the human sequence to help identify genes. "We thought we'd sequence a chimp that we'd call Jim—for historical purposes—but we don't think anyone would have noticed."

Venter closed by returning to one of his favorite themes: the importance of the environment in shaping human behavior. The 'colon cancer' gene is no such thing, he argued. It is in fact a gene for repairing DNA, and although mutation carriers are at higher risk of the disease, the cancer is triggered by environmental factors. Cystic fibrosis is another example, mutations in a single gene giving rise to a broad spectrum of illnesses, including sterility, asthma, and sinusitis, in addition to the overt lung disease in cystic fibrosis. The genome is finally putting to rest the simplistic notion of one gene for one disorder.

THE CHALLENGE FACING VENTER in the weeks and months following the genome celebration was to convince new clients to sign up for the Celera database, in spite of the freely available human sequence. Some universities felt obliged to subscribe because of their need to stay competitive, but others easily justified the expense. According to one respected geneticist, $250,000 for a Celera subscription could easily be justified if it led to discoveries that would lead to the funding of a couple of new federal grants. By the summer of 2001, Celera had signed up more than fifty database subscribers including, somewhat ironically, the National Cancer Institute, the largest of the National Institutes of Health, and a close partner of the genome center. The Wellcome Trust, by contrast, refused to sanction any subscriptions from its grantees.

The key for Celera to winning new subscribers was to stay one step ahead of the publicly available data by adding value to the human sequence. In this regard, providing the near-complete genome sequence of three strains of laboratory mice was a notable achievement. Celera's sequencing strategy also won tacit approval from the US government, in the form of a $20-million grant to sequence the rat genome in collaboration with one of its former rivals, the Baylor College of Medicine genome center, as well as part of the mosquito genome. Among the many other species queuing up to be sequenced is the dog. Venter has already signed up his first canine volunteer—his own giant poodle Shadow. "I thought we'd keep it in the family," he says with a wink.

When *Time* named Venter "Scientist of the Year," it ran a striking full-page photograph of the Celera chief executive, with one arm in a sleek black Italian suit, the other in a white lab coat. Whether Venter can successfully juggle his business and academic ambitions will be fascinating to watch during the first few years of the postgenome era. His stiffest challenge is to convince investors that Celera will be as big a player in the new era of proteomics as it was in genomics, particularly as Celera is still losing vast sums of money (more than $200 million in 2001). Celera aims to generate one million protein sequences a day that will reveal the protein composition of blood, spinal fluid and tumors, information that cannot be extracted from DNA sequence alone. But Celera will face formidable competition. For example, Hitachi, Oracle, and Myriad Genetics have forged a $500 million alliance to sequence the full complement of proteins from a variety of human tissues.

In the summer of 2001, Venter finally conceded what the biotech industry had been predicting for months, even years. Celera declared that it was moving into the drug discovery business, acquiring the biotech company Axys Pharmaceuticals and embarking on a new therapeutic program in cancer. The notion that Celera could flourish exclusively as an information company is history.

IT WILL BE YEARS, PROBABLY DECADES, before the full impact of the human genome sequence is felt. But it is not too early for the sequencing pioneers to be lauded for their efforts. Collins, Lander, Sulston and Venter have already won several prestigious prizes, but will one or more win the big one?

In February 2001, a few days after the genome announcement,

Venter and Collins held a joint press conference in San Francisco attended by hundreds of reporters. After they had dutifully recapped the highlights of the sequence, the final question was left to a Swedish journalist, who voiced what many of the reporters were doubtless wondering: did the sequencing of the human genome warrant the Nobel Prize?

Collins, without a moment's hesitation, leant into the microphone and replied, "I think it would have to be given to abou t 3,492 people to properly recognize all the people that have put an effort into this."

The heads of the press corps shifted in unison toward Venter. But for once, the man who had galvanized the race to sequence the human genome had nothing to say.

NOTES

253   "Scientist of the Year": M. D. Lemonick, "Gene Mapper," *Time,* December 25, 2000, pp.110–113. The coveted title of "Person of the Year" narrowly went to President George W. Bush.

254   "confused about the purpose of scientific publication": G. Kolata, "Celera to charge other companies to use its genome data," *New York Times,* December 8, 2000.

254   "the only intellectually honest position": M. Ashburner, *Genome Technology,* February 2001.

254   the critics of *Science* were being "short-sighted": Kolata, *New York Times.*

255   The next day, McKie drafted his newspaper's lead story: R. McKie, "Revealed: the secret of human behaviour," the *Observer,* February 11, 2001.

255   "but flies have two pairs of wings and can fly," M. Bobrow, Wellcome Trust press conference, London, February 12, 2001.

256   "Celera's assembly is remarkably similar to the HGP's assembly": Wellcome Trust press release.

256   "It didn't work," Lander insisted: J. Gillis, "The Gene Map and Celera's Detour," *Washington Post,* February 12, 2001.

256   "He's playing with half-truths and innuendos": Gillis, *Washington Post.*

257   testimonials from pharmaceutical clients: S. Hensley, "Celera's genome anchors it atop biotech," *Wall Street Journal,* February 12, 2001.

257   Under the striking headline: C. Abraham, "Private sector wins genetic-code race," *Globe and Mail,* February 12, 2001.

257   "This undoubtedly was—and is—a large step for mankind": Editorial, "Burying the hatchet," *Nature Biotechnology,* 19 (2001): 181.

257   one of the most enthralling scientific reports ever published: International Human Genome Sequencing Consortium. "Initial sequencing and analysis of the human genome," *Nature* 409 (2001): 860–921.

257   "to explain periodically to my three wonderful children": E. S. Lander, "After Deciphering the Map, the Next Task Is a Guidebook for the Human Genome," *New York Times,* September 12, 2000.

258   "the consortium published a correction": International Human Genome Sequencing Consortium, "Initial sequencing and analysis of the human genome (correction)," *Nature* 412, (2001): 565–566.

258   The Celera article was co-authored by 274 scientists: J. C. Venter et al., "The sequence of the human genome," *Science* 291 (2001): 1304–1351.

259   with apologies to David Letterman: F. Collins, address at the AAAS meeting, San Francisco, February 17, 2001.

259   "If it wasn't there before, it's there now!": J. C. Venter, address at the AAAS meeting, San Francisco, February 18, 2001.

261   "We thought we'd sequence a chimp": Venter, AAAS, February 18, 2001.

262   "I thought we'd keep it in the family": Venter, BioVision, Lyon, February 9, 2001.

262   a $500 million alliance: A. Pollack, "3 companies will try to identify all human proteins," *New York Times,* April 6, 2001.

263   "I think it would have to be given to about 3,492 people": F. Collins, AAAS press conference, February 17, 2001.

Notes

INTRODUCTION

1   "nature's finest midwife, interpreter and namesake": S. J. Gould, Introduction to *A Bedside Nature: Genius and Eccentricity in Science, 1869–1953*, ed. W. Gratzer (New York: Macmillan, 1996).

1   "the chic place for scientists to disport themselves": W. Gratzer, Foreword to *A Bedside Nature.*

2   the astounding claims of Jacques Benveniste: E. Davenas et al., "Human Basophil Degranulation Triggered by Very Dilute Antiserum Against IgE," *Nature* 333 (1988): 816–818; J. Maddox, J. Randi, and W. W. Stewart, "'High-Dilution' Experiments a Delusion," *Nature* 334 (1988): 287–290.

2   We wish to suggest a structure for the salt: J. D. Watson and F. H.C. Crick, "A Structure for Deoxyribose Nucleic Acid," *Nature* 171 (1953): 737–738.

3   mapping of the gene for an inherited form of Lou Gehrig's disease: T. Siddique et al., "Linkage of a Gene Causing Familial Amyotrophic Lateral Sclerosis to Chromosome 21 and Evidence of Genetic-Locus Heterogeneity," *New England Journal of Medicine* 3243 (1991): 1381–1384. *Nature* passed on publishing the report describing the localization of the ALS gene. The discovery of the mutated gene, ironically, the first gene ever isolated from chromosome 21, appeared in the journal two years later: D. R. Rosen et al., "Mutations in Cu/Zn Superoxide Dismutase Gene Are Associated with Familial Amyotrophic Lateral Sclerosis." *Nature* 362 (1993): 59–62.

3   the start of an ambitious international program: J. D. Watson, "The Human Genome Project: Past, Present, and Future," *Science* 248 (1990): 44–51.

3   the entertainment consisted of twenty of the most inspiring scientists: A full meeting report appears in *Nature Genetics* 4 (1993): 1–4.

4   the distinction of publishing the first article: F. S. Collins, "Positional Cloning—Let's Not Call It Reverse Anymore," *Nature Genetics* 1 (1992): 3–6.

4   Collins identified the gene for cystic fibrosis: K. Davies and M. White,

Breakthrough: The Race to Find the Breast Cancer Gene (New York: Wiley, 1996).

4   a high-profile collaboration between two genetics superstars: Ibid.

4   a revolutionary method for identifying the thousands of genes: M. D. Adams et al., "Complementary DNA Sequencing: Expressed Sequence Tags and Human Genome Project," *Science* 252 (1991): 1651–1656.

5   Venter dramatically changed the course of the Human Genome Project: L. Belkin, "Splice Einstein and Sammy Glick. Add a Little Magellan," *New York Times Magazine,* August 23, 1998, p. 26; L. Roberts, "The Gene Hunters," *U.S. News & World Report,* January 3–10, 2000, pp. 34–38; R. Preston, "The Genome Warrior," *New Yorker,* June 12, 2000, pp. 66–83.

6   "Intense competitors sometimes trade a little trash talk": D. Kennedy, "Coming Aboard," *Science* 288 (2000): 1745.

6   "Today, we are learning the language in which God created life": N. Wade, "Scientists Complete Rough Draft of Human Genome," *New York Times,* June 27, 2000; F. Golden and M. D. Lemonick, "The Race Is Over," *Time,* July 3, 2000, pp. 18–23.

7   "This is just halftime for genetics": E. Lander, Washington, D.C., June 26, 2000. The quote is adapted from J. F. Crow, "Two Centuries of Genetics: A View from Halftime," *Annual Review of Genomics and Human Genetics* 1 (2000).

7   Mendel had established the fundamental rules of play: G. Mendel, "Versuche ueber Pflanzen-Hybriden," *Verhandlungen Des Naturforschenden Vereines, Abnahndlungen, Brunn* 4 (1866): 3–47; V. Orel, *Gregor Mendel: The First Geneticist* (Oxford: Oxford University Press, 1996); R. Marantz Henig, *The Monk in the Garden* (Boston: Houghton Mifflin, 2000).

7   Sir Archibald Garrod proposed that a disease called alkaptonuria: A. E. Garrod, "The Incidence of Alkaptonuria: A Study in Chemical Individuality," *Lancet* II (1902): 1616–1620.

8   estimates have ranged wildly from 40,000 to over 100,000: S. Aparicio, "How to Count . . . Human Genes," *Nature Genetics* 25 (2000): 129–130.

9   a 3-billion-year-old Fortran code:.E. Lander, interview, Chevy Chase, Md., May 1999.

10   the childhood of the human race is about to come to an end: C. De Lisi and D. Galas, "The Legacy of the Human Genome Project: Some Lessons and Implications," August 11, 2000. Available at: http://www.Doubletwist.com

CHAPTER ONE: *KNIGHTS OF THE DOUBLE HELIX*

James D. Watson's *The Double Helix* (New York: Norton, 1980)—particularly this edition edited by Gunther Stent—is a remarkable account of the events leading up to the discovery of the structure of DNA, complete with original reviews and papers. Francis Crick's *What Mad Pursuit* (New York:

Basic Books, 1988) is less outspoken but provides excellent insights into the discovery of the genetic code. The indispensable history of this period is Horace Freeland Judson's *The Eighth Day of Creation* (Cold Spring Harbor, N.Y.: Cold Spring Harbor Laboratory Press, 1996). Another good historical overview is *The Path to the Double Helix* by Robert Olby (New York: Dover, 1994). The best book about the politics behind the launch of the Human Genome Project is *The Gene Wars* by Robert Cook-Deegan (New York: Norton, 1994). Many books cover the science that led up to the launch, none better than *Genome* by Jerry Bishop and Michael Waldholz (New York: Simon & Schuster, 1990).

11 Its significance would be comparable: R. Dulbecco, "A Turning Point in Cancer Research: Sequencing the Human Genome," *Science* 231 (1986): 1055–1056.

12 "The total human sequence is the grail": Cook-Deegan, *The Gene Wars*.

12 For physicists, the grail is the identification: P. Regan, *Daily Telegraph*, October 9, 1999.

12 Biophysicists consider the grail: G. Petsko, "The Grail Problem," *Genome Biology* 1 (2000).

12 "It is a sure sign of their alienation": R. C. Lewontin, "The Dream of the Human Genome," *New York Review of Books*, May 28, 1992, pp. 31–40.

12 Indeed, *human blueprint*: R. Shapiro, *The Human Blueprint* (New York: St. Martin's Press, 1991).

12 "Genes are not like engineering blueprints": I. Stewart, *Life's Other Secret* (New York: John Wiley, 1998).

12 "an immense book, a recipe of extravagant length": M. Ridley, *Genome* (London: Fourth Estate, 1999).

13 The principal cartographer of the chemical kingdom: Biographical details taken from D. Q. Posin, *Mendeleyev* (New York: Whittlesey House, 1948) and P. W. Atkins, *The Periodic Kingdom* (New York: Basic Books, 1995).

13 One night in February 1869: O. Sacks, "Everything in Its Place," *New York Times Magazine*, April 18, 1999, pp. 126–129.

13 Mendeleyev tumbled the sovereigns: Reminiscence by the treasurer of the Chemical Society, T. E. Thorpe, *Nature* 75 (1907): 373.

14 "The Human Genome Project aims to produce biology's periodic table": E. S. Lander, "The New Genomics: Global Views of Biology," *Science* 274 (1996): 536–539.

14 there is one potentially important difference: Crick, *What Mad Pursuit*.

14 "having a parts list doesn't tell you how to put it together": E. S. Lander, CNN, June 1, 2000.

15 "DOE's program for unemployed bomb-makers": Cook-Deegan, *The Gene Wars*.

16 during the late 1980s, the proportion of grants: B. D. Davis and colleagues, "The Human Genome and Other Initiatives," *Science* 249 (1990): 342–343.

16 one of the founding fathers: N. Wade, "A Founder of Modern Biology

Shapes the Genome Era, Too," *New York Times,* March 7, 2000. In September 2000, Brenner was named one of the winners of the Lasker Award (in prestige second only to the Nobel Prize).

16   I am not one who believes: S. Brenner, "The Human Genome: The Nature of the Enterprise," *Ciba Foundation Symposium* 149 (1990): 6–17.

17   "would simultaneously reassure Congress": J. D. Watson, "The Human Genome Project: Past, Present, and Future," *Science* 248 (1990): 44–49.

17   "no sane person would have contemplated": M. Perutz, *I Wish I'd Made You Angry Earlier* (Cold Spring Harbor, N.Y.: Cold Spring Harbor Laboratory Press, 1998).

18   The inducing substance, on the basis of its chemical and physical properties: O. T. Avery, C. M. MacLeod, and M. McCarty, "Studies on the Chemical Nature of the Substance Inducing Transformation of Pneumococcal Types. Induction of Transformation by a Desoxyribonucleic Acid Fraction Isolated from Pneumococcus Type III," *Journal of Experimental Medicine* 79 (1944): 137–158.

19   It was a chilling quote: J. B. Leathes, "Function and Design," *Science* 64 (1926): 187–194.

19   "If we are right," Avery wrote: M. McCarty, *The Transforming Principle* (New York: Norton, 1985); Judson, *The Eighth Day of Creation.*

19   "The evidence presented supports the belief": McCarty, *The Transforming Principle.*

19   "the most interesting and portentous biological experiment of the 20th century": Ibid.

20   "striking, but perhaps meaningless": Olby, *The Path to the Double Helix.*

21   Watson be played by Woody Allen: S. Brenner, "How the Quest Was Won," *Current Biology* 7 (1997): R596.

21   "tall, gawky, scraggly": F. Jacob, *The Statue Within* (Cold Spring Harbor, N.Y.: Cold Spring Harbor Laboratory Press, 1995).

21   "they shared the sublime arrogance": Perutz, *I Wish I'd Made You Angry Earlier.*

22   "two pitchmen in search of a helix": Judson, *The Eighth Day of Creation.*

22   "Though the odds still appeared against us": Watson, *The Double Helix.*

22   "Thus by the time I had cycled back to the gate": Ibid.

23   "winged into the Eagle to tell everyone": Ibid.

23   "the most exciting day of my life": S. Brenner, *Outstanding Papers in Biology* (London: Current Biology, 1993).

23   "the most famous event in biology": Watson, *The Double Helix.*

23   Watson and Crick launched the era of molecular biology: J. D. Watson and F. H.C. Crick, "A Structure for Deoxyribose Nucleic Acid," *Nature* 171 (1953): 737–738.

23   a second, more elaborate analysis: J. D. Watson and F. H.C. Crick, "Genetical Implications of the Structure of Deoxyribonucleic Acid," *Nature* 171(1953): 964–967.

23   "the greatest achievement of science": Watson, *The Double Helix.*

24 *Homage to Crick and Watson:* I. Gibson, *The Shameful Life of Salvador Dali* (London: Faber & Faber, 1997).

24 the authorship of *The Double Helix:* L. Roberts, "The Gene Hunters," *U.S. News & World Report,* January 3, 2000, pp. 34–38..

24 Two best-selling authors warmly welcomed Watson: Watson, *The Double Helix.*

24 "the Sylvia Plath of molecular biology": B. Maddox, "The Dark Lady of DNA?" *Observer,* March 5, 2000.

25 "I've just had the first good night's sleep in 20 years": C. Holden, "Award to DNA Sleuth," *Science* 287 (2000): 579.

25 Crick drafted a form letter to handle the deluge: H. Zuckerman, *Scientific Elite* (New York: Transaction, 1995).

26 the list of authors read "Alpher, Berthe and Gamow": S. W. Hawking, *A Brief History of Time* (New York: Bantam Books, 1988).

26 "The sequence of bases determines in a unique way": Judson, *The Eighth Day of Creation.*

27 "The Sequence Hypothesis": Ibid.

27 the "comma-free code": J. Maynard Smith, "Too Good To Be True," *Nature* 400 (1999): 223; see also Judson, *The Eighth Day of Creation.* Crick's argument went as follows: There are sixty-four possible triplet arrangements of four letters. Crick discounted triplets of the same letter (e.g., CCC, GGG) because they could be read out of phase, which left sixty permutations. For any permissible triplet (e.g., ATG), two alternative orders would be disallowed (GAT, TGA). That allowed for twenty possible triplets, equal to the number of amino acids.

27 "It seemed so pretty, almost elegant": Crick, *What Mad Pursuit.*

27 Crick eventually published a formal account: F. H.C. Crick, J. S. Griffith, and L. E. Orgel, "Codes Without Commas," *Proceedings of the National Academy of Sciences U.S.A.* 43 (1957): 416–421.

27 "the most elegant biological theory": Judson, *The Eighth Day of Creation.*

28 Nirenberg first presented his results: Judson, *The Eighth Day of Creation;* Crick, *What Mad Pursuit.*

28 In a stunning series of bacterial experiments: F. H.C. Crick, S. Brenner, et al., "General Nature of the Genetic Code for Proteins," *Nature* 192 (1961): 1227–1232.

28 "the most elegant paper *Nature* ever published": J. Maddox, Preface to Judson's, *The Eighth Day of Creation.*

29 "Only once," he later recalled: J. D. Watson, "The Human Genome Project: Past, Present, and Future."

29 "We need no more vivid reminders": Ibid.

29 Watson wrote back sarcastically: J. D. Watson, "Too Many Naughts," *Nature* 350 (1991): 550.

30 "The magnification is wrong," Davis complained: B. D. Davis et al., "The Human Genome and Other Initiatives," *Science* 249 (1990): 342–343.

30   "in the position of a nonmusician": J. S. Jones, "Songs in the Key of Life," *Nature* 354 (1991): 323.

30   Watson's former Ph.D adviser, Salvador Luria: R. Wright, "Achilles' Helix," *New Republic*, July 9–16, 1990, pp. 21–31; also see S. E. Luria, "Human Genome Program," *Science* 246 (1989): 873.

30   the "Human *Gnome* Initiative": C. Ezzell, "The Business of the Human Genome," *Scientific American*, July 2000, pp. 48–49.

30   Responding to charges of "Japan bashing": R. Lewin, "In the Beginning Was the Genome," *New Scientist*, July 21,1990, pp. 34–38; Cook-Deegan, *The Gene Wars*.

31   Watson abruptly quit on April 10, 1992: L. Roberts, "Why Watson Quit as Project Head," *Science* 256 (1992): 301–302.

31   Watson fired one last shot: C. Anderson, "Watson Resigns, Genome Project Open to Change," *Nature* 356 (1992): 549.

32   Walter Gilbert had an almost preternatural vision: W. Gilbert, "Towards a Paradigm Shift in Biology," *Nature* 349 (1991): 99.

CHAPTER TWO: *READING THE BOOK OF LIFE*

Matt Ridley's *Genome* (London: Fourth Estate, 1999) and Steve Jones's *The Language of Genes* (New York: Anchor Books, 1993) are both marvelous guides to the role of genes in health and disease. *Human Molecular Genetics*, 2nd ed., by Tom Strachan and Andrew Read (New York: John Wiley, 1999) is as good a textbook as any other in the field.

33   If the sequence were written: This analogy was adapted from one given in Strachan and Read, *Human Molecular Genetics*.

33   An average gene would run about 5 pages: The coding sequence of most genes does not consist of one contiguous run of bases, but is divided into segments, or exons, separated by stretches of garbled DNA called *introns*. These introns are spliced out of the final message.

33   "Could such a collection of unbelievably dull reading": Strachan and Read, *Human Molecular Genetics*.

34   Suppose we magnified a typical cell about 300,000 times: This analogy is adapted from Boyce Rensberger, *Life Itself* (New York: Oxford University Press, 1996).

35   switching on telomerase in cells: X. R. Jiang et al., "Telomerase Expression in Human Somatic Cells Does Not Induce Changes Associated with a Transformed Phenotype," *Nature Genetics* 21 (1999): 111–114; C. P. Morales et al., "Absence of Cancer-Associated Changes in Human Fibroblasts Immortalized with Telomerase," *Naure Genetics* 21 (1999): 115–118.

35   The aglet at the end of a human chromosome: DNA sequences have a directionality, based on the orientation of certain chemical groups in each

base. Sequences are traditionally read in the 5' to 3' direction (hence purists might quibble with the suggestion that G is the first letter).

36   the tip of the telomere folds back: J. D. Griffith et al., "Mammalian Telomeres End in a Large Duplex Loop," *Cell* 97 (1999): 503–514.

36   "deep in conversation, battering away at each other": R. Shapiro, *The Human Blueprint* (New York: St. Martin's Press, 1991).

36   "the best idea I have ever had": F. Sanger, "Sequences, Sequences and Sequences," *Annual Review of Biochemistry* 57 (1988): 1–28.

37   by sequencing the 5,375 bases of a virus: F. Sanger et al., "Nucleotide Sequence of Bacteriophage øX174 DNA," *Nature* 265 (1977): 687–695.

37   "I like to call Fred Sanger's virus": F. J. Dyson, *The Sun, the Genome, and the Internet* (New York: Oxford University Press, 1999).

37   Sanger's recipe: F. Sanger, S. Nicklen, and A. R. Coulson. "DNA Sequencing with Chain-Terminating Inhibitors," *Proceedings of the National Academy of Sciences U.S.A.* 74 (1977): 5463–5467.

37   Sanger was invited to write a brief memoir: Sanger, "Sequences, Sequences and Sequences."

38   "My respect for [*Jurassic Park* scientist] Dr. Wu's scientific ability": M. S. Boguski, "A Molecular Biologist Visits Jurassic Park," *BioTechniques* 12 (1992): 668–669.

38   a segment of DNA for the sequel: M. Crichton, *The Lost World* (New York: Knopf, 1995). Bioinformatics expert Will Gilbert has created a simple online test of Crichton's dino DNA sequence, which shows how to conduct a search for sequences homologous to that presented in *The Lost World*. In addition to identifying the source material that provided the sequence used in the book, users can also convert the DNA sequence into the corresponding protein sequence, and in so doing reveal the hidden message. See http://bioinformatics.unh.edu/SMART/

40   estimates varied from about 30,000: S. Aparicio, "How to Count . . . Human Genes," *Nature Genetics* 25 (2000): 129–130.

40   Ewan Birney has started an international betting contest: R. L. Hotz, "Betting on the Genes in a Human," *Los Angeles Times,* July 30, 2000. The winner of the Genesweep lottery will probably be announced in 2003, and also gets a signed, leather-bound copy of James Watson's *The Double Helix*. Details are available at http://www.ensembl.org/genesweep.html

41   The sites that vary most commonly: N. Wade, "In the Hunt for Useful Genes, a Lot Depends on 'Snips'," *New York Times,* August 11, 1998.

42   The Newcastle group checked the sequence of this gene: P. Valverde et al., "Variants of the Melanocyte Stimulating Hormone Receptor Gene Are Associated with Red Hair and Fair Skin in Humans," *Nature Genetics* 11 (1995): 328–330.

42   changes in a single gene have a profound effect: E. Healy et al., "Melanocortin-1-Receptor Gene and Sun Sensitivity in Individuals Without Red Hair," *Lancet* 355 (2000): 1072–1073; R. A. Sturm, N. F. Box,

and M. Ramsay, "Human Pigmentation Genetics: The Difference Is Only Skin Deep," *BioEssays* 20 (1998): 712–721. In general, variations in the melanocortin-1 receptor correlate with an inability to tan and increased susceptibility to sunburn.

42   "variation at this locus": K. Owens and M-C. King, "Genomic Views of Human History," *Science* 286 (1999): 451–453.

42   Evolutionary biologists are searching for clues: Ridley, *Genome.* See chromosome 15.

43   researchers have catalogued more than 5,000 distinct genetic disorders: V. A. McKusick, *Mendelian Inheritance in Man,* 12th ed. (Baltimore: Johns Hopkins University Press, 1998). OMIM, the online, updated version of the catalogue, is available at: http://www.ncbi.nlm.nih.gov/omim

43   In more than 1,000 cases: S. E. Antonarakis and V. A. McKusick, "OMIM Passes the 1,000-Disease-Gene Mark," *Nature Genetics* 25 (2000): 11.

43   DNA alterations that have a frequency of less than 2 percent: The definition of a mutation is found at: http://www.nhgri.nih.gov/DIR/VIP/glossary

44   eggs from young female models: T. Weingarten and M. Hosenball, "A Fertile Scheme," *Newsweek,* November 8, 1999, pp. 78–79.

44   the ratio was less than 2 to 1: N. B. Bohossian, H. Skaletsky, and D. C. Page, "Unexpectedly Similar Rates of Nucleotide Substitution Found in Male and Female Hominids," *Nature* 406 (2000): 622–625.

44   The double helix is under constant bombardment: M. Gaines. "Inside Science: Radiation and Risk," *New Scientist* 18, March 2000, pp. 1–4.

45   In *The Odyssey:* A. Mandelbaum, *The Odyssey of Homer* (Berkeley: University of California Press, 1990), book 9, pp. 171–192.

45   a condition called holoprosencephaly: Maximillian Muenke, interview, July 14, 2000.

45   a single-letter change in a gene called *sonic hedgehog:* E. Roessler et al., "Mutations in the Human *Sonic Hedgehog* Gene Cause Holoprosencephaly," *Nature Genetics* 14 (1996): 357–360. *Sonic hedgehog* was named after the popular computer game character, and is related to a fruit fly gene called *hedgehog,* so named because flies with mutations in this gene develop a bristled appearance.

46   using this compound to treat basal cell carcinoma: A. E. Bale, "Sheep, Lilies and Human Genetics," *Nature* 406 (2000): 944–945.

46   The alteration of an A to a G: W. Ahmad et al., "Alopecia Universalis Associated with a Mutation in the Human *Hairless* Gene," *Science* 279 (1998): 720–724.

46   Inherited defects in the energy-producing enzyme: Pythagoras, the famous Greek philosopher, warned his followers, *"A fabis abstinete"*—eat no beans. The prevalence of carriers for glucose-6-phosphate dehydrogenase deficiency due to malaria resistance also causes marked susceptibility to toxins in undercooked fava beans, resulting in anemia. B. N. Ames, "Dietary Carcinogens and Anticarcinogens," *Science* 221 (1983): 1256.

47 warding off the *Salmonella typhi* bacteria: G. B. Pier et al. *"Salmonella typhi*
Uses CFTR to Enter Intestinal Epithelial Cells," *Nature* 393 (1998): 79–82.

47 Mäntyranta carried a mutation: G. D. Longmore, "Erythropoietin Recep-
tor Mutations and Olympic Glory," *Nature Genetics* 4 (1993): 108–109; A.
de la Chapelle, A-L. Traskelin, and E. Juvonen, "Truncated Erythropoietin
Receptor Causes Dominantly Inherited Benign Human Erythrocytosis,"
Proceedings of the National Academy of Sciences U.S.A. 90 (1993): 4495–4499;
C. Aschwanden, "Gene Cheats," *New Scientist,* January 15, 2000, pp. 25–29.

48 a corrupt version of the *CCR5* gene: R. Liu et al., "Homozygous Defect in
HIV-1 Coreceptor Accounts for Resistance of Some Multiply-Exposed
Individuals to HIV-1 Infection," *Cell* 86 (1996): 367–377; M. Samson et al.,
"Resistance to HIV-1 Infection in Caucasian Individuals Bearing Mutant
Alleles of the CCR-5 Chemokine Receptor Gene," *Nature* 382 (1996):
722–725; M. Dean et al., "Genetic Restriction of HIV-1 Infection and
Progression to AIDS by a Deletion Allele of the CKR5 Structural Gene,"
Science 273 (1996): 1856–1862.

48 They possess a variant form of apolipoprotein A1: C. Cox, "Italian Village
Gives World the Secret of Long Life," *Daily Telegraph,* January 23, 2000.

49 the sudden reappearance of a trait: B. K. Hall, "Atavisms and Atavistic Mutations," *Nature Genetics* 10 (1995): 126–127.

49 In the small town of Zacadecas in Mexico: N. Angier, *Woman* (Boston:
Houghton Mifflin, 1999).

49 has traced the responsible gene to a region of the X chromosome: L. E.
Figuera, M. Pandolfo, P. W. Dunne, J. Cantu, and P. I. Patel, "Mapping of the
Congenital Generalized Hypertrichosis Locus to Chromosome Xq24–
q27.1," *Nature Genetics* 10 (1995): 202–207. The decision we took to publish this paper in 1995 on the grounds of unusual general interest was
borne out by worldwide media interest, not all of it positive. The *New York
Times* called me a "carnival barker" guilty of "indulging in a bit of sensationalism"; curiously, the *Times* piece ran beneath a picture of Lon Chaney,
Hollywood's original werewolf.

50 about four new mutations that change the sequence: A. Eyre-Walker and
P. D. Keightley, "High Genomic Deleterious Mutation Rates in Hominids," *Nature* 397 (1999): 344–347; J. Crow, "The Odds of Losing at Genetic Roulette," *Nature* 397 (1999): 293–294.

50 headaches, stomach upsets, poor vision, and other ailments: J. F. Crow, "The
High Spontaneous Mutation Rate: Is It a Health Risk?" *Proceedings of the
National Academy of Sciences U.S.A.* 94 (1997): 8380–8386.

50 such variations hold the key: A full account of Botstein's discovery is given
in J. E. Bishop and M. Waldholz, *Genome* (New York: Simon & Schuster,
1990). See also K. Davies and M. White, *Breakthrough* (New York: Wiley,
1996)

CHAPTER THREE: *The Eye of The TIGR*

52   *Time* magazine's list of the top 100 greatest minds: *Time,* March 29, 1999.

52   Craig Venter was born on October 14, 1946: L. Belkin, "Splice Einstein and Sammy Glick. Add a Little Magellan." *New York Times Magazine,* August 23, 1998, p. 26; J. Shreeve, "The Code Breaker," *Discover* (May 1998): 45–51; N. Wade, "The Genome's Combative Entrepreneur," *New York Times,* May 18, 1999; T. Anton, *Bold Science: Seven Scientists Who Are Changing Our World* (New York: W. H. Freeman, 2000).

53   "First Lyndon Johnson messed up my surfing career": B. Appleyard, "Is This Man the Enemy of Science? Or Is He the Saviour of Mankind?" *Sunday Times Magazine* [London], June 11, 2000, pp. 14–21.

53   "You can't live through a situation like [Vietnam]": J. C. Venter, *CNN Newstand,* June 3, 1999.

53   "I felt very fortunate to survive the year": Ibid.

55   Venter requested funding from Watson's genome center: R. Cook-Deegan, *The Gene Wars* (New York: Norton, 1994).

55   Venter's team sequenced approximately 60,000 bases of DNA: A. Martin-Gallardo et al., "Automated DNA Sequencing and Analysis of 106 Kilobases from Human Chromosome 19q13.3," *Nature Genetics* 1 (1992): 34–39; W. R. McCombie et al., "Expressed Genes, *Alu* Repeats and Polymorphisms in Cosmids Sequenced from Chromosome 4p16.3," *Nature Genetics* 1 (1992): 348–353.

55   "informative polymorphisms will be present near [any] gene": McCombie et al., "Expressed Genes."

58   In 1983, Paul Schimmel and two colleagues at MIT: S. D. Putney, W. C. Herlihy, and P. Schimmel, "A New Troponin T and cDNA Clones for 13 Different Muscle Proteins, Found by Shotgun Sequencing," *Nature* 302 (1983): 718–821.

58   "It cannot be evaded": S. Brenner, "The Human Genome: The Nature of the Enterprise," *Ciba Foundation Symposium* 149 (1990): 6–17.

58   "The development of methods of cloning and sequencing DNA": Ibid.

58   By 1988, Brenner's group was sampling cDNAs: S. Brenner, "Loose Ends," *Current Biology* 4 (1994): 384.

59   In April 1991, Venter withdrew his stalled proposal: Cook-Deegan, *The Gene Wars.*

59   Venter presented his new EST strategy in a landmark article: M. D. Adams et al., "Complementary DNA Sequencing: Expressed Sequence Tags and Human Genome Project," *Science* 252 (1991): 1651–1656.

59   That prediction was borne out six years later: A. Joutel et al., *"Notch3* Mutations in CADASIL, a Hereditary Adult-Onset Condition Causing Stroke and Dementia," *Nature* 383 (1996): 707–710.

60   "In our own laboratory": Adams et al., "Complementary DNA Sequencing."

60   Venter's pronouncements upset several senior figures: L. Roberts, "Gambling on a Shortcut to Genome Sequencing," *Science* 252 (1991): 1618–1619.

60   in February 1992, Venter was in the news again: M. D. Adams et al., "Sequence Identification of 2,375 Human Brain Genes," *Nature* 355 (1992): 632–634.

61   Criticism flared briefly that some of the ESTs: T. R. Burglin and T. M. Barnes, "Introns in Sequence Tags," *Nature* 357 (1992): 367.

61   While scientists pondered the merits of EST sequencing: J. Maddox, *Nature* 357 (1992): 13.

61   "I didn't want to miss the boat": L. Thompson, "NIH's Rush to Patent Human Genes," *Washington Post,* October 28, 1991.

62   Venter disclosed it during a Senate meeting: Cook-Deegan, *The Gene Wars.*

62   He challenged the "non-obviousness": Thompson, "NIH's Rush."

62   "could be run by monkeys": Cook-Deegan, *The Gene Wars.*

62   "I compare this information": R. Herman, "The Great Gene Gold Rush," *Washington Post,* June 16, 1992.

62   buyers of uncut diamonds: Brenner, "The Human Genome."

63   "It makes a mockery of what most people feel": L. Roberts, "NIH Gene Patents, Round Two," *Science* 255 (1992): 912–913.

63   Vice President Al Gore charged that the patent applications: Cook-Deegan, *The Gene Wars.*

63   "a patent should not be granted": P. Aldhous, "MRC Follows NIH on Patents," *Nature* 356 (1992): 98.

63   "a little silly . . . a tempest in a teapot": Thompson, "NIH's Rush."

63   "The rationale," Healy explained later: B. Healy, "On Patenting Genes," *New England Journal of Medicine* 327 (1992): 664–668.

63   "vague, indefinite, misdescriptive": Cook-Deegan, *The Gene Wars.*

63   Particularly damaging to the NIH case: C. Anderson, "NIH cDNA Patent Rejected; Backers Want to Amend Law," *Nature* 359 (1992): 263.

64   "Craig wasn't just turned down": J. Gillis, "One Man's Race to Map Genetic Code," *Washington Post,* August 22, 1998.

64   Venter spurned a $70 million offer: J. Carey, "The Gene Kings," *Business Week,* May 8, 1995, pp. 72–78.

64   Steinberg's plan, outlined in a fifteen-minute meeting with Venter: S. Sugawara, "A Healthy Vision," *Washington Post,* November 16, 1992.

64   "It's really remarkable," he exclaimed: Ibid.

64   "By the year 2000": Ibid.

65   "when we got married, I had made millions too!": L. Weeks, "Mr. Green Genes," *Washington Post,* February 17, 1998.

65   Steinberg invited Haseltine down to Washington: Ibid.

65   Venter set about creating the world's largest DNA sequencing institute: M. D. Adams et al., "A Model for High-Throughput Automated DNA Sequencing and Analysis Core Facilities," *Nature* 368 (1994): 474–475.

66   "He has never invented anything": Carey, "The Gene Kings."
67   *Business Week* dubbed Venter and Haseltine: Ibid.
67   "I had a radical idea, it worked": Ibid.
67   Venter's financial stake in HGS: L. M. Fisher, "Mining the Genome: Big Science as Big Business," *New York Times,* January 30, 1994.
67   Venter jokingly told a *Washington Post* reporter: A. Phillips, "He Leaves His Body to Science, His Heart to Sailing," *Washington Post,* November 24, 1996.
68   captained *Sorcerer* to victory in the Atlantic Cup Challenge: Ibid.

CHAPTER FOUR: *LOADING THE BASES*

The scientific breakthroughs in gene mapping and disease gene identification are well told in *Genome* (New York: Simon & Schuster, 1990) by Jerry Bishop and Michael Waldholz, as well as Waldholz's *Curing Cancer* (New York: Simon & Schuster, 1997). Further information about Francis Collins's life and career is contained in *Breakthrough: The Race to Find the Breast Cancer Gene* (New York: Wiley, 1996) by Kevin Davies and Michael White.

69   "My whole career has been spent training for this job": S. Veggeberg, "Scientists Express Relief as Francis Collins Is Named New Director of NIH Genome Project," *Scientist,* May 3, 1993.
70   Born in April 1950: Davies and White, *Breakthrough.*
70   Bruce Bryer was born in Washington State: Bishop and Waldholz, *Genome.*
71   a clever scheme called "positional cloning": S. H. Orkin, "Reverse Genetics and Human Disease," *Cell* 47 (1986): 845–850; F. S. Collins, "Positional Cloning: Let's Not Call It Reverse Anymore," *Nature Genetics* 1 (1992): 3–6.
71   "It helps me to remember what matters": G. Kolata, "Unlocking the Secrets of the Genome," *New York Times,* November 30, 1993.
71   they had snared the cystic fibrosis gene: J. M. Rommens et al., "Identification of the Cystic Fibrosis Gene: Chromosome Walking and Jumping," *Science* 245 (1989): 1059–1065; J. R. Riordan et al., "Identification of the Cystic Fibrosis Gene: Cloning and Characterization of Complementary DNA," *Science* 245 (1989): 1066–1072; B.-S. Kerem et al., "Identification of the Cystic Fibrosis Gene: Genetic Analysis," *Science* 245 (1989): 1073–1080; see also K. Davies, "The Search for the Cystic Fibrosis Gene," *New Scientist,* October 21, 1989, pp. 54–58.
72   Some judicious lobbying by Max Cowan: W. M. Cowan, interview, September 14, 2000.
72   both articles were published on the same day: L. Roberts, "Down to the Wire for the NF Gene," *Science* 249 (1990): 236–238.
72   "appreciating something that up until then": F. Collins, "Faith and Reason," PBS, September 1998.

73   "Driving with my father through a wooded road": A. Wexler, *Mapping Fate* (Berkeley: University of California Press, 1995).

73   "the complete ruin of a human being": Ibid.

74   all of the Huntington's patients tested: Bishop and Waldholz, *Genome*.

74   It's just like reading a great mystery novel: Wexler, *Mapping Fate*.

75   a telltale difference in the DNA from chromosome 4: Huntington's Disease Collaborative Research Group, "A Novel Gene Containing a Trinucleotide Repeat That Is Expanded and Unstable on Huntington's Disease Chromosomes," *Cell* 72 (1993): 971–983.

76   the first patchy framework maps: H. Donis-Keller et al., "A Genetic Linkage Map of the Human Genome," *Cell* 50 (1987): 319–337.

76   "Good God," thought Barataud: P. Rabinow, *French DNA* (Chicago: University of Chicago Press, 1999).

76   "a turning point in the history of humanity": Ibid.

77   Eighteen months later, in June 1994: G. Gyapay et al., "The 1993–1994 Généthon Human Genetics Linkage Map," *Nature Genetics* 7 (1994): 246–339.

77   he anticipated the future "amelioration of our genetic patrimony": Rabinow, *French DNA*.

78   rapidly building mouse maps: W. F. Dietrich et al., "A Genetic Map of the Mouse with 4,006 Simple Sequence Length Polymorphisms," *Nature Genetics* 7 (1994): 220–245.

78   On March 7, 1994, Froguel appealed in writing: Rabinow, *French DNA*.

79   "I don't want to be alive": "Fingering Fibroblast Growth Factor Receptors," *Nature Genetics* 8 (1994): 1–2.

80   Collins said the technique had moved "from perditional to traditional": F. S. Collins, "Positional Cloning Moves from Perditional to Traditional," *Nature Genetics* 9 (1995): 347–350.

80   "Gone were the opportunities for brief flirtations with different markers": B. Sykes, "Bone Disease Cracks Genetics," *Nature* 348 (1990): 18–20.

81   It is announced that molecular biologists have discovered: H. F. Judson, *The Eighth Day of Creation* (Cold Spring Harbor, N.Y.: Cold Spring Harbor Laboratory Press, 1996).

82   The crucial breakthrough came in 1990: Davies and White, *Breakthrough*.

82   James Watson counseled her: Waldholz, *Curing Cancer*.

83   It was only when Weber made a timely call: Davies and White, *Breakthrough*.

83   Myriad had won the race: Y. Miki et al., "A Strong Candidate for the Breast and Ovarian Cancer Susceptibility Gene *BRCA1*," *Science* 266 (1994): 66–71.

83   "a genetic trophy so ferociously coveted and loudly heralded": N. Angier, "Scientists Identify a Mutant Gene Tied to Hereditary Breast Cancer," *New York Times*, September 15, 1994.

83   She closed with an emotional appeal for her colleagues to fight on: Davies and White, *Breakthrough*.

84   The result was positive: G. Kolata, "Breaking Ranks, Lab Offers Test to As-sess Risk of Breast Cancer," *New York Times,* April 1, 1996.

84   "as a small token of our concern": R. Weiss, "Genetic Testing's Human Toll," *Washington Post,* July 21, 1999.

84   a large team headed by geneticist Yosef Shiloh: K. Savitsky et al., "A Single Ataxia Telangiectasia Gene with a Product Similar to PI-3 Kinase," *Science* 268 (1995): 1749–1753.

85   the genetic origins of a subset of acute myeloid leukemia: P. Liu et al., "Fusion Between Transcription Factor CBFß/PEBP2ß and a Myosin Heavy Chain in Acute Myeloid Leukemia," *Science* 261 (1993): 1041–1044.

85   One unusually conscientious reviewer: C. Macilwain, " 'Ambition and Im-patience' Blamed for Fraud," *Nature* 384 (1996): 6–7.

85   A subsequent inquiry: "Findings of Scientific Misconduct," *NIH Guide* 26, July 18, 1997.

86   Calling it "the most painful experience of my professional career": P. Recer, "Doctoral Candidate Faked Experiments," *Seattle Times,* October 30, 1996.

86   any hope of sequencing the human genome: F. Collins and D. Galas, "New Five-Year Plan for the U.S. Human Genome Project," *Science* 262 (1993): 43–46.

86   dense maps of genetic markers: C. Dib et al., "A Comprehensive Genetic Map of the Human Genome Based on 5,264 Microsatellites," *Nature* 380 (1996): 152–154. *Nature* published an extended reprint of the Généthon linkage map, featuring William Blake's painting of a nude Isaac Newton perched on a rock, intently measuring a line on a scroll of paper with a compass.

86   However, these clones were notoriously unstable: C. Anderson, "Genome Shortcut Leads to Problems," *Science* 259 (1993): 1685–1687.

86   In October 1995, Olson penned a commentary: M. V. Olson, "A Time to Sequence," *Science* 270 (1995): 394–396.

87   "All human genome sequence information should be freely available": Sanger Centre and Washington University Genome Center, "Toward a Complete Human Genome Sequence," *Genome Research* 8 (1998): 1097–1108.

87   By early 1998, only two of the pilot centers: E. Pennisi, "DNA Sequencers' Trial by Fire," *Science* 280 (1998): 814–817.

87   only 2 to 3 percent of the DNA in the human genome had been se-quenced: L. Rowen, G. Mahairas, and L. Hood, "Sequencing the Human Genome," *Science* 278 (1997): 605–607.

87   "I don't know how we're going to do that": Pennisi, "DNA Sequencers' Trial by Fire." Adams's quote appeared in the May 8, 1998, issue of *Science,* the same day that Venter and Hunkapiller announced the creation of their new genome company, with Adams playing a central role.

88   In April 1996, a coalition of 633 scientists based in one hundred different

laboratories: A. Goffeau et al., "The Yeast Genome Directory," *Nature* 387 (suppl.) (1997): 1–105.

88   "the security council of model genetic organisms": G. Fink, "Anatomy of a Revolution," *Genetics* 149 (1998): 473–477. Membership of the Security Council includes the lambda phage; two bacteria (*Bacillus subtilis* and *Escherichia coli*); a fungus *(Saccharomyces cerevisiae)*; a unicellular alga *(Chlamydomonas reinhardtii); a* plant *(Arabidopsis); a* worm *(Caenorhabditis elegans); an* insect *(Drosophila melanogaster); a* mammal *(Mus musculus);* and of course humans (although Fink insists the latter "smacks of self-interest on the part of the selection committee").

88   "life had become subtle enough for the hierarchy of genes": N. Calder, *Timescale* (New York: Viking Press, 1983).

88   "As a more long term possibility": W. B. Wood et al., *The Nematode* (Cold Spring Harbor, N. Y.: Cold Spring Harbor Laboratory Press, 1988).

89   Brenner already had a simple organism in mind: Ibid.

89   "I have a weakness for grandiose, meaningless projects": M. Chalfie, "The Worm Revealed," *Nature* 396 (1998): 620–621.

89   Their success attracted the attention of American venture capitalist Frederick Bourke: L. Roberts, "Sequencing Venture Sparks Alarm," *Science* 255 (1992): 677–678.

90   "The work with the nematode illuminated the path forward": A. Klug, Wellcome Trust press release, June 28, 2000.

90   In December 1998, Sulston and Waterston finally completed their 15-year, $45 million endeavor: C. *elegans* Sequencing Consortium, "Genome Sequence of the Nematode *C. elegans:* A Platform for Investigating Biology," *Science* 282 (1998): 2012–2018.

91   profound effects on the longevity and social interactions of worms: G. Ferry, "The Human Worm," *New Scientist,* December 5, 1998, pp. 32–35.

91   "We have come to realize humans are more like worms": N. Wade, "Animal's Genetic Program Decoded, in a Science First," *New York Times,* December 11, 1998.

91   "totally immoral and disgusting": K. Perry, "John Sulston: Altruist or Moralist?" *Guardian,* June 26, 2000.

91   "I find it a terrible shame that this important moment": "The Map Maker," *Guardian,* April 19, 1999.

92   Genome information should not be patented, he argues: J. Sulston, "Forever Free," *New Scientist,* April 1, 2000, pp. 46–47.

92   Venter wants "to establish a monopoly position on the human sequence": Perry, "John Sulston: Altruist or Moralist?"

92   "Galileo, a great hero of mine": "The Map Maker."

92   "Genetic engineering is no more than a branch of surgery": J. Gerard, "Putting the Monsters to the Back of His Mind," *Sunday Times* [London], July 2, 2000.

92   "a disaster—no cohesion, no focus, no game plan": D. Birch, "Daring Sprint to the Finish," *Baltimore Sun,* April 13, 1999.

93   "When it was the human genome initiative, it was really great": S. Brenner, interview, June 24, 2000.

93   it became "politically imperative" to persevere with existing sequencing technology: F. Dyson, *The Sun, the Genome, and the Internet* (New York: Oxford University Press, 1999).

93   derisively called the leaders of the sequencing centers the "Liars' Club": N. Wade, *New York Times*, March 10, 1998.

CHAPTER FIVE: *THE CIRCLE OF LIFE*

94   By 1994, TIGR was operating at full capacity: M. Adams et al. "A Model for High-Throughout Automated DNA Sequencing and Analysis Core Facilities," *Nature* 368 (1994): 474–475.

94   TIGR became the prototype for the genomics revolution: J. Carey, "Gene Hunters Go for the Big Score," *Business Week,* August 16, 1993.

95   By the early 1990s, Johns Hopkins University's Bert Vogelstein: M. Waldholz, *Curing Cancer* (New York: Simon & Schuster, 1997).

96   Vogelstein espoused his multistep model for colon cancer: B. Vogelstein et al., "Genetic Alterations During Colorectal-Tumor Development," *New England Journal of Medicine* 319 (1988): 525–532.

96   Finnish researchers brought to his attention: Waldholz, *Curing Cancer.*

96   Vogelstein and veteran Finnish geneticist Albert de la Chapelle: P. Peltomaki et al., "Genetic Mapping of Locus Predisposing to Human Colorectal Cancer," *Science* 260 (1993): 812–816.

97   Kolodner had already submitted his paper: R. Fishel et al., "The Human Mutator Gene Homolog MSH2 and Its Association with Hereditary Nonpolyposis Colon Cancer," *Cell* 75 (1993): 1027–1038.

97   Vogelstein rushed his paper to the same journal: F. S. Leach et al., "Mutations of a MutS Homolog in Hereditary Nonpolyposis Colon Cancer," *Cell* 75 (1993): 1215–1225.

97   Vogelstein's partner Ken Kinzler telephoned Venter: R. F. Service, "Stalking the Start of Colon Cancer," *Science* 263 (1994): 1559–1560; Waldholz, *Curing Cancer.*

98   the second gene was reported simultaneously: C. E. Bronner et al., "Mutation in the DNA Mismatch Repair Gene Homologue *hMLH1* Is Associated with Hereditary Non-Polyposis Colon Cancer," *Nature* 368 (1994): 258–261; N. Papadopoulos et al., "Mutation of a *mutL* Homolog in Hereditary Colon Cancer," *Science* 263 (1994): 1625–1629.

99   Tanzi and Schellenberg published the discovery of a single misspelling: E. Levy-Lahad, W. Wasco, et al., "Candidate Gene for the Chromosome 1 Familial Alzheimer's Disease Locus," *Science* 269 (1995): 973–977.

100   "They're coming after us with torches and pitchforks": E. Marshall, "A Showdown over Gene Fragments," *Science* 266 (1994): 208.

100 That year, he flew home from Europe for an emergency operation: J. Shreeve, "The Code Breaker," *Discover* (May 1998): pp. 44–51.

100 At an October 1994 conference: Marshall, "A Showdown over Gene Fragments": 208–210.

100 all of Merck's sequences would be made publicly available: D. Dickson, " 'Gene Map' Plan Highlights Dispute over Public vs. Private Interests," *Nature* 371 (1994): 365–366.

101 "If you publish this Venter stuff": J. Maddox, "Directory to the Human Genome," *Nature* 376 (1995): 459–460.

101 *Nature* finally published "The Genome Directory": M. D. Adams et al., "Initial Assessment of Human Gene Diversity and Expression Patterns Based Upon 83 Million Nucleotides of cDNA Sequence," *Nature* 377 (suppl.) (1995): 3–174.

102 The final prediction estimate for the total number of human genes: C. Fields, M. D. Adams, O. White, and J. C. Venter, "How Many Genes in the Human Genome?" *Nature Genetics* 7 (1994): 345–346.

103 "What the hell do you think you're doing": J. Cohen, "How Many Genes Are There?" *Science* 275 (1995): 769.

103 Estimates range from about 40,000 to well over 120,000: S. Aparicio, "How to Count . . . Human Genes," *Nature Genetics* 25 (2000): 129–130.

103 "People like to have a lot of genes": Cohen, "How Many Genes Are There?"

103 systematically mapping the location of the ESTs: G. D. Schuler et al., "The Human Transcript Map," *Science* 274 (1996): 547–562.

103 "Once we have the sequence data": J. C. Venter, "Genome Directory" press pack, September 28, 1995.

104 Venter was attending a genetics conference in Bilbao: D. Birch, "The Seduction of a Scientist," *Baltimore Sun,* April 12, 1999.

104 "I didn't know there was another Hamilton Smith at Hopkins": D. Birch, "Hamilton Smith's Second Chance," *Baltimore Sun,* April 11, 1999.

105 In the aftermath of the Nobel Prize: Birch, "The Seduction of a Scientist."

105 "My guess is, we both wish": Ibid.

106 Venter liked it almost immediately: Ibid.

106 Venter finally received a pink sheet: N. Wade, "Bacterium's Full Gene Makeup Is Decoded," *New York Times,* May 26, 1995.

107 The accolades flew thick and fast: N. Wade, "First Sequencing of Cell's DNA Defines Basis of Life," *New York Times,* August 1, 1995.

107 Venter's team published the first bacterial genome sequence in *Science:* R. D. Fleischmann et al., "Whole-Genome Random Sequencing and Assembly of *Haemophilus influenzae* Rd," *Science* 269 (1995): 496–512.

107 "Better to call a spade a spade": K. M. Devine and K. Wolfe, "Bacterial Genomes: A TIGR in the Tank," *Trends in Genetics* 11 (1995): 429–431.

108 Venter's team began sequencing the DNA of *Mycoplasma genitalium:* C. M. Fraser, "The Minimal Gene Complement of *Mycoplasma genitalium,*" *Science* 270 (1995): 397–403.

109   That left about 350 genes in *M. genitalium:* C. A. Hutchison III et al., "Global Transposon Mutagenesis and a Minimal Mycoplasma Genome," *Science* 286 (1999): 2165–2169.

110   "One way to identify a minimal gene set for self-replicating life": Ibid.

110   "The key point," says Fraser: C. Fraser, "Life: What Exactly Is It?" *HMS Beagle,* June 25, 1999. Available at: http://www.biomednet.com.

110   "That vital spark from inanimate matter to animate life": R. Fortey, *Life* (New York: Knopf, 1997).

110   When Bill Haseltine was asked once whether he was playing God: L. Weeks, "Mr. Green Genes," *Washington Post,* February 17, 1998.

110   "This is a divide which takes us into a brave new world": "The Mysteries of Creation," BBC Online, December 10, 1999.

110   The panel concluded that experiments minimizing the number of genes: M. K. Cho, D. Magnus, A. L. Caplan, D. McGee, and the Ethics of Genomics Group, *Science* 286 (1999): 2087–2090.

111   *Alvin* combed the floor of the Pacific Ocean: V. Morell, "Life's Last Domain," *Science* 273 (1996): 1043–1045.

112   Woese had to rely on an indirect strategy: V. Morell, "Microbiology's Scarred Revolutionary," *Science* 276 (1997): 699–702.

112   a "sulphorous surrogate of Hades": Fortey, *Life.*

113   "was greeted with wrath and ridicule": R. A. Garrett, *"Methanococcus jannaschii* and the Golden Fleece," *Current Biology* 6 (1996): 1377–1380.

113   Venter and Woese were not the first to embark on sequencing archaea: Ibid.

114   the results appeared in *Science* in August 1996: C. J. Bult et al., "Complete Genome Sequence of the Methanogenic Archaeon, *Methanococcus jannaschii," Science* 273 (1996): 1058–1073.

114   "It tells us things about life on this planet": "Proof of Third Class of Life Sheds New Light on Possible Life in the Universe," UPI, August 22, 1996.

114   "We are one tribe with bacteria": Fortey, *Life.*

115   Venter's group sequenced the genome of *Thermotoga maritima:* K. E. Nelson et al., "Evidence for Lateral Gene Transfer Between Archaea and Bacteria from Genome Sequence of *Thermotoga maritima," Nature* 399 (1999): 323–329.

115   "It is as if we have failed at the task that Darwin set for us": W. F. Doolittle, "Uprooting the Tree of Life," *Scientific American* (February 2000): 90–95.

115   publication of the sixth archaea sequence: D. M. Faguy and W. Ford Doolittle, "Lessons from the *Aeropyrum pernis* Genome," *Current Biology* (1999): R883–R886.

115   more than thirty microbial genomes have been published: C. M. Fraser, J. A. Eisen, and S. L. Salzberg, "Microbial Genome Sequencing," *Nature* 406 (2000): 799–803. For a complete list of current microbial sequencing projects, see http://www.tigr.org/tbd/mdb/mdb.html

117   In August 2000, Fraser and her TIGR colleagues celebrated the completion: J. F. Heidelberg et al., "DNA Sequence of Both Chromosomes of the

Cholera Pathogen *Vibrio cholerae,*" *Nature* 406 (2000): 477–483. See also T. Radford, "Life in Time of Cholera," *Guardian,* August 3, 2000.

117 tales of dissent and disagreement: B. Berselli, "Gene Split," *Washington Post,* July 7, 1997.

117 Venter and Haseltine were suddenly dissolving their partnership: N. Wade, "Genome Project Partners Go Their Separate Ways," *New York Times,* June 24, 1997.

118 "raised the ante worldwide for sequencing the human genome": R. Nowak, "Homing in on the Human Genome," *Science* 269 (1995): 269.

118 "Sequence, sequence, sequence": D. Birch, "Daring Sprint to the Summit," *Baltimore Sun,* April 13, 1999.

CHAPTER SIX: *TREASURES OF THE LOST WORLDS*

120 Zamel first became aware of the Tristanians in 1961: N. Zamel, "In Search of the Genes of Asthma on the Island of Tristan da Cunha," *Canadian Respiratory Journal* 2 (1995): 18–22.

120 The first serious case of asthma on the island was reported in 1910: Ibid.

121 Zamel began a thorough clinical evaluation: N. Zamel et al., "Asthma on Tristan da Cunha: Looking for the Genetic Link," *American Journal Respiratory Crit. Care Medicine* 153 (1996): 1902–1906.

122 Sequana refused to reveal its asthma discovery: G. Vogel, "A Scientific Result Without the Science," *Science* 276 (1997): 1327.

122 The identity of the candidate genes: N. Zamel, interview, July 6, 2000.

122 A Canadian organization called the Rural Advancement Foundation: T. Monmaney, "Gene Sleuths Seek Asthma's Secrets on Remote Island," *Los Angeles Times,* April 30, 1997.

123 In 1961, a five-year-old boy was referred to a well-known physician: H. H. Hobbs and D. J. Rader, "ABC1: Connecting Yellow Tonsils, Neuropathy, and Very Low HDL," *Journal of Clinical Investigation* 104 (1999): 1015–1017.

124 In 1999, three teams of researchers simultaneously identified the Tangier disease gene: S. G. Young and C. J. Fielding, "The ABCs of Cholesterol Efflux," *Nature Genetics* 22 (1999): 316–318; A. Brooks-Wilson. et al., "Mutations in *ABC1* in Tangier Disease and Familial High-Density Lipoprotein Deficiency," *Nature Genetics* 22 (1999): 336–345; M. Bodzioch et al., "The Gene Encoding ATP-Binding Cassette Transporter 1 Is Mutated in Tangier Disease," *Nature Genetics* 22 (1999): 347–351; S. Rust et al., "Tangier Disease Is Caused by Mutations in the Gene Encoding ATP-Binding Cassette Transporter 1," *Nature Genetics* 22 (1999): 352–355.

124 boosts production of the Tangier protein: J. J. Repa et al., "Regulation of Absorption and ABC1-Mediated Efflux of Cholesterol by RXR Heterodimers," *Science* 289 (2000): 1524–1529.

125 From the frozen tundra of western Finland: V. C. Sheffield, E. M. Stone, and R. Carmi, "Use of Isolated Inbred Human Populations for Identification

of Disease Genes," *Trends in Genetics* 14 (1998): 391–396; L. Belkin, "The Clues Are in the Blood," *New York Times Magazine,* April 26, 1998, p. 46.

126   the Finns feel they live "at the edge of the inhabitable world": L. Peltonen, A. Jalanko, and T. Varilo, "Molecular Genetics of the Finnish Disease Heritage," *Human Molecular Genetics* 8 (1999): 1913–1923.

126   geneticists recognized that Finland was unusually fertile ground: A. de la Chapelle and F. A. Wright, "Linkage Disequilibrium Mapping in Isolated Populations: The Example of Finland Revisited," *Proceedings of the National Academy of Sciences U.S.A.* 96 (1998): 12416–12423.

127   "Although our populations are human": J. Hästbacka, A. de la Chapelle, I. Kaitila, P. Sistonen, A. Weaver, and E. Lander, "Linkage Disequilibrium Mapping in Isolated Founder Populations: Diastrophic Dysplasia in Finland," *Nature Genetics* 2 (1992): 204–211.

127   the result of Lander's calculations was that the *DTD* gene: Ibid.

127   Lander and colleagues soon identified the gene: J. Hästbacka et al., "The Diastrophic Dysplasia Gene Encodes a Novel Sulfate Transporter: Positional Cloning by Fine-Structure Linkage Disequilibrium Map," *Cell* 78 (1994): 1073–1087; J. Hästbacka et al., "Identification of the Finnish Founder Mutation for Diastrophic Dysplasia (DTD)," *European Journal of Human Genetics* 7 (1999): 664–670.

127   This couple has the dubious distinction of having introduced a dominant mutation: T. Jenkins, "The South African Malady," *Nature Genetics* 13 (1996): 7–9; P. N. Meissner et al., "A R59W Mutation in Human Protoporphyinogen Oxidase Results in Decreased Enzyme Activity and Is Prevalent in South Africans with Variegate Porphyria," *Nature Genetics* 13 (1996): 95–97.

128   Suspicion as to the root cause of the madness of King George: Jenkins, "South African Malady"; S. Jones, *The Language of Genes* (New York: Anchor, 1994).

128   The challenge is to see whether the methods: G. Thomson and M. S. Esposito, "The Genetics of Complex Diseases," *Trends in Genetics* 15 (1999): M17–M20; N. Risch, "Searching for Genetic Determinants in the New Millennium," *Nature* 405 (2000): 847–856.

129   Glaxo Wellcome has pinpointed three genes using SNPs: M. Waldholz and E. Tanouye, "Glaxo to Report It's Closing In on Genes Linked to 3 Diseases," *Wall Street Journal,* October 19, 1999.

129   the Polynesian predisposition to cardiovascular disease: J. Hinde, "Genetic Bounty," *New Scientist,* May 6, 2000, p. 23.

129   Friedman has initiated a collaboration with the Department of Health: *Research in Progress 2000* (Chevy Chase, Md.: Howard Hughes Medical Institute, 2000).

130   recently identified what they suspect is the key gene: Y. Horikawa et al., "Genetic Variation in the Gene Encoding Calpain-10 Is Associated with Type 2 Diabetes Mellitus," *Nature Genetics* 26 (2000): 163–175.

130 "The men looked robust but clumsy": J. Verne, *Journey to the Centre of the Earth* (London: Penguin Books, 1965).

132 trace virtually every instance of breast cancer on the island: S. Thorlacius et al., "A Single *BRCA2* Mutation in Male and Female Breast Cancer Families from Iceland with Varied Cancer Phenotypes," *Nature Genetics* 13 (1996): 117–119.

132 "The single, ancient *BRCA2* mutation was a golden key": M. Specter, "Decoding Iceland," *New Yorker,* January 18, 1999, pp. 40–51.

132 One of the most famous characters in Icelandic lore: D. Roberts and J. Krakauer, *Iceland: Land of the Sagas* (New York: Villard, 1990).

132 a direct descendant of Egil has returned to his native land: R. Kunzig, "Blood of the Vikings," *Discover* (December 1998): 90–99; Specter, "Decoding Iceland"; I. Wickelgren, "Decoding a Nation's Genes," *Popular Science,* January 2000.

133 "I had forgotten how magical the summer is": Kari Stefánsson, interview, February 19, 2000.

134 "I thought it would be interesting to coerce 'big pharma' ": Ibid.

134 the pinnacle of medieval literature: *The Complete Sagas of Icelanders* (Reykjavik: Leifure Eiriksson Publishing, 1997).

135 "we can all trace our ancestry back to Egil!": Thordur Kristjansson, interview in Reykjavik, August 22, 2000.

135 Opponents were appalled by the prospect: S. Lyall, "A Country Unveils Its Gene Pool and Debate Flares," *New York Times,* February 16, 1999; R. C. Lewontin, "People Are Not Commodities," *New York Times,* January 23, 1999; B. Palsson and S. Thorgeirsson, "Decoding Developments in Iceland," *Nature Biotechnology,* 17 (1999): 407; P. Billings, "Iceland, Blood and Science," *American Scientist* 87 (1999): 199–200; G. Páalsson and P. Rabinow, "Iceland: The Case of a National Human Genome Project," *Anthropology Today,* October 1999.

136 Stefánsson points out that the penalty: E. Mahsood, "Gene Warrior," *New Scientist,* July 15, 2000, pp. 42–45.

136 including unsavory allegations of bribery: K. Philipkoski, "Genetics Scandal Inflames Iceland," *Wired News,* March 20, 2000; Mahsood, "Gene Warrior."

136 "There are those that claim that this constitutes coercion": Stefánsson interview, February 19, 2000.

136 a rival biotechnology company called UVS: M. Enserink, "Start-Up Claims Piece of Iceland's Gene Pie," *Science* 287 (2000): 951.

136 "This is just an attempt to cash in on our sort of success": Stefánsson interview, February 19, 2000.

137 "The people still support me, and the doctors think I'm the devil": Specter, "Decoding Iceland."

137 "it would be unethical not to try to use": Lyall, "A Country Unveils Its Gene Pool."

137   "We have the potential to have": Kari Stefánsson, interview in Reykjavik, August 22, 2000.

137   "The Delaware registration was simply": Ibid.

137   "of particular note is the ambitious $60 million plan": R. McKie and P. Wintour, "Mass Gene Bank Launched," *Observer,* February 13, 2000.

138   "It's the first ever mapping of common stroke": Stefánsson interview, February 19, 2000.

139   "Jeff is incontinent with excitement over the work with osteoporosis": Stefánsson interview, August 22, 2000.

139   "Not only is longevity familial": Stefánsson interview, February 19, 2000.

139   "What Celera and the Human Genome Project are doing": Ibid.

CHAPTER SEVEN: *PRIZE FIGHT*

141   mapping genes that cause inherited disease: J. L. Weber, "Know Thy Genome," *Nature Genetics* 7 (1994): 343–344.

141   Myers's chief claim to fame: S. F. Altschul, W. Gish, W. Miller, E. W. Myers, and D. J. Lipman, "Basic Local Alignment Search Tool," *Journal of Molecular Biology* 215 (1990): 403–410.

142   "They trounced him," says Myers: R. Preston, "The Genome Warrior," *New Yorker,* June 12, 2000, pp. 66–83.

142   "Part of the opposition," he recalls: Jim Weber, interview, June 15, 2000.

142   The Human Genome Project needed a radically new approach: J. L. Weber and E. W. Myers, "Human Whole-Genome Shotgun Sequencing," *Genome Research* 7 (1997): 401–409.

143   Green exposed a series of potential scientific and economic flaws: P. Green, "Against a Whole-Genome Shotgun," *Genome Research* 7 (1997): 410–417.

143   "It is not clear how one would deal with hiring": Ibid.

143   Venter and colleagues proposed that the tedious chore : J. C. Venter, H. O. Smith, and L. Hood, "A New Strategy for Genome Sequencing," *Nature* 381 (1996): 364–366. The idea began with a "seed" clone, the ends of which would be sequenced. Overlapping DNA fragments containing those sequences would then be identified and sequenced in turn, producing an expanding, overlapping tile of DNA clones spreading in both directions along the chromosome which would be mapped and sequenced simultaneously.

144   "there would be no genome field at all": A. Knox, "Hot Rod of a Sequencer Pushed Genome Effort," *Philadelphia Inquirer,* June 27, 2000.

144   Hood's prototype machine had many advantages: L. M. Smith et al., "Fluorescence Detection in Automated DNA Sequence Analysis," *Nature* 321 (1984): 674–679; R. Cook-Deegan, *The Gene Wars* (New York: Norton, 1994); P. G. Gosselin and P. Jacobs, "DNA Device's Heredity Scrutinized by U.S.," *Los Angeles Times,* May 14, 2000.

145   Hunkapiller and Hood had toyed with a radical idea: C. Wills, *Exons, Introns and Talking Genes* (New York: Basic Books, 1991).

146   "We didn't have skills in-house": Q. Hardy, "Riches from a Genetic Gold Rush," *Forbes*, February 21, 2000, pp. 102–103.

146   "we spent a few days working through the math": N. Wade, "Beyond Sequencing of Human DNA," *New York Times*, May 12, 1998.

147   In his story, Wade outlined how the new company: N. Wade, "Scientist's Plan: Map All DNA Within 3 Years," *New York Times*, May 10, 1998.

148   "the full Monty": J. Carey, "The Duo Jolting the Gene Business," *Business Week*, May 25, 1998, pp. 71–72.

148   they would not be held hostage: J. Gillis and R. Weiss, "Private Firm Aims to Beat Government to Gene Map," *Washington Post*, May 12, 1998.

148   "If successful," Wade suggested: Wade, "Scientist's Plan."

148   Wade's article drew a swift rebuttal from Varmus: H. Varmus, "Progress On All Fronts in Race to Map Genes," *New York Times*, May 17, 1998.

149   "On your marks": A. Coghlan and L. Kleiner, "On Your Marks . . . ," *New Scientist*, May 23, 1998, p. 4.

149   "Consider it modern biology's equivalent": J. Travis, "Another Human Genome Project," *Science News*, May 23, 1998, pp. 334–335.

149   "the definitive source of genomic and related medical information": Tony White, teleconference, May 11, 1998.

150   "the takeover of the human-genome project": Wade, "Beyond Sequencing of Human DNA."

150   he compared Venter's assault on the genome project: Preston, "The Genome Warrior."

151   "It was coincidental," Morgan explains: Michael Morgan, interview, London, August 17, 2000.

151   "To leave this to a private company": N. Wade, "International Project Gets Lift," *New York Times*, May 17, 1998.

151   "They weren't cheering me!" Morgan interview, August 17, 2000.

151   "I really don't see this as being any great advance": Wade, "International Project Gets Lift."

151   "If I were on the other side of this": P. Smaglik, "Privatizing the Human Genome," *The Scientist*, June 8, 1998.

152   "It's really Craig Venter going after the Nobel Prize": Carey, "The Duo Jolting the Gene Business."

152   "The era of government-sponsored big science": W. A. Haseltine, "Gene-Mapping, Without Tax Money," *New York Times*, May 21, 1998.

152   "Leaving aside the egomania": Sydney Brenner, interview, June 24, 2000.

152   "An essential feature of the business plan": J. C. Venter, M. D. Adams, G. G. Sutton, A. R. Kerlavage, H. O. Smith, and M. Hunkapiller, "Shotgun Sequencing of the Human Genome," *Science* 280 (1998): 1540–1542.

153   It is our hope that this program is complementary: Ibid.

153   the House Subcommittee on Energy and the Environment convened a meeting: Testimony of Maynard Olson, Craig Venter, and Francis Collins

at U.S. House of Representatives Committee on Science meeting, June 17, 1998. Available at: http://www.house.gov/science

154   an issue of *Mad* magazine: T. Friend, "The Book of Life: Twin Efforts Will Attempt to Write It," *USA Today,* June 9, 1998.

155   For more than 25 years I have looked at the little fly: C. Stern, "Two or Three Bristles," *American Scientist* 42 (1954): 213.

155   When Morgan visited his wife in hospital: M. M. Green, "The 'Genesis of the White-Eyed Mutant' in *Drosophila melanogaster:* A Reappraisal," *Genetics* 142 (1996): 329–331. Most books on Morgan maintain that *white* was not the first mutant fruit fly Morgan observed, but Green's article makes a compelling argument in favor.

156   That summer, Morgan published his white-eye results: T. H. Morgan, "Sex Limited Inheritance in *Drosophila,*" *Science* 32 (1910): 120–122.

156   "They may throw some further light on the process of heredity": G. Allen, *Thomas Hunt Morgan* (Princeton, N.J.: Princeton University Press, 1978).

156   Morgan's fly room contained: R. E. Kohler, *Lords of the Fly* (Chicago: University of Chicago Press, 1994); J. Weiner, *Time, Love, Memory* (New York: Knopf, 1999).

157   the locations of five fruit fly traits: Ibid.

157   the first map of genes on a single chromosome: Ibid.

157   "I beg to inform you that the Nobel Prize": Allen, *Thomas Hunt Morgan.*

158   "It is often difficult to distinguish what is Morgan's work": Text of Nobel Prize presentation available at: http://www.nobel.se

158   Morgan's students continued to make spectacular discoveries: G. M. Rubin and E. B. Lewis, "A Brief History of *Drosophila's* Contributions to Genome Research," *Science* 287 (2000): 2216–2218.

158   In 1980, they published a classic paper: C. Nüsslein-Volhard and E. Wischaus, "Mutations Affecting Segment Number and Polarity in *Drosophila,*" *Nature* 287 (1980): 795–801.

159   "an offer I could not refuse": G. Taubes, *"Drosophila* Genome Sequence Completed," *HHMI Bulletin,* July 2000, pp. 4–5.

160   By making the sequence of the entire human genome available: J. C. Venter, U.S. House of Representatives Committee on Science meeting, June 17, 1998.

160   "Entrepreneurs have a vital role in any economy": "The Gene Machine: Research Which Must Not Be Privatized," *Guardian,* September 21, 1999.

160   "We would have to refuse some women": J. Burn, "You Can't Put a Price on a Human Goldmine," *Daily Mail,* June 27, 2000.

161   royalties for a gene test for Canavan's disease: J. Peres, "Genetic Tests Reduce Neighborhood's Grief," *Chicago Tribune,* September 12, 1999; G. Kolata, "Who Owns Your Genes?" *New York Times,* May 15, 2000.

161   the successful screening for Tay-Sachs disease: G. Kolata, "Nightmare or the Dream of a New Era in Genetics?" *New York Times,* December 7, 1993; L. Wingerson, *Unnatural Selection* (New York: Bantam, 1998).

161   The most notorious landlords of the genome: R. T. King Jr., "Code Green:

Mapping Human Genome Will Bring Glory to Some, But Incyte Prefers Cash," *Wall Street Journal,* February 10, 2000; M. Compton, "Lean Green Gene-Counting Machine," *Salon,* April 24, 2000; A. Pollack, "Is Everything for Sale?" *New York Times,* June 27, 2000.

161   "when the Human Genome Project sequence is completed": Haseltine, "Gene-Mapping, Without Tax Money."

161   Donna MacLean, a British waitress and poet: J. Meek, "Poet Attempts the Ultimate in Self-Invention—Patenting Her Own Genes," *Guardian,* February 29, 2000; E. Goodman, "Our Genes for Sale—Get 'em While They're Hot," *Boston Globe,* March 8, 2000.

162   "big and ambitious, even audacious": M. Wadman, "Human Genome Deadline Cut by Two Years," *Nature* 395 (1998): 207; E. Marshall, "NIH to Produce a 'Working Draft' of the Genome by 2001," *Science* 281 (1998): 1774–1775.

162   "The best service to the scientific community": Wadman, "Human Genome Deadline."

163   Funneling over $80 million over five years to just three centers: N. Wade, "One of 2 Teams in Genome-Map Race Sets an Earlier Deadline," *New York Times,* March 16, 1999; M. Wadman, "Human Genome Project Aims to Finish 'Working Draft' Next Year," *Nature* 398 (1999): 177–178.

163   "putting humanity in a Waring blender": N. Wade, "In Genome Race, Government Vows to Move Up Finish," *New York Times,* September 14, 1998.

163   "Cliff Notes version of the genome": Friend, "The Book of Life."

163   "operating manager and field marshal": E. Marshall, "Sequencers Endorse Plan for a Draft in 1 Year," *Science* 284 (1999): 1439–1440.

164   His early interest was in mathematics: C. Cookson, "Cartographer of Life," *Financial Times,* December 22, 1995; P. J. Hilts, "Love of Numbers Leads to Chromosome 17," *New York Times,* September 10, 1996; A. Zitner, "The DNA Detective," *Boston Globe Sunday Magazine,* October 10, 1999, p. 17.

165   appropriate criteria for forensic DNA fingerprinting: E. S. Lander and B. Budowle, "DNA Fingerprinting Dispute Laid to Rest," *Nature* 371 (1994): 735–738.

165   mapping genes for complex traits: E. Lander and L. Kruglyak, "Genetic Dissection of Complex Traits: Guidelines for Interpreting and Reporting Linkage Results," *Nature Genetics* 11 (1995): 241–247.

165   comprehensive genetic map of the mouse: W. Dietrich et al., "A Comprehensive Genetic Map of the Mouse Genome," *Nature* 380 (1996): 149–152.

165   Venter set up his new corporate headquarters: K. Jegalian, "The Gene Factory," *Technology Review* (March–April 1999): 64–68; J. Gillis, "A Gene Dream," *Washington Post,* September 27, 1999.

166   "What are stock options?" J. C. Venter, Washington D.C. press conference, June 26, 2000.

167   "The unanimous opinion of the international sequencing community":

F. S. Collins, A. Patrinos, et al., "New Goals for the U.S. Genome Project, 1998–2003," *Science* 282 (1998): 682.

167   "The day we announced Celera": P. E. Ross, "Gene Machine," *Forbes,* February 21, 2000, pp. 98–104.

CHAPTER EIGHT: *THE STORY OF US*

169   In 1975, geneticist Mary-Claire King published: M. C. King and A. C. Wilson, "Evolution at Two Levels in Humans and Chimpanzees," *Science* 188 (1975): 107–116.

169   "I kept thinking the project was a disaster": T. Bass, "The Gene Detective," *Omni* (July 1993): 68.

169   King went on to make her name: K. Davies and M. White, *Breakthrough* (New York: Wiley, 1996); M. Waldholz, *Curing Cancer* (New York: Simon & Schuster, 1997); P. Bock, "Mission Possible," *Seattle Times,* May 31, 1998.

170   physical and behavioral differences between *Homo sapiens* and *Pan troglodytes:* A. Gibbons, "Which of Our Genes Makes Us Human?" *Science* 281 (1998): 1432–1434; K. Hopkin, "The Greatest Apes," *New Scientist,* May 15, 1999, pp. 26–30.

170   "Knowing the genetic basis of these changes": J. Maddox, "The Unexpected Science to Come," *Scientific American* (December 1999): 62–67.

170   this chromosomal junction marks the location of the genes for the soul: M. Ridley, *Genome* (London: Fourth Estate, 1999).

171   GenoPlex scientists believe that the cognitive differences: N. Wade, "Human or Chimp? 50 Genes Are the Key," *New York Times,* October 20, 1998.

171   only one significant biochemical difference between humans and chimpanzees: H. H. Chou et al., "A Mutation in Human CMP-Sialic Acid Hydroxylase Occurred After the *Homo-Pan* Divergence," *Proceedings of the National Academy of Sciences U.S.A.* 95 (1998): 11751–11756.

172   a rational explanation for the failure of kidney transplants: Hopkin, "Greatest Apes."

172   "Maybe their mice will speak," says Varki dryly: Gibbons, "Which of Our Genes."

172   launching a human genome evolution project: E. McConkey and M. A. Goodman, "Human Genome Evolution Project Is Needed," *Trends in Genetics* (September 1997): 350–351.

172   A primate genome project would help: E. H. McConkey and A. Varki, "A Primate Genome Project Deserves High Priority," *Science* 289 (2000): 1295–1296; A. Varki, "A Chimpanzee Genome Project Is a Biomedical Imperative," *Genome Research* 10 (2000): 1065–1070.

173   the extinction of Miss Waldron's red colobus: A. C. Revkin, "West African Monkey Is Extinct, Scientists Say," *New York Times,* September 12, 2000.

NOTES   291

173 Collins has already embarked on studies of chimpanzee DNA polymor-
     phisms: J. G. Hacia et al., "Determination of Ancestral Allele for Human
     Single-Nucleotide Polymorphisms Using High-Density Oligonucleotide
     Arrays," *Nature Genetics* 22 (1999): 164–167.
173 But Bob Waterston sees problems: Lecture at American Society of Human
     Genetics, Philadelphia, October 6, 2000.
173 the origins and global migration patterns of human populations: K. Owens
     and M-C. King, "Genomic Views of Human History," *Science* 286 (1999):
     451–453; S. Pääbo, "Human Evolution," *Trends in Genetics* 15 (1999):
     M13–M16.
173 mitochondrial DNA provides a pristine record of the maternal line of de-
     scent: D. C. Wallace, M. D. Brown, and M. T. Lott, "Mitochondrial DNA
     Variation in Evolution and Disease," *Gene* 238 (1999): 211–230. Some re-
     cent evidence leaves open the possibility that some paternal mitochondrial
     DNA can be inherited: see P. Awadalla, A. Eyre-Walker, and J. Maynard
     Smith, "Linkage Disequilibrium and Recombination in Hominid Mito-
     chondrial DNA," *Science* 286 (1999): 2524–2525; E. Straus, "mtDNA
     Shows Signs of Paternal Influence," *Science* 286 (1999): 2436.
174 The other staple of human evolutionary studies is the Y chromosome: M.
     A. Jobling and C. Tyler-Smith, "Fathers and Sons: The Y Chromosome
     and Human Evolution," *Trends in Genetics* 11 (1995): 449–456; M. Jensen,
     "All About Adam," *New Scientist,* July 11, 1998: 35–39.
174 retraced the incredible evolutionary journey: B. T. Lahn and D. C. Page,
     "Four Evolutionary Strata on the Human X Chromosome," *Science* 286
     (1999): 964–967; N. Wade, "Earliest Divorce Case: X and Y Chromo-
     somes," *New York Times,* October 29, 1999.
174 the male sex-determination pathway: A. Swain and R. Lovell-Badge,
     "Mammalian Sex Determination: A Molecular Drama," *Genes & Develop-
     ment* 13 (1999): 755–767.
174 a family of genes required for sperm development: B. T. Lahn and D. C.
     Page, "Functional Coherence of the Y Chromosome," *Science* 278 (1997):
     675–680.
174 the Y has been subjected to waves of gene rearrangements: Lahn and Page,
     "Four Evolutionary Strata."
174 In 1856, a major archaeological discovery: R. McKie, *Ape Man: The Story of
     Human Evolution* (London: BBC Worldwide, 2000).
175 Art, symbols, music, notation, language: I. Tattersall, *Becoming Human* (New
     York: Harcourt Brace, 1998).
175 "The sounds made a picture in his head": W. Golding, *The Inheritors* (New
     York: Harcourt Brace, 1998).
175 it remains unclear whether they were evicted by modern humans: I. Tat-
     tersall and J. H. Schwartz, "Hominids and Hybrids: The Place of Nean-
     derthals in Human Evolution," *Proceedings of the National Academy of Sciences
     U.S.A.* 96 (1999): 7117–7119.
175 DNA exposed to the elements would decompose within about 10,000

years: T. Lindahl, "Instability and Decay of the Primary Structure of DNA," *Nature* 362 (1993): 709–715.

175 Having rehearsed the extraction of ancient DNA: G. Poinar Jr., "Ancient DNA," *American Scientist* 87 (1999): 446–457.

176 Pääbo found that there were twenty-seven differences: M. Krings, A. Stone, R. W. Schmitz, H. Krainitzki, M. Stoneking, S. Pääbo, "Neandertal DNA Sequences and the Origin of Modern Humans," *Cell* 90 (1997): 19–30. (Both spellings—Neandertal and Neanderthal—are acceptable.)

176 indispensable verification of Pääbo's findings: I. V. Ovchinniko et al., "Molecular Analysis of Neanderthal DNA from the Northern Caucasus," *Nature* 404 (2000): 490–493; M. Höss, "Neanderthal Population Genetics," *Nature* 404 (2000): 453–454.

176 the skeleton of an unusual four-year-old child in Portugal: C. Duarte, J. Mauricio, P. B. Pettitt, P. Souto, E. Trinkaus, H. van der Plicht, and J. Zilhao, "The Early Upper Paleolithic Human Skeleton from the Abrigo do Lagar Velho (Portugal) and Modern Human Emergence from Iberia," *Proceedings of the National Academy of Sciences U.S.A.* 96 (1999): 7604–7609. See also R. Kunzig, "Learning to Love Neanderthals," *Discover* (August 1999), 68–75; K. Wong, "Who Were the Neanderthals?" *Scientific American* (April 2000): 99–107; and McKie, *Ape Man*.

176 Tattersall praised the Portuguese study as "brave and imaginative [and] courageous": Tattersall and Schwartz, "Hominids and Hybrids."

177 Wilson conducted a thorough comparison of mitochondrial DNA: R. L. Cann, M. Stoneking, and A. C. Wilson, "Mitochondrial DNA and Human Evolution," *Nature* 325 (1987): 31–36. "The popular evocations of Eve in her African Eden," recalled *Nature* senior editor Henry Gee, "were such that one would have been forgiven for thinking that she had been interviewed in person for her comments on the *Nature* paper": H. Gee, "Statistical Cloud over African Eden," *Nature* 355 (1992): 583.

177 "All these mitochondrial DNAs stem from one woman": Cann, Stoneking, and Wilson, "Mitochondrial DNA."

178 "The earliest divergences in the current mitochondrial gene pool occurred in Africa": Pääbo, "Human Evolution."

178 "Y-chromosome Adam" lived somewhat later than "mitochondrial Eve": L. L. Cavalli-Sforza, *Genes, Peoples, and Languages* (New York: North Point Press, 2000).

178 the likelihood that some Asians returned to Africa: M. Hammer et al., "Out of Africa and Back Again," *Molecular Biology & Evolution* 15 (1998): 427.

178 mitochondrial Eve had eighteen "daughters": N. Wade, "The Human Family Tree: 10 Adams and 18 Eves," *New York Times,* May 2, 2000; Wallace, Brown, and Lott, "Mitochondrial DNA Variation." There are nine European mitochondrial lineages (H, I, J, K; T, U, V, W, X); eight Asian lineages (A, B, C, D, E, F, G, M); and the major African lineage (L, subdivided into L1, L2, L3). Lineages A to D are found in American Indians, as is the rare European haplotype X.

179   this population represents one of the most ancient African populations: Y.
      S. Chen et al., "mtDNA Variation in the South African Kung and Khwe—
      and Their Genetic Relationships to Other African Populations," *American
      Journal of Human Genetics* 66 (2000): 1362–1383.

179   there is a much greater degree of variation of Y-chromosome markers: M.
      T. Seielstad, E. Minch, and L. L. Cavalli-Sforza, "Genetic Evidence for a
      Higher Female Migration Rate," *Nature Genetics* 20 (1998): 278–280; M.
      Stoneking, "Women on the Move," *Nature Genetics* 20 (1998): 219–220;
      Cavalli-Sforza, *Genes, Peoples, and Languages.*

179   For example, Pääbo's group has sequenced more than 10,000 bases: H.
      Kaessmann, F. Heissig, A. von Haeseler, and S. Pääbo, "DNA Sequence
      Variation in a Non-Coding Region of Low Recombination on the
      Human X Chromosome," *Nature Genetics* 22 (1999): 78–81. The calcu-
      lated times back to a common ancestor will be three to four times longer
      in studies based on genes on the X chromosome or autosomes, compared
      with mitochondrial DNA or Y chromosome markers. This is because a
      breeding pair of humans can potentially transmit three X chromosomes
      and four autosomes but only one Y chromosome and a single pool of mi-
      tochondrial DNA (from the mother).

180   the mitochondrial sequence of Cheddar Man matched that of Adrian Tar-
      gett: S. O'Neill, "Cheddar Man Is My Long-Lost Relative," *Daily Telegraph,*
      March 8, 1997.

180   as he fondly calls them, the Seven Daughters of Eve: A. Ahuja, "So God
      Created Woman," *Times,* April 19, 2000. The descendants of Ursula (more
      formally, the bearers of mitochondrial haplotype U) account for 15 per-
      cent of European mitochondrial DNAs, and are also found among African
      Bantu, suggesting this is the founding European lineage. The other fre-
      quencies are Helena (H), 40 percent; Tara (T), 15 percent; Jasmine (J), 11
      percent; Katrine (K), 9 percent; Xenia (X), 7 percent; and Valda (V), 5 per-
      cent. In addition to the seven daughters of Eve, there are two additional
      minor European haplotypes, I and W.

180   Ursula, the oldest, lived in northern Greece: Ibid.

181   Sykes has set up a company: http://www.oxfordancestors.com

181   "It's really quite low": Ahuja, "So God Created Woman."

181   an exhaustive study of nearly one hundred markers: L. L. Cavalli-Sforza, P.
      Menozzi, and A. Piazza, *The History and Geography of Human Genes*
      (Princeton, N.J.: Princeton University Press, 1994); Cavalli-Sforza, *Genes,
      Peoples and Languages.*

182   Japan's current population is a mixture of two cultures: M. F. Hammer and
      S. Horai, "Y Chromosomal DNA Variation and the Peopling of Japan,"
      American Journal of Human Genetics 56 (1995): 951–962; J. Travis, "Jomon
      Genes," *Science News* 151 (1997): 106–107.

182   half of the Jewish priests shared the same genetic signature: K. Skorecki et
      al., "Y Chromosomes of Jewish Priests," *Nature* 385 (1997): 32; M. G.
      Thomas et al., "Origins of Old Testament Priests," *Nature* 394 (1998):

138–140; J. Travis, "The Priest's Chromosome?" *Science News* 154 (1998): 218–220. The prevalent genetic signature in both Ashkenazic and Sephardic *cohanim* is called the Cohen modal haplotype—a specific set of variations at six locations on the Y chromosome.

182   the Welsh Indiana Jones: J. Karp, "Seeking Lost Tribes of Israel in India, Using DNA Testing," *Wall Street Journal,* May 11, 1998.

182   Parfitt traces the Lembas' long journey to southern Africa: T. Parfitt, *Journey to the Vanished City* (New York: Vintage, 2000). Parfitt's travels were featured in a PBS/NOVA documentary, which produced an excellent accompanying web site: http://cgi.pbs.org/wgbh/nova/israel/

183   a significant portion of the Lemba Y chromosomes: M.G. Thomas et al., "Y Chromosomes Traveling South: The Cohen Modal Haplotype and the Origins of the Lemba—the 'Black Jews of Southern Africa,' " *American Journal of Human Genetics* 66 (2000): 674–686; N. Wade, "Group in Africa Has Jewish Roots, DNA Indicates," *New York Times,* May 9, 1999.

183   surprising similarities in the Y chromosomes of Jews, Palestinians, and other Arab populations: M. F. Hammer et al., "Jewish and Middle Eastern Non-Jewish Populations Share a Common Pool of Y-Chromosome Biallelic Haplotypes," *Proceedings of the National Academy of Sciences U.S.A.* 97 (2000): 6769–6774; N. Wade, "Y Chromosome Bears Witness to Story of the Jewish Diasporas," *New York Times,* May 9, 2000.

183   Kittles offered to compare an individual's DNA with the database: B. Macintyre, "DNA to Unlock the Past of Black American Slaves," *Times,* April 26, 2000.

183   "We're talking about American slavery": A. Allen, "Flesh and Blood and DNA," *Salon,* May 12, 2000.

183   the Human Genome Diversity Project: L. L. Cavalli-Sforza et al., "Call for a Worldwide Survey of Human Genetic Diversity: A Vanishing Opportunity for the Human Genome Project," *Genomics* 11 (1991): 490–498.

183   the plans have stirred anger and resentment: P. Salopek, "Unlocking the Rainbow," *Chicago Tribune,* April 27, 1997; R. W. Wallace, "The Human Genome Diversity Project: Medical Benefits Versus Ethical Concerns," *Molecular Medicine Today* 4 (1998): 59–62.

184   Genetic fingerprinting was invented: A. J. Jeffreys, V. Wilson, and S. L. Thein, "Hypervariable 'Minisatellite' Regions in Human DNA," *Nature* 314 (1985): 67–73.

184   In 1987, Dawn Ashworth was raped and strangled: J. Wambaugh, *The Blooding* (New York: Morrow, 1989).

185   Jeffreys declared a complete match: E. Hagelberg, I. C. Gray, and A. J. Jeffreys, "Identification of the Skeletal Remains of a Murder Victim by DNA Analysis," *Nature* 352 (1991): 427–429.

185   he was in fact the Nazi "Angel of Death," Josef Mengele: A. J. Jeffreys, M. J. Allen, E. Hagelberg, and A. Sonnberg, "Identification of the Skeletal Remains of Josef Mengele by DNA Analysis," *Forensic Science International* 56 (1992): 65–76.

185   "When she went back to her grandparents' house": Bass, "The Genetic Detective."

185   three hundred years of Romanov rule in Russia were brutally terminated: R. K. Massie, *The Romanovs: The Last Chapter* (New York: Random House, 1995).

186   "inappropriate to carry the Russian Imperial family": Ibid.

186   The mitochondrial DNA sequences were identical except for position 16169: P. Gill et al., "Identification of the Remains of the Romanov Family by DNA Analysis," *Nature Genetics* 6 (1994): 130–135.

186   Incredibly, Georgij's mitochondrial DNA also revealed heteroplasmy: P. Ivanov et al., "Mitochondrial DNA Sequence Heteroplasmy in the Grand Duke of Russia Georgij Romanov Establishes the Authenticity of the Remains of Tsar Nicholas II," *Nature Genetics* 12 (1996): 417–420. Georgij's mitochondrial DNA had a T:C ratio of 62:38 at position 16169, compared with 28:72 for Czar Nicholas.

187   but the claims of the last pretender: P. Gill et al., "Establishing the Identity of Anna Anderson Manahan," *Nature Genetics* 9 (1995): 9–10.

187   "The longest-running mini-series in American history": E. S. Lander and J. J. Ellis, "Founding Father," *Nature* 396 (1998): 13–14.

187   The Jefferson saga aroused the interest of Eugene Foster: B. Murray and B. Duffy, "Jefferson's Secret Life," *U.S. News & World Report*, November 9, 1998, pp. 58–63.

187   a complete match: E. A. Foster et al., "Jefferson Fathered Slave's Last Child," *Nature* 396 (1998): 27–28.

187   "reappears to remind us of a truth that should be self evident": Lander and Ellis, "Founding Father."

188   the Federal Bureau of Investigation (FBI) informed independent counsel Kenneth Starr: C. R. Babcock, "The DNA Test," *Washington Post*, September 22, 1998; J. Toobin, *A Vast Conspiracy* (New York: Random House, 2000).

188   K39 and Q3243 were identical "to a reasonable degree of scientific certainty": *The Starr Report* (New York: Public Affairs, 1998). The FBI report of August 6, 1998, contained the following DNA typing results:

Specimen	LDLR	GYPA	HBGG	D7S8	Gc	DOA1	D1S80
Q3243–1	BB	BB	AB	AB	AC	1.1, 1.2	24, 24
Q3243–2	BB	BB	AB	AB	AC	1.1, 1.2	24, 24
K39	BB	BB	AB	AB	AC	1.1, 1.2	24, 24

The report stated that "the source of specimen K39 [Bill Clinton] is included as a potential contributor of the DNA obtained from specimens Q3243–1 and Q3243–2 [two semen stains removed from specimen Q3243]. The probability of selecting an unrelated individual at random having the same [DNA] types as detected in the questioned specimens is approximately . . . 1 in 43,000 in the Caucasian population." A subsequent test examined seven additional DNA markers.

188   "be empowered to bring posthumous impeachment charges": E. Zimiles, "A Job for Starr," *New York Times*, November 4, 1998.

188   the royal physician smuggled out the boy's heart: A. Swardson, "A Telltale Heart Finds Its Place in History," *Washington Post*, April 20, 2000.

189   the two labs were able to compare mitochondrial DNA sequence: Ibid.

189   The death of Napoleon Bonaparte in 1821: A. Sage, "Napoleon DNA Test Demanded," *Times*, February 3, 2000.

189   suspiciously reminiscent of Marfan's syndrome: V. McKusick, "Abraham Lincoln and Marfan Syndrome," *Nature* 352 (1991): 280.

189   it would be feasible to test Lincoln's DNA: P. R. Reilly, *Abraham Lincoln's DNA* (Cold Spring Harbor, N.Y.: Cold Spring Harbor Laboratory Press, 2000).

189   the case of French artist Toulouse-Lautrec: J. B. Frey, "What Dwarfed Toulouse-Lautrec?" *Nature Genetics* 10 (1995): 128–130.

190   identified the gene mutation in families with this disorder: B. D. Gelb, G. P. Shi, H. A. Chapman, and R. J. Desnick, "Pycnodysostosis, a Lysosomal Disease Caused by Cathepsin K Deficiency," *Science* 273 (1996): 1236–1238.

190   forensic DNA analysis has enormous potential: J. Kluger, "DNA Detectives," *Time*, January 11, 1999, pp. 62–63; J. Alter, "The Death Penalty on Trial," *Newsweek*, June 12, 2000, pp. 24–34.

190   "the gold standard of innocence": M. Chapman, "The Gift of Life," *Guardian*, June 7, 2000.

190   1 in 57 billion: B. S. Weir, "DNA Statistics in the Simpson Matter," *Nature Genetics* 11 (1995): 365–368.

190   a national DNA database contains some 700,000 profiles: Chapman, "The Gift of Life."

190   profiles of over 250,000 criminals: Kluger, "DNA Detectives."

191   DNA fingerprinting of house pets: M. A. Menotti-Raymond, V. A. David, and S. J. O'Brien, "Pet Cat Hair Implicates Murder Suspect," *Nature* 386 (1997): 774.

CHAPTER NINE: *THE CROESUS CODE*

192   the one-billionth letter: D. Dickson and C. Macilwain, " 'It's a G': the One-Billionth Nucleotide," *Nature* 402 (1999): 331.

192   Transatlantic festivities took place a few days later: Ibid.; J. Gillis, "Genome Project Reaches Milestone," *Washington Post*, November 23, 1999.

193   He had invited forty-five leading *Drosophila* scientists: E. Pennisi, "Ideas Fly at Gene-Finding Jamboree," *Science* 287 (2000): 2182–2184.

194   "The first human chromosome sequence": I. Dunham et al., "The DNA Sequence of Human Chromosome 22," *Nature* 402 (1999): 489–495. In response to charges that *Nature*'s use of Christian religious imagery on its

cover was inadvertently supporting creationism, *Nature's* editor, Phillip Campbell, said that the image "combined iconic symbolism with the science without implying that the Bible is true or that evolution is not the key to making sense of biology": *Nature* 403 (2000): 242.

194   a timely validation of the mapping strategy: B. Appleyard, "The Race for the Croesus Code," *Sunday Times,* December 5, 1999.

194   "as important an accomplishment": S. Connor, "The Secret of Life," *Independent,* December 1, 1999.

194   the invention of the wheel: T. Radford, "Cracking the Code of Human Life," *Guardian,* December 2, 1999.

194   "A new era has dawned": N. Wade, "After 10 Years' Effort, Genome Mapping Team Achieves Sequence of a Human Chromosome," *New York Times,* December 2, 1999.

194   "I don't often pick up a scientific paper": Ibid.

194   "seeing the surface or the landscape": R. Weiss, "Scientists Decode Layout of Human Chromosome," *Washington Post,* December 2, 1999.

195   the assembly of more than 33,400,000 letters of DNA: Dunham et al., "The DNA Sequence." The chromosome 22 sequence is available on the Internet at: http://www.sanger.ac.uk/HGP/Chr22 and www.genome.ou.edu/Chr22.html

195   the long arm of the chromosome contained a dozen gaps: Ibid. Some DNA fragments, when readied for sequencing, may "poison" the bacteria intended to propagate them, while other sequences are too repetitive for the sequencing enzyme to read through.

195   "As a result": Ibid.

196   the CEO of Incyte Genomics, Roy Whitfield: M. Compton, "Lean, Green Gene-Counting Machine," *Salon,* April 24, 2000.

196   Richard Durbin dismisses these inflated estimates as "bogus": Richard Durbin, interview, Cambridge, August 18, 2000.

197   DiGeorge syndrome, a relatively common congenital disorder: P. J. Scambler, "Engineering a Broken Heart," *Nature* 401 (1999): 335–337.

197   researchers have mapped putative schizophrenia genes: B. Riley and R. Williamson, "Sane Genetics for Schizophrenia," *Nature Medicine* 6 (2000): 253–255.

197   a thinly veiled jibe at Celera: Dunham et al., "The DNA Sequence."

197   "unbridled exaltation, yes—but also, solemn reflection": Editorial, *Guardian,* December 2, 1999.

198   "As 1999 draws to a close and we approach the third millennium": P. Little, "The Book of Genes," *Nature* 402 (1999): 467–468.

198   the Motley Fool announced that it would invest heavily in Celera: J. Surowiecki, "When Motley Fool Speaks, (Some) People Listen," *Slate,* December 28, 1999.

198   "It will not be good if the public effort is seen to lose": N. Wade, "Talk o f Collaboration on Decoding of the Genome," *New York Times,* November 14, 1999.

198   Collins led a delegation of scientists: J. Gillis, "Gene Map Alliance Hopes Fade," *Washington Post,* March 6, 2000.

199   a draft statement of "Shared Principles": Letter from Collins, Varmus, Bobrow, and Waterston to Venter, White, Levine, and Gilman, February 28, 2000. The letter was released to the press by the Wellcome Trust on March 5, 2000: Gillis, "Gene Map Alliance"; J. Martinson, "Protocol on Gene Research at Risk as Firm Demands Exclusive Rights," *Guardian,* March 7, 2000.

200   The meeting ended with no agreement: J. Gillis, "Gene-Mapping Controversy Escalates," March 7, 2000.

201   *Time* tagged the public genome project "the proverbial tortoise": D. Thompson, "The Gene Machine," *Time,* January 24, 2000, p. 58.

201   more than 60 percent of the human disease and cancer genes: G. M. Rubin et al., "Comparative Genomics of the Eukaryotes," *Science* 287 (2000): 2204–2215.

202   This involves a contraption called an inebriometer: H. J. Bellen, "The Fruit Fly: A Model Organism to Study the Genetics of Alcohol Abuse and Addiction?" *Cell* 93 (1998): 909–912.

202   *Science* devoted almost half its regular issue to the *Drosophila* genome sequence: M. D. Adams et al., "The Genome Sequence of *Drosophila melanogaster,*" Science 287 (2000): 2185–2195, and other papers.

202   "The *Drosophila* sequence is a critical resource": T. Kornberg and M. Krasnow, "The *Drosophila* Genome Sequence: Implications for Biology and Medicine," *Science* 287 (2000): 2218–2220.

203   Venter said he still wanted to print out one copy of the *Drosophila* sequence: J. C. Venter, press conference, Celera, March 30, 2000.

203   "People who live in glass houses shouldn't throw stones": P. G. Gosselin and P. Jacobs, "Biotech Firm's Mix-up on Fly Genome Creates a Stir," *Los Angeles Times,* April 20, 2000.

203   Finally, on February 28, Collins faxed a "confidential" letter: Letter from Collins et al., February 28, 2000.

204   the Wellcome Trust released the letter to the media: Gillis, "Gene-Mapping Controversy."

204   White angrily attacked the British charity's move as "slimy" and "dumb": Ibid.

204   "The Human Genome Project, supposedly one of mankind's noblest undertakings": Ibid.

204   "we continue to be interested in pursuing good-faith discussions toward collaboration": J. Gillis, "Celera Leaves Door Open to Genetic Research Deal," *Washington Post,* March 8, 2000.

204   the 2 billionth letter of DNA: S. Begley, "Decoding the Human Body," *Newsweek,* April 10, 2000, pp. 52–57.

205   To realize the full promise of this research: Joint statement by President Clinton and Prime Minister Tony Blair, White House press release, March 14, 2000.

205   Celera's share price plummeted: A. Berenson and N. Wade, "A Call for Sharing of Research Causes Gene Stocks to Plunge," *New York Times*, March 15, 2000. Just two weeks after Celera raised almost $1 billion in a follow-up public offering of more than 4 million shares, its share price fell sharply from more than $270 to $100 after the Clinton-Blair statement.

205   Tony Blair had initiated discussions with the White House: D. Hencke, R. Evans, and T. Radford, "Blair and Clinton Push to Stop Gene Patents," *Guardian*, September 20, 1999.

206   Venter penned an op-ed article: J. C. Venter, "Clinton and Blair Shouldn't Destroy Our Research," *Wall Street Journal*, March 21, 2000.

206   "We are not in the business of patenting humanity": W. A. Haseltine, "21st Century Genes," *Washington Post*, March 28, 2000.

206   Haseltine had been thrust into the center: M. Waldholz, "AIDS Discovery Spurs Some to Challenge a Patent Filing That Boosted HGS Stock," *Wall Street Journal*, March 16, 2000; A. Clark and J. Meek, "Drug Firms Laying Claim to Our Genes," *Guardian*, February 18, 2000. The value of Human Genome Sciences' stock rose 40 percent on news of the *CCR5* gene patent.

207   The furor provided ammunition: M. Ensenrink, "Patent Office May Raise the Bar on Gene Claims," *Science* 287 (2000): 1196–1197; R. F. Harris, "Patenting Genes: Is It Necessary and Is It Evil?" *Current Biology* 10 (2000): R174–R175; A. Allen, "Who Owns Your DNA?" *Salon*, March 7, 2000. In December 1999, the U.S. Patent Office decided to adopt more stringent requirements for demonstrating gene utility.

207   "If patents have the word '-like' in the title": Venter, press conference, March 30, 2000. Incyte and Human Genome Sciences maintain that they apply for patents only on well characterized genes.

207   "What lies ahead is a battle": Editorial, *Guardian*, March 16, 2000.

207   "It is vital that all researchers have access to the full genome": B. Alberts and A. Klug, *Nature* 404 (2000): 325.

208   "People think this is a biotech company": Venter, press conference, March 30, 2000.

209   "I am happy to again show the Subcommittee": J. C. Venter, U.S. House of Representatives Committee on Science, Subcommittee on Energy and Environment, April 6, 2000.

210   Dr. Venter and others are responsible for speeding up the sequencing: K. Calvert, "Celera's Role in Opening Up New Frontiers," *Nature* 405 (2000): 613.

210   "We're sequencing the genome for the world for free": J. C. Venter, *Newshour with Jim Lehrer*, PBS, April 6, 2000.

211   Waterston estimated that Celera's modest coverage: P. Smaglik, "Critics Challenge Celera's Claims over Human Genome Sequence," *Nature* 404 (2000): 691–692.

211   "You should not take at face value": "HGP Head Refutes Celera's Claim," *Wired News*, April 10, 2000.

211   the Joint Genome Institute had sequenced about 300 million bases: N. Wade, "Energy Department Experts Open Way to Gene Search in Genome Project," *New York Times,* April 14, 2000.

212   the sequencing of 33,546,361 bases of the smallest human chromosome: The Chromosome 21 Mapping and Sequencing Consortium, "The DNA Sequence of Human Chromosome 21," *Nature* 405 (2000): 311–318.

212   The colorful language of the press announcement: Press invitation, German Human Genome Project, Berlin, April 30, 2000.

213   "Observations on an Ethnic Classification of Idiots": A. P. Mange and E. J. Mange, *Genetics: Human Aspects* (Philadelphia: Saunders, 1980).

214   "How many IQ points would it take": P. Jacobs, "Team Decodes Down Syndrome Chromosome," *Los Angeles Times,* May 9, 2000.

214   a series of class action lawsuits: B. Barker, "A Calm Look at Class Actions and Celera," Motley Fool, May 3, 2000.

215   "a real validation of the public sequencing project": A. Coghlan, "Life's Encyclopedia," *New Scientist,* May 13, 2000, p. 11.

215   "The whole finish-line mentality is silly": "Avoid Wacky (Genome) Races," *Nature* 405 (2000): 103.

215   "So far, good sense and manners": Ibid.

CHAPTER TEN: *THE EIGHTH DAY*

Several books examine the potential impact of genetic technology on life and medicine. Good places to start are Lee Silver's *Remaking Eden* (New York: Avon Books, 1997) and Phillip Kitcher's *The Lives to Come* (New York: Touchstone, 1997) . The history of gene therapy is well told in Larry Thompson's *Correcting the Code* (New York: Simon & Schuster, 1994) and *Altered Fates* by J. Lyon and P. Gorner (New York: Norton, 1996).

216   Within 10 years, every baby born in a hospital: J. C. Venter, press conference, Celera, March 30, 2000.

216   In 10 years, we should be able to make predictions: F. S. Collins, *The Charlie Rose Show,* PBS Television, June 20, 2000.

217   "There is no disease, except some cases of trauma": Ibid.

217   "When the whole human genome is available on CD-ROMs": C. Cookson, "Cartographer of Life," *Financial Times,* December 22, 1995.

218   the powerful technology of DNA microarrays: D. J. Lockhart, E. A. Winzeler, "Genomics, Gene Expression and DNA Arrays," *Nature* 405 (2000): 827–836; "The Chipping Forecast," *Nature Genetics* 21 (suppl.) (1999): 1–60.

218   The Affymetrix chips resemble a microscopic chessboard: R. J. Lipzhutz, S. P. A. Fodor, T. R. Gingeras, and D. J. Lockhart, "High Density Synthetic Oligonucleotide Arrays," *Nature Genetics* 21 (suppl.) (1999): 20–24.

218   pioneered by Patrick Brown and colleagues at Stanford University: Details of Brown's DNA microarray system are freely available on his web site: http://cmgm.stanford.edu/pbrown/index.html

219   a survey of the activity of 8,000 genes in dozens of breast tumors: C. M. Perou et al., "Molecular Portraits of Human Breast Tumours," *Nature* 406 (2000): 747–752.

219   the most common form of non-Hodgkin's lymphoma: A. A. Alizadeh et al., "Distinct Types of Diffuse Large B-Cell Lymphoma Identified by Gene Expression Profiling," *Nature* 403 (2000): 503–511.

219   There are on average four to eight SNPs in every gene: N. Risch, "Searching for Genetic Determinants in the New Millennium," *Nature* 405 (2000): 847–856.

219   identify thousands of critical genetic variations in an individual's DNA: M. Cargill et al., "Characterization of Single-Nucleotide Polymorphisms in Coding Regions of Human Genes," *Nature Genetics* 22 (1999): 231–238; M. K. Halushka et al., "Patterns of Single-Nucleotide Polymorphisms in Candidate Genes for Blood-Pressure Homeostasis," *Nature Genetics* 22 (1999): 239–247.

219   assessing your genetic constitution: G. J. B. van Ommen, E. Bakker, and J. T. den Dunnen, "The Human Genome Project and the Future of Diagnostics, Treatment, and Prevention," *Lancet* 354 (suppl.) (1999): 5–10.

220   pharmacogenomics, one of the most expectant areas: A. D. Roses, "Pharmacogenetics and the Practice of Medicine," *Nature* 405 (2000): 857–865.

220   Although Collins is not in the business of making rash predictions: F. S. Collins, "Shattuck Lecture—Medical and Societal Consequences of the Human Genome Project," *New England Journal of Medicine* 341 (1999): 28–37.

220   By 2010, he predicts, tests for twenty-five common diseases: L. K. Altman, "Genomic Chief Has High Hopes, and Great Fears, for Genetic Testing," *New York Times,* June 27, 2000.

220   we can already predict what some of the first commonly available tests: A. G. Motulsky, "If I Had a Gene Test, What Would I Have and Who Would I Tell," *Lancet* 354 (suppl.) (1999): 35–37.

221   "All of us carry probably four or five really fouled-up genes": F. S. Collins, "Evolution of a Vision: Genome Project Origins, Present and Future Challenges, and Far-Reaching Benefits," *Human Genome News* 7 (September–December 1995), p. 3.

221   No other area of research has seen such high expectations and bitter disappointments: T. Friedmann, ed., *The Development of Human Gene Therapy* (Cold Spring Harbor, N.Y., Cold Spring Harbor Laboratory Press, 1999).

222   the first officially sanctioned trial for gene therapy: Thompson, *Correcting the Code;* Lyon and Gorner, *Altered Fates.*

222   the therapy had not led to permanent cure for Ashanthi: R. M. Blaese et al., "T Lymphocyte-Directed Gene Therapy for ADA-SCID: Initial Trial Results after 4 Years," *Science* 270 (1995): 475–480. Some success was reported

by an Italian group: C. Bordignon et al., "Gene Therapy in Peripheral Blood Lymphocytes and Bone Marrow for ADA-Immunodeficient Patients," *Science* 270 (1995): 470–475.

222   Wilson devised a radical ex vivo therapy: M. Grossman et al., "Successful Ex Vivo Gene Therapy Directed to Liver in a Patient with Familial Hypercholesterolaemia," *Nature Genetics* 6 (1994): 335–341.

223   Gelsinger developed a severe inflammatory reaction: R. Weiss and D. Nelson, "Teen Dies Undergoing Experimental Gene Therapy," *Washington Post,* September 29, 1999.

223   "This tragedy has rocked the field down to its toes": E. Marshall, E. Pennisi, and L. Roberts, "In the Crossfire: Collins on Genomes, Patents, and 'Rivalry,' " *Science* 287 (2000): 2396–2398.

223   the first tangible success for this heavily touted field: M. Cavazzana-Calvo et al., "Gene Therapy of Human Severe Combined Immunodeficiency (SCID)-X1 Disease," *Science* 288 (2000): 669–672.

223   researchers are eagerly exploring a number of alternative strategies: E. Licking, "Gene Therapy: One Family's Story," *Business Week,* July 12, 1999, pp. 94–104.

223   The result is a virus that "is easy to make": L. Jaroff, "Fixing the Genes," *Time,* January 11, 1999, pp. 68–73.

224   Kmiec is pursuing a technique called chimeraplasty: E. Kmiec, "Gene Therapy," *American Scientist* 87 (1999): 240–247; T. Gura, "Repairing the Genome's Spelling Mistakes," *Science* 285 (1999): 316–318. Kimeragen has merged with the French biotech firm ValiGen, and Kmiec has moved from Penn to the University of Delaware.

224   studies in animal models have been encouraging: B. T. Kren et al., "Correction of the UDP-Glucuronosyltransferase Gene Defect in the Gunn Rat Model of Crigler-Najjar Syndrome Type I with a Chimeric Oligonucleotide," *Proceedings of the National Academy of Sciences U.S.A.* 96 (1999): 10349–10354.

224   Clinical trials are being considered for Mennonite children: D. Grady, "At Gene Therapy's Frontier, the Amish Build a Clinic," *New York Times,* June 29, 1999.

225   "Dare we be entrusted with improving upon the results": J. Watson, "All for the Good," *Time,* January 11, 1999, p. 91.

225   "The prospect of a 'product recall' ": E. S. Lander, "In Wake of Genetic Revolution, Questions About Its Meaning," *New York Times,* September 12, 2000.

225   Preimplantation genetic diagnosis works like this: A. H. Handyside and J. D. Delhanty, "Preimplantation Genetic Diagnosis: Strategies and Surprises," *Trends in Genetics* 13 (1997): 270–275.

226   In the first case in 1990: A. H. Handyside, E. H. Kontogianni, K. Hardy, and R. M. Winston, "Pregnancies from Biopsied Human Preimplantation Embryos Sexed by Y-Specific DNA Amplification," *Nature* 344 (1990): 768–770.

226 Hughes began a transatlantic collaboration: A. H. Handyside, J. G. Lesko, J. J. Tarin, R. M. Winston, and M. R. Hughes, "Birth of a Normal Girl After *in vitro* Fertilization and Preimplantation Diagnostic Testing for Cystic Fibrosis," *New England Journal of Medicine* 327 (1992): 905–909.

226 Hughes ran afoul of those government restrictions: J. Crewdson, "U.S. Fires Researcher over Use of Embryonic DNA," *Chicago Tribune,* January 9, 1997; E. Marshall, "Varmus Grilled over Breach of Embryo Research Ban," *Science* 276 (1997): 1963.

226 tests were also being performed in Hughes's NIH lab: J. Crewdson, "Gene Test That Went Awry Was Concealed," *Chicago Tribune,* March 9, 1997.

226 successfully arranged the births of twenty-five babies spared the Huntington's gene: M. Hughes, interview, July 27, 2000.

226 the first successful preimplantation diagnosis for sickle cell anemia: K. Xu, Z. M. Shi, L. L. Veeck, M. R. Hughes, and Z. Rosenwaks, "First Unaffected Pregnancy Using Preimplantation Genetic Diagnosis for Sickle Cell Anemia," *Journal of the American Medical Association* 281 (1999): 1701–1706.

227 Despite a $20,000 price tag in the United States: F. Golden, "Good Eggs, Bad Eggs," *Time,* January 11, 1999, pp. 56–59.

227 he is helping give rise to "designer babies": N. Boyce, "Designing a Dilemma," *New Scientist,* December 11, 1999.

227 The decision to have a baby: On August 29, 2000, in the United States, Lisa Nash gave birth to a baby boy Adam, one of 15 embryos conceived artificially. Using PGD, Adam's cells were found not only to have been spared the gene for the fatal genetic disorder Fanconi anaemia, but also to provide a match for the Nash's affected daughter, Molly. One month after birth, Adam's umbilical cord cells were transplanted into his six-year-old sister, Molly, to replenish her bone marrow.

228 "If we could honestly promise young couples": J. D. Watson, *A Passion for DNA* (Cold Spring Harbor, N.Y.: Cold Spring Harbor Laboratory Press, 2000).

228 another film set in the "not-too-distant-future": *GATTACA,* DVD, Columbia–TriStar, 1998.

229 none is beyond the realm of possibility: Silver, *Remaking Eden.*

229 A particularly provocative cover of *Time:* M. D. Lemonick, "Smart Genes?" *Time,* September 13, 1999, pp. 54–58.

229 his supermice could store and recall information: Y. P. Yang et al., "Genetic Enhancement of Learning and Memory in Mice," *Nature* 401 (1999): 63–69.

229 a big hit at Tsien's son's "show-and-tell" day: J. Z. Tsien, "Building a Brainier Mouse," *Scientific American,* April 2000, pp. 62–68.

229 the notion of a single gene's controlling IQ: N. Wade, "Scientist Creates Smarter Mouse: Work on Formation of Memory May Someday Help People," *New York Times,* September 2, 1999; N. Wade, "Of Smart Mice

and an Even Smarter Man," *New York Times,* September 7, 1999; R. M. Sapolsky, "Genetic Hyping," *The Sciences* (March/April 2000): 12–15.

229   the "not-in-our-genes" crowd: R. C. Lewontin, S. Rose, and L. J. Kamin, *Not in Our Genes* (New York: Pantheon, 1984): R. Hubbard and E. Wald, *Exploding the Gene Myth* (Boston: Beacon Press, 1993).

230   the results of the largest twin study of cancer on record: P. Lichtenstein et al., "Environmental and Heritable Factors in the Causation of Cancer— Analyses of Cohorts of Twins from Sweden, Denmark, and Finland," *New England Journal of Medicine* 343 (2000): 78–85.

230   they can be classified into six classes: D. Hanahan and R. Weinberg, "The Hallmarks of Cancer," *Cell* 100 (2000): 57–70.

231   John Colapinto describes the tragic case of a young boy: J. Colapinto, *As Nature Made Him* (New York: HarperCollins, 2000).

231   the gene was appropriately named Hephaestin: C. D. Vulpe et al., "Hephaestin, a Ceruloplasmin Homologue Implicated in Intestinal Iron Transport, Is Defective in the *sla* Mouse," *Nature Genetics* 21 (1999): 195–199; J. Kaplan and J. P. Kushner, "Mining the Genome for Iron," *Nature* 403 (2000): 711–713.

231   a cartoon map of the Y chromosome: "Why Map Y?" *Science* 281 (1993): 675.

231   she has demonstrated that perfect pitch is a highly hereditary trait: S. Baharloo, S. K. Service, N. Risch, J. Gitschier, and N. B. Freimer, "Familial Aggregation of Absolute Pitch," *American Journal of Human Genetics* 67 (2000): 755–758. See http://www.perfectpitchstudy.org

231   many human behavioral traits are at least partially influenced: M. McGue and T. J. Bouchard Jr., "Genetic and Environmental Influences on Human Behavioral Differences," *Annual Review of Neuroscience* 21 (1998): 1–24.

231   the behavior of a highly antisocial Dutch family: H. G. Brunner et al., "Abnormal Behavior Associated with a Point Mutation in the Structural Gene for Monoamine Oxidase A," *Science* 262 (1993): 578–580.

232   "Are those people who go to work in suits happier": D. Lykken and A. Tellegen, "Happiness Is a Stochastic Phenomenon," *Psychological Science* 7 (1996): 186–189.

232   "About half of your sense of well-being": D. Goleman, "Forget Money: Nothing Can Buy Happiness, Some Researchers Say," *New York Times,* July 16, 1996; J. Adler, "The Happiness Meter," *Newsweek,* July 29, 1996, p. 78; S. Begley, "Born Happy?" *Newsweek,* October 14, 1996, pp. 78–80.

232   "How you feel right now is about equally genetic and circumstantial": D. Hamer, "The Heritability of Happiness," *Nature Genetics* 14 (1996): 125–126.

233   Hamer made international headlines in 1993: D. H. Hamer, S. Hu, V. L. Magnuson, N. Hu, and A. M. L. Pattatucci, "A Linkage Between DNA Markers on the X Chromosome and Male Sexual Orientation," *Science* 261 (1993): 321–326.

233 variation in the length of the gene for the dopamine D4 receptor: D. Hamer and P. Copeland, *Living with Our Genes* (New York: Doubleday, 1998); J. M. Nash, "The Personality Genes," *Time,* April 27, 1998, pp. 60–61.

234 one positive benefit of having the "high-anxiety" form: Hamer and Copeland, *Living with Our Genes.*

234 "*g* is one of the most reliable and valid measures": R. Plomin, "Genetics and General Cognitive Ability," *Nature* 402 (suppl.) (1999): C25-C29; R. Plomin and J. C. DeFries, "The Genetics of Cognitive Abilities and Disabilities," *Scientific American* (May 1998): 62–69; S. Begley, "A Gene for Genius?" *Newsweek,* May 25, 1998.

234 Patients with Williams's syndrome are missing several genes: H. M. Lenhoff, P. P. Wang, F. Greenberg, and U. Bellugi, "Williams Syndrome and the Brain," *Scientific American* (December 1997): pp. 68–73.

234 the location of a gene for female intuition: D. H. Skuse et al., "Evidence from Turner's Syndrome of an Imprinted X-Linked Locus Affecting Cognitive Function," *Nature* 387 (1997): 705–708.

235 Take speed, for example: J. Entine, *Taboo* (New York: Public Affairs Press, 2000); A. Anthony, "White Men Can't Run," *Observer,* June 4, 2000.

235 "I think black sportsmen": R. Williams, "Time for French Lessons," *Guardian,* June 22, 2000.

235 the heritability of facial features: See http://www.gene.ucl.ac.uk/face

CHAPTER ELEVEN: *THE LANGUAGE OF GOD*

236 "Today we are learning the language in which God created life": The White House, Office of the Press Secretary, June 26, 2000. See http://www.whitehouse.gov/WH/New/html/20000626.html

236 "Today is the day that we hand over the gift": T. Radford, "Scientists Revel in a Day of Glory," *Guardian,* June 27, 2000.

238 "fix it . . . make these guys work together": F. Golden and M. D. Lemonick, "The Race Is Over," *Time,* July 23, 2000, pp. 18–23.

238 "I don't think I've ever seen them as tense": Ibid.

238 posed for the cover of *Time* magazine: Ibid.

238 "They're trying to say it's not a race, right?": R. Preston, "The Genome Warrior," *New Yorker,* June 12, 2000, pp. 66–83.

238 "The journalistic feeding frenzy has served no one": A. Regaldo, "Riding the DNA Railroad," *Technology Review* (July–August 2000): 94–98.

239 "Competition is an essential part of science": E. Lander, Washington D.C. press conference, June 26, 2000.

239 "Your goal is not to foresee the future": F. Collins, Washington D.C. press conference, June 26, 2000.

240 There has been considerable speculation: Preston, "Genome Warrior."

240   Venter attributed much of the success to a "secret weapon": S. Hensley, "New Race Heats Up to Turn Gene Information into Drugs Discoveries," *Wall Street Journal,* June 26, 2000.

240   "I asked Gene [Myers] what that really was": J. C. Venter, Washington D.C. press conference, June 26, 2000.

241   On the day that Celera made history: J. Gillis, "Gene Mapper's Stock Tumbles," *Washington Post,* June 28, 2000.

241   "I can well imagine technology making the wheel obsolete": Radford, "Scientists Revel."

241   "Along with Bach's music": N. Hawkes, "Benefits 'Some Distance in Future,' " *Times,* June 27, 2000.

241   "a phenomenal day": J. C. Venter, *Nightline,* ABC News, June 26, 2000.

241   "on behalf of our species, thank you!": R. Krulwich, *Nightline,* ABC News, June 26, 2000.

241   "I'm a little confused on timing": "The Coming Revolution," *New Scientist,* July 1, 2000, p. 3.

241   Schubert's Unfinished Symphony could be declared "finished": "First Draft of Genome Sets New Industry Standards," *Nature Biotechnology* 18 (2000): 803. ("The announcement," wrote editor Andy Marshall, "like the celebrations of the millennium, preceded the event by 12 months.")

242   baseball games would be called after eight innings: A. Frew, "Washington Diary, " *New Scientist,* July 29, 2000, p. 61.

242   The reason for the carefully orchestrated announcement: "The Genetic Starting-Line," *Economist,* July 1, 2000, pp. 21–22.

242   the only time that a gap could be found: M. Morgan, interview, London, August 17, 2000.

242   "It's like a private company in 1967": A. Pollack, "Finding Gold in Scientific Pay Dirt," *New York Times,* June 28, 2000.

242   "The race at this point is not for the DNA": Ibid.

242   "the beginning of thousands of races": K. Philipkoski, "Genome Mappers to Make Amends," *Wired News,* June 22, 2000.

242   "The idea that this is a tremendous scientific accomplishment": S. Brenner, interview, June 24, 2000.

242   "personality, profit, passion—and pizza!": D. Sawyer, *Good Morning America,* ABC, June 27, 2000.

243   "Now scientists are concentrating": C. Kilborne, *The Late Show,* CBS, June 27, 2000.

243   John Malkovich would star as Venter: "First Draft," *Nature Biotechnology,* (August 2000).

244   Christiano continues her search: A. Christiano, interview, May 15, 2000.

244   The mouse sequence: In October 2000, the NIH, the Wellcome Trust, SmithKline Beecham, Merck and Affymetrix announced that they were forming a consortium to sequence the mouse genome in six months, at a cost of $58 million. The consortium selected a strain called "Black 6" not

being studied by Celera. Unveiling the new public–private partnership, Francis Collins joked with reporters, "This is not a race—write this down, this is not a race!"

244   Venter caused a major stir: A. Saegusa, "US Firm's Bid to Sequence Rice Genome Causes Stir in Japan," *Nature* 398 (1999): 245.

244   The dog genome is also an attractive potential target: J. Groopman, "Pet Scan," *New Yorker,* May 10, 1999, p. 47; R. K. Wayne and E. A. Ostrander, "Origin, Genetic Diversity, and Genome Structure of the Domestic Dog," *BioEssays,* 21 (1999): 247–257; E. A. Ostrander, F. Gailbert, and D. F. Patterson, "Canine Genetics Comes of Age," *Trends in Genetics,* 16 (2000): 117–124. Perhaps as long as 100,000 years ago, humans began to domesticate dogs. Today, there are some 300 breeds, all descendants of the gray wolf, exhibiting a rich variety of behavioral and physical traits ranging from those of the Great Dane (80 kg) to the Chihuahua (0.5 kg). Extensive inbreeding over the past few centuries has resulted in the manifestation of nearly 400 documented recessive diseases, including hemophilia, blindness, various cancers, and narcolepsy (narcolepsy being the first important disease-related gene identified initially in dogs).

244   they might one day share the Nobel Prize: D. Birch, "Daring Sprint to the Summit," *Baltimore Sun,* April 13, 1999.

245   "It's only by understanding protein function": A. Coghlan and M. Le Page, "Spot the Difference," *New Scientist,* July 22, 2000, p. 14.

245   This technique could be used to detect abnormal proteins: A. Pandey and M. Mann, "Proteomics to Study Genes and Genomes," *Nature* 405 (2000): 837–846.

245   "I am certain that a century from now": J. Gillis, "IBM to Put Genetics on Fast Track," *Washington Post,* June 3, 2000.

245   the new buzzwords of the postgenome era: C. Ezzell, "Beyond the Human Genome," *Scientific American* (July 2000): 64–69.

245   "The difference between physiology and functional genomics": "Up the Function," *Nature Genetics* 16 (1997): 111–112.

246   a prototype protein microarray: G. MacBeath and S. L. Schreiber, "Printing Proteins as Microarrays for High-Throughput Function Determination," *Science* 289 (2000): 1760–1763.

246   partners for more than 1,000 yeast proteins: P. Uetz et al., "A Comprehensive Analysis of Protein–Protein Interactions in *Saccharomyces cerevisiae,*" *Nature* 403 (2000): 623–627.

246   "The first look at a new protein structure": A. Sali and J. Kuriyan, "Macromolecular Complexes and Structural Genomics," *Trends in Biochemical Sciences* 24 (1999): M20–M24.

246   leading the call for a structural genomics initiative: S. Burley et al., "Structural Genomics: Beyond the Human Genome Project," *Nature Genetics* 23 (1999): 151–157.

247   an international structural genomics consortium: D. Butler, "Wellcome

Discusses Structural Genomics Effort with Industry," *Nature* 406 (2000): 923–924.

247   determine the three-dimensional structures of 5,000 proteins in five years: K. Garber, "The Next Wave of the Genomics Business," *Technology Review* (July–August 2000): 46–56; A. Pollack, "The Next Chapter in the Book of Life: Structural Genomics," *New York Times,* July 4, 2000.

247   "If you can input 200,000 or 400,000 compounds": D. Rotman, "The Bell Labs of Biology," *Technology Review* (March–April 2000): 94–98.

248   hailed as biology's latest holy grail: J. Meek and M. Ellison, "On the Path of Biology's Holy Grail," *Guardian,* June 5, 2000; J. Bowen, "Blue Gene," *Salon,* December 9, 2000; Gillis, "IBM to Put Genetics on Fast Track."

248   "When all of this [genome project] mania dies down": Brenner interview, June 24, 2000.

248   the warp-speed evolution of the universal genetic code: A. B. Martin and P. G. Schultz, "Opportunities at the Interface of Chemistry and Biology," *Trends in Biochemical Sciences* 24 (1999): M24–M28.

248   "Why are there only four bases in our DNA": U. Kher, "Doing It Nature's Way," *Time,* August 7, 2000, pp. 68–77.

249   To incorporate an unnatural amino acid into proteins: T. Hesman, "Code Breakers: Scientists Are Altering Bacteria in a Most Fundamental Way," *Science News* 157(2000): 360–362; D. Rotman, "The Bell Labs of Biology," *Technology Review* (March–April 2000): 94–98.

249   Researchers are exploring ways to expand the genetic lexicon: Martin and Schultz, "Opportunities at the Interface."

249   "There haven't been any significant changes": N. Farndale, "A Brief History of the Future," *Daily Telegraph,* January 4, 2000.

250   "Genetics *per se* can never be evil": J. Watson, *A Passion for DNA* (Cold Spring Harbor, N.Y.: Cold Spring Harbor Laboratory Press, 2000).

250   the British zoologist William Bateson settled into his seat: R. Marantz Henig, *The Monk in the Garden* (Boston: Houghton Mifflin, 2000).

251   The isolation of human embryonic stem cells: N. Wade, "Scientists Cultivate Cells at Root of Human Life," *New York Times,* November 6, 1998; "Embryo Cell Research: A Clash of Values," *New York Times,* July 2, 1999; S. J. Wright, "Human Embryonic Stem-Cell Research: Science and Ethics," *American Scientist* 87 (1999): 352–361.

251   the birth of Dolly the sheep: G. Kolata, *Clone* (New York: William Morrow, 1999); I. Wilmut, K. Campbell, and C. Tudge, *The Second Creation* (New York: Farrar Straus & Giroux, 2000).

251   The genome sequence can be considered a genetic palimpsest: S. Eddy, interview, July 17, 2000.

Acknowledgments

IN FEBRUARY 1676, Isaac Newton penned a letter to his critic Robert Hooke that included the famous line, "If I have seen further it is by standing on ye shoulders of Giants." Originally a subtle reference to Hooke's stooped posture, the quotation has since become a popular means of crediting the essential scientific foundations built by one's predecessors. It was a phrase heard frequently during the widespread celebration of the completion of the Human Genome Project.

Since the project's launch in 1990, I have enjoyed the friendship and support of countless scientists around the world, while playing a small part in communicating their discoveries through the pages of *Nature* and *Nature Genetics*. From my unique vantage point as the founding editor of *Nature Genetics*, I had the great pleasure of working closely with the major figures in the story, including Craig Venter, Francis Collins, Eric Lander, Daniel Cohen, Mary-Claire King, and too many others to mention by name. Quite simply, this book would not have been possible without the trust and support of the thousands of dedicated geneticists with whom I worked during those years. Special thanks go to Mark Boguski, Sydney Brenner, James Crow, Jane Gitschier, Michael Morgan, Kári Stefánsson, James Watson, and Noé Zamel.

I shared this thrilling vantage point with a small band of extraordinarily talented editors—Adrian Ivinson, Laurie Goodman, Barbara Cohen, and Bette Phimister (the current editor of *Nature Genetics*)—

whom I cannot thank enough. I also thank my colleagues for their support at the beginning, especially Sir John Maddox, Barbara Culliton, Mary Waltham, and Nicky Byam-Shaw. Thanks also to many other good friends at *Nature* who helped in various ways during this project, including Carina Dennis, David Dickson, Adam Felsenfeld, Tim Lincoln, Colin Macilwain, and David Swinbanks.

The coverage of the Human Genome Project in the media has been subject to criticism in some quarters for sensationalizing the "race." However, I am deeply indebted to the writers who have covered the project in the lay and scientific press on both sides of the Atlantic. Particular thanks go to Nicholas Wade of the *New York Times,* whose superb coverage of all aspects of the story has been invaluable.

My former colleagues at the Howard Hughes Medical Institute graciously allowed me the freedom to research and work on this book. Particular thanks go to all of the HHMI investigators working; Cathy Harbert, Marianne Beane, and the staff of the Purnell Choppin Library; and Laura Bonetta, David Jarmul, Jim Keeley, and Joe Perpich. Most of all, I extend my appreciation to Max Cowan, from whom I learned so much.

This book could not have happened without the extraordinary support of my agents and editors. I cannot begin to thank my agents, Jennifer Gates Hayes and Esmond Harmsworth at Zachary Shuster Harmsworth, who have been so supportive. Special thanks are also due to my editors, Stephen Morrow at The Free Press and Peter Tallack at Weidenfeld & Nicolson, who were behind this book from the beginning and have helped in so many ways to make it a reality. Edith Lewis and Beverly Miller worked wonders with the rough text. And thanks to Michael White, as always.

P. G. Wodehouse once dedicated a book to his daughter, "without whose never-failing sympathy and encouragement this book would have been finished in half the time." I could not have written this book without the love and occasional distraction of my two wonderful children, Kyle and Fiona. And, most of all, I thank my amazing wife, Susan, for everything.

Rockville, Maryland
October 2000

Index